Magnetic Sensors and Magnetometers

For a listing of related titles from *Artech House*,
turn to the back of this book.

Magnetic Sensors and Magnetometers

Pavel Ripka

Editor

Artech House
Boston • London
www.artechhouse.com

Library of Congress Cataloging-in-Publication Data
Magnetic sensors and magnetometers / Pavel Ripka, editor.
   p. cm.
  Includes bibliographical references and index.
  ISBN 1-58053-057-5 (alk. paper)
    1. Engineering instruments. 2. Magnetic instruments. 3. Magnetometer. 4. Detectors.
  I. Ripka, Pavel. II. Title.

TA165 .M34 2000   00-050816
621.34—dc21   CIP

British Library Cataloguing in Publication Data
Magnetic sensors and magnetometers.
  1. Magnetic instruments
  I. Ripka, Pavel
  621.3'4

ISBN 1-58053-057-5

Cover design by Gary Ragaglia

© 2001 ARTECH HOUSE, INC.
685 Canton Street
Norwood, MA 02062

All rights reserved. Printed and bound in the United States of America. No part of this book may be reproduced or utilized in any form or by any means, electronic or mechanical, including photocopying, recording, or by any information storage and retrieval system, without permission in writing from the publisher.
   All terms mentioned in this book that are known to be trademarks or service marks have been appropriately capitalized. Artech House cannot attest to the accuracy of this information. Use of a term in this book should not be regarded as affecting the validity of any trademark or service mark.

International Standard Book Number: 1-58053-057-5
Library of Congress Catalog Card Number: 00-050816

10 9 8 7 6 5 4 3 2 1

Contents

	Preface	**xv**
	Acknowledgments	*xviii*
	References	*xviii*
1	**Basics**	**1**
1.1	Magnetic Quantities and Units	1
1.1.1	Roots of Magnetism	1
1.1.2	Magnetic Field and Matter	4
1.1.3	Units and Magnitudes of Magnetic Fields	8
1.2	Magnetic States of Matter	9
1.2.1	Weakly Magnetic Materials	10
1.2.2	Ferromagnetism and Ferrimagnetism	11
1.3	Magnetic Materials for Sensor Applications	28
1.3.1	Soft Magnetic Materials	29
1.3.2	Hard Magnetic Materials	31
1.3.3	Magnetostrictive Materials	33
1.3.4	Thin Films	33
1.4	Sensor Specifications	39
1.4.1	Full-Scale Range, Linearity, Hysteresis, and Temperature Coefficient of Sensitivity	39

1.4.2	Offset, Offset Temperature Coefficient, and Long-Term Stability	40
1.4.3	Perming	40
1.4.4	Noise	41
1.4.5	Resistance Against Environment (Temperature, Humidity, Vibrations)	42
1.4.6	Resistance Against Perpendicular Field and Field Gradient	43
1.4.7	Bandwidth	43
1.4.8	Other Parameters	43
	References	44
2	**Induction Sensors**	**47**
2.1	Air Coils	48
2.1.1	Voltage Sensitivity at Low Frequencies	51
2.1.2	Thermal Noise	52
2.1.3	The Influence of the Parasitic Capacitances	53
2.1.4	Current Output (Short-Circuited Mode)	55
2.2	Search Coils With a Ferromagnetic Core	57
2.2.1	Voltage Output Sensitivity	57
2.2.2	Thermal Noise of the Cored Induction Sensor (Voltage Output)	62
2.2.3	The Equivalent Circuit for Cored Coils	62
2.2.4	Cored Coils With Current Output	63
2.3	Noise Matching to an Amplifier	64
2.4	Design Examples	65
2.5	Other Measuring Coils	65
2.5.1	Rotating Coil Magnetometers	65
2.5.2	Moving-Coils, Extraction Method	69
2.5.3	Vibrating Coils	69
2.5.4	Coils for Measurement of H	70
	References	72

3	**Fluxgate Sensors**	**75**
3.1	Orthogonal-Type Fluxgates	78
3.2	Core Shapes of Parallel-Type Fluxgates	79
3.2.1	Single-Rod Sensors	80
3.2.2	Double-Rod Sensors	80
3.2.3	Ring-Core Sensors	81
3.2.4	Race-Track Sensors	82
3.3	Theory of Fluxgate Operation	83
3.3.1	The Effect of Demagnetization	85
3.4	Core Materials	88
3.5	Principles of Fluxgate Magnetometers	90
3.5.1	Second-Harmonic Analog Magnetometer	90
3.5.2	Digital Magnetometers	93
3.5.3	Nonselective Detection Methods	94
3.5.4	Auto-Oscillation Magnetometers	96
3.6	Excitation	97
3.7	Tuning the Output Voltage	97
3.8	Current-Output (or Short-Circuited) Fluxgate	100
3.8.1	Broadband Current Output	101
3.8.2	Tuning the Short-Circuited Fluxgate	104
3.9	Noise and Offset Stability	105
3.9.1	Zero Offset	108
3.9.2	Offset From the Magnetometer Electronics	109
3.9.3	Other Magnetometer Offset Sources	110
3.10	Crossfield Effect	110
3.11	Designs of Fluxgate Magnetometers	111
3.11.1	Portable and Low-Power Instruments	111
3.11.2	Station Magnetometers	111
3.12	Miniature Fluxgates	112
3.13	ac Fluxgates	114

3.14	Multiaxis Magnetometers	115
3.14.1	Three-Axial Compensation Systems	116
3.14.2	Individually Compensated Sensors	117
3.15	Fluxgate Gradiometers	119
	References	120

4	**Magnetoresistors**	**129**
4.1	AMR Sensors	130
4.1.1	Magnetoresistance and Planar Hall Effect	130
4.1.2	Magnetoresistive Films	134
4.1.3	Linearization and Stabilization	136
4.1.4	Sensor Layout	144
4.2	GMR Sensors	150
4.2.1	Introduction	150
4.2.2	Spin Valve Effect Basics	152
4.2.3	Sensor Construction	163
4.2.4	Applications	166
	References	169

5	**Hall-Effect Magnetic Sensors**	**173**
5.1	Basics of the Hall Effect and Hall Devices	175
5.1.1	The Hall Effect	175
5.1.2	Structure and Geometry of a Hall Device	179
5.1.3	Main Characteristics of Hall Magnetic Field Sensors	180
5.1.4	Other Problems	183
5.2	High Electron Mobility Thin-Film Hall Elements	184
5.2.1	Introduction to Thin-Film Hall Elements	184
5.2.2	Highly Sensitive InSb Hall Elements	185
5.2.3	InAs Thin-Film Hall Elements by MBE	192
5.2.4	InAs Deep Quantum Wells and Application to Hall Elements	198

5.2.5	Conclusion	200
5.3	Integrated Hall Sensors	201
5.3.1	Historical Perspective	201
5.3.2	CMOS Hall Elements	205
5.3.3	Hall Offsets	206
5.3.4	Excitation	210
5.3.5	Amplification	213
5.3.6	Geometry Considerations	215
5.3.7	Vertical Hall Elements	218
5.3.8	Packaging for Integrated Hall Sensors	219
5.3.9	Trimming Methods and Limitations	221
5.3.10	Applications and Trends	222
5.4	Nonplatelike Hall Magnetic Sensors	223
5.4.1	Introduction	224
5.4.2	Vertical Hall Devices	225
5.4.3	Cylindrical Hall Devices	230
5.4.4	Two-Axis Vertical Hall Devices	232
5.4.5	Three-Axis Hall Devices	237
	References	240
6	**Magneto-Optical Sensors**	**243**
6.1	Faraday and Magneto-Optical Kerr Effects	244
6.1.1	Faraday Effect	244
6.1.2	Magneto-Optical Kerr Effect	247
6.2	Sensors of Magnetic Fields and Electric Currents	248
6.2.1	Polarimetric Measurements	249
6.2.2	Magneto-Optical Current Transformers Based on Diamagnets	251
6.2.3	MOCTs Based on Transparent Ferromagnets	256
6.2.4	MOCTs With Direct Registration of the Domain Wall Positions	260

6.3	Geometric Measurements	263
	References	264

7 Resonance Magnetometers — 267

7.1	Magnetic Resonance	267
7.1.1	Historical Overview	268
7.1.2	Absolute Reproducibility of Magnetic Field Measurements	270
7.2	Proton Precession Magnetometers	271
7.2.1	Mechanical Gyroscopes	271
7.2.2	Classic Proton-Free Precession Magnetometer	274
7.2.3	Overhauser-Effect Proton Magnetometers	289
7.3	Optically Pumped Magnetometers	294
7.3.1	Metastable He^4 Magnetometers	294
7.3.2	Alkali Metal Vapor Self-Oscillating Magnetometers	298
	References	301

8 Superconducting Quantum Interference Devices (SQUIDs) — 305

8.1	Introduction	305
8.1.1	Superconductivity	305
8.1.2	Meissner Effect	307
8.1.3	Flux Quantization	308
8.1.4	Josephson Effect	308
8.1.5	SQUIDs	309
8.2	SQUID Sensors	311
8.2.1	Materials	312
8.3	SQUID Operation	313
8.3.1	RF SQUIDs	314
8.3.2	dc SQUIDs	316
8.3.3	Noise and Sensitivity	317

8.3.4	Control Electronics	321
8.3.5	Limitations on SQUID Technology	322
8.4	Input Circuits	323
8.4.1	Packaging	323
8.4.2	The SQUID as a Black Box	324
8.4.3	Sensitivity	324
8.4.4	Detection Coils	326
8.4.5	Gradiometers	327
8.4.6	Electronic Noise Cancellation	329
8.5	Refrigeration	331
8.5.1	Dewars	331
8.5.2	Closed-Cycle Refrigeration	333
8.6	Environmental Noise (Noise Reduction)	334
8.6.1	Gradiometers for Noise Reduction	334
8.6.2	Magnetic Shielding	336
8.7	Applications	337
8.7.1	Laboratory Applications	337
8.7.2	Geophysical Applications	342
8.7.3	Nondestructive Test and Evaluation	342
8.7.4	Medical Applications	343
	References	345
9	**Other Principles**	**349**
9.1	Magnetoimpedance and Magnetoinductance	350
9.1.1	Materials	355
9.1.2	Sensors	357
9.2	Magnetoelastic Field Sensors	358
9.2.1	Fiber-Optic Magnetostriction Field Sensors	359
9.2.2	Magnetostrictive-Piezoelectric Sensors	361
9.2.3	Shear-Wave Magnetometers	362

9.3	Biological Sensors	362
9.3.1	Magnetotactic Bacteria	363
9.3.2	Magnetic Orientation in Animals	364
	References	365
10	**Applications of Magnetic Sensors**	**369**
10.1	Biomagnetic Measurements	369
10.2	Navigation	371
10.3	Military and Security	380
10.3.1	UXO	380
10.3.2	Target Detection and Tracking	382
10.3.3	Antitheft Systems	382
10.4	Automotive Applications	383
10.5	Nondestructive Testing	383
10.6	Magnetic Marking and Labeling	384
10.7	Geomagnetic Measurements: Mineral Prospecting, Object Location, and Variation Stations	385
10.8	Space Research	391
10.8.1	Deep-Space and Planetary Magnetometry	391
10.8.2	Space Magnetic Instrumentation	392
10.8.3	Measurement of Magnetic Fields Onboard Spacecraft	394
	References	399
11	**Testing and Calibration Instruments**	**403**
11.1	Calibration Coils	405
11.1.1	Field Compensation Systems	410
11.2	Magnetic Shielding	410
11.2.1	Magnetic Shielding Theory	410
11.2.2	Transverse Magnetic Shielding	412
11.2.3	Axial Magnetic Shielding	413

11.2.4	Flux Distribution	419
11.2.5	Annealing	420
11.2.6	Demagnetizing	421
11.2.7	Enhancement of Magnetic Shielding by Magnetic Shaking	421
	References	422

12 Magnetic Sensors for Nonmagnetic Variables — 425

12.1	Position Sensors	425
12.1.1	Sensors With Permanent Magnet	426
12.1.2	Eddy-Current Sensors	429
12.1.3	Linear Transformer Sensors	433
12.1.4	Rotation Transformer Sensors	437
12.1.5	Magnetostrictive Position Sensors	438
12.1.6	Wiegand Sensors	438
12.1.7	Magnetic Trackers	442
12.2	Proximity and Rotation Detectors	444
12.3	Force and Pressure	446
12.4	Torque Sensors	448
12.5	Magnetic Flowmeters	453
12.6	Current Sensors	454
12.6.1	dc/ac Hall and MR Current Sensors	455
12.6.2	Current Clamps	455
12.6.3	Magnetometric Measurement of Hidden Currents	455
12.7	Sensors Using Magnetic Liquids	455
	References	456

Magnetic Sensors, Magnetometers, and Calibration Equipment Manufacturers — 459

List of Symbols and Abbreviations — 467

About the Authors 473

Index 477

Preface

In recent years, several valuable books and review papers on various types of magnetic sensors, magnetometers, and their applications have appeared, but there is no up-to-date text fully covering this important, complex, and exciting field.

The most comprehensive of the existing sources is, undoubtedly, *Magnetic Sensors*, edited by R. Boll and K. J. Overshott [1]. Although it was published over a decade ago, many parts of that book are still very valuable; it covers most of the sensor types (except for resonance sensors and SQUIDs), but not many of the problems of magnetometers and their applications. *Solid State Magnetic Sensors*, written by Roumenin [2], covers only semiconductor sensors and SQUIDs. That book is oriented toward interesting but often rather impractical devices, such as magnetotransistors and carrier-domain magnetometers. *Hall Effect Devices*, written by R. S. Popovic [3], is still valuable but very specialized.

Recent—and the most valuable—review papers were written by Gibbs and Squire [4], Heremans [5], Rozenblat [6], Lenz [7], and Popovic et al. [8].

Although magnetic sensors are usually only briefly mentioned in books on sensors in general, books such as [9–11] give important comparisons to the properties of other sensor types. They also discuss many problems common to various sensor groups, for example, sensor construction, thermal stability, interfacing, and signal processing.

We wanted to make this book useful as a tool and thus have tried to keep a realistic, comprehensive, and practical approach.

- *The realistic approach*. This book does not concentrate on exotic methods and techniques, which often appear in conference papers, but have little or no practical impact. We compare the performances of individual sensors and magnetometers (both reported laboratory prototypes and commercially available devices) in relation to the various application areas.
- *The comprehensive approach*. This book covers not only sensors but also magnetometers, including multichannel and gradiometric systems. We discuss special issues, such as crosstalk and crossfield sensitivity, and cover problems of testing and calibrating magnetic sensors.
- *The practical approach*. Theory serves for understanding the working principles of real devices. This book gives information that will help in the selection of an existing sensor suitable for a particular application, or in the development of a customized sensor, if the available sensors are not suitable.

This book will explain the basic principles, available device parameters, and application rules and give extensive reference for further reading to the book's audience:

- University teachers and students (including postgraduates) mainly in the field of physics, geophysics, electrical engineering, and measurement and instrumentation;
- Instructors and participants in military courses (navigation, bomb location, weapon and vehicle detection);
- Users and designers of industrial sensors;
- Marketers and consultants in the field of sensor systems and industrial automation;
- Integrators and programmers of systems that contain sensors.

We assume that the reader has a general education in physics (at an undergraduate level).

Chapter 1 could have been entitled "Basics of Magnetism Revisited." The text does not replicate the introductory chapters on magnetism that can be found on the bookshelves of every physicist and electrical engineer. Instead, its aim is to reframe information that is—or once was—familiar to the reader in a new light. For more conventional and detailed introductions to magnetism, electromagnetics, and magnetic materials, we recommend several excellent textbooks [12–14].

This book may look inconsistent; in fact, the world of magnetic sensors itself is not very consistent. While some sensors (e.g., fluxgates) have been in development for more than half a century, others (e.g., magneto-optical and GMI sensors) are fresh. Although it is tricky, we decided also to cover these fast-developing areas. We even discuss sensors that we consider not particularly prospective, with the aim just to show their principles and point out their disadvantages. Some sensors (like magnetoresistors and Hall sensors) are available on the market, and users have a lot of support from manufacturers in the form of reference brochures and application manuals. In those cases, this book explains the principles and applications rather than the details of the manufacturing process. Induction and fluxgate sensors, on the other hand, usually are sold only as a part of magnetometers; it makes sense to develop those sensors custom-made for specific applications; this book may partly serve as a "cookbook" for that purpose.

The term *magnetometer* is used with two meanings: (1) a device for the measurement of the magnetic field (the more common meaning) and (2) an instrument for the measurement of magnetic moment (e.g., a rotating or vibrating sample magnetometer). There is no danger of confusion, because the type of instrument usually is specified (e.g., a proton magnetometer measures magnetic field).

The term *magnetic sensor* also has two meanings: (1) sensors that work on magnetic principles (the wider definition) and (2) sensors that measure the magnetic field. Although this book concentrates on magnetic field sensors and magnetometers, magnetic sensors for measurement of nonmagnetic variables are covered in Chapter 12.

One of the most important applications of precise magnetometers is in geophysics. For further reading, we recommend the excellent basic book *Applied Geophysics*, written by Telford et al. [15], and an extremely useful handbook on magnetic measurements and observatories by Jankowski and Sucksdorff [16].

Finally, we briefly mention two trends in the development of magnetic sensors: miniaturization and the use of new materials. Microtechnologies are already in use for fabrication of nonsemiconductor sensors: fluxgates, induction position sensors, and so forth. High-aspect ratio structures such as multilayer micromachined coils can be made using current micromachining technologies [17]. The applications of amorphous tapes and wires to all kinds of magnetic sensors are reviewed in [18]. The use of multilayers in GMR magnetoresistors is discussed in Chapter 4.

Acknowledgments

We thank our colleagues and professional friends for reading parts of the manuscript and for their valuable comments. We express particular gratitude to O. V. Nielsen, J. Hochreiter, and K. Záveta for their comments and suggestions. We thank our families for their support and patience.

Pavel Ripka
Prague, December 2000

References

[1] Boll, R., and K. J. Overshott (eds.), *Magnetic Sensors*, Vol. 2 of *Sensors*, Veiden, Germany: VCH, 1989.

[2] Roumenin, S., *Solid State Magnetic Sensors*, Lausanne, Switzerland: Elsevier, 1994

[3] Popovic, R. S., *Hall Effect Devices*, Bristol, England: Adam Hilger, 1991.

[4] Gibbs M. R. J., and P. T. Squire, "A Review of Magnetic Sensors," *Proc. IEEE*, Vol. 78, 1990, pp. 973–989.

[5] Heremans, J., "Solid State Magnetic Sensors and Applications," *J. Phys. D*, Vol. 26, 1993, pp. 1149–1168.

[6] Rozenblat, M. A., "Magnetic Sensors—State-of-the-Art and Trends," *Automation and Remote Control*, Vol. 56, No. 6, 1995, pp. 771–809.

[7] Lenz, J. E., "Review of Magnetic Sensors," *Proc. IEEE*, Vol. 78, 1990, pp. 973–989.

[8] Popovic, R. S., J. A. Flanagan, and P. A. Besse, "The Future of Magnetic Sensors," *Sensors and Actuators A*, Vol. 56, 1996, pp. 39–55.

[9] Norton, H. A., *Sensor and Analyzer Handbook*, London: Prentice Hall, 1982.

[10] Fraden, J., *AIP Handbook on Modern Sensors*, New York: American Institute of Physics, 1993.

[11] Pallas-Areny, R., and J. G. Webster, *Sensors and Signal Conditioning*, New York: Wiley, 1991.

[12] Jiles, D., *Introduction to Magnetism and Magnetic Materials*, London: Chapman & Hall, 1998.

[13] Kraus, J. D., *Electromagnetics*, 2nd ed., New York: McGraw-Hill, 1984.

[14] Crangle, J., *Solid State Magnetism*, New York: Van Nostrand Reinhold, 1991.

[15] Telford, W. M., L. P. Geldart, and R. E. Sheriff, *Applied Geophysics*, Cambridge, England: Cambridge University Press, 1990.

[16] Jankowski, J., and C. Sucksdorff, *Guide for Magnetic Measurements and Observatory Practice*, Warsaw: IAGA, 1996.

[17] Seidemann, V., M. Ohnmacht, and S. Büttgenbach, "Microcoils and Microrelays—An Optimized Multilayer Fabrication Process," *Sensors and Actuators A*, Vol. 83, Iss. 1–3, 2000, pp. 124–129.

[18] Meydan, T., "Application of Amorphous Materials to Sensors," *J. Magn. Magn. Mater.*, Vol. 133, 1995, pp. 525–532.

1

Basics
Hans Hauser and Pavel Ripka

This chapter begins with a phenomenology of magnetism and describes the physical quantities and their units. After a short overview about matter and magnetism, the emphasis is on magnetically ordered media, including anisotropy, nonlinearity, and hysteresis. Both the fundamental properties and the performance characterize selected materials that are important for sensor applications. The last section of the chapter examines the basics of sensor specifications.

1.1 Magnetic Quantities and Units

Magnetism is a comparatively new subject of science research. The main general reviews have been published since about the mid-twentieth century [1–12]. The first part of this chapter is based mainly on those references, which are well suited for further reading.

1.1.1 Roots of Magnetism

This section describes a phenomenology of magnetism, tracing the historical development as an introduction. The mathematical dualism between "vortex ring and double layer" (*Wirbelring und Doppelschicht*) leads to two different approaches to explaining the magnetic properties of matter. The units of the physical quantities develop consequently from empirically found electromagnetic laws.

1.1.1.1 Magnetic Force and Charges

First we consider magnetic forces based on dipoles. The magnetic behavior of matter also can be explained by moving electrical charges on an atomic level, and later we use that explanation for further considerations.

When one is dealing with magnetic fields, the basic physical experience is a fundamental phenomenon: the force **F** between magnetic poles. The Coloumb interaction for two pointlike poles with the pole strength $Q_{m,1,2}$ is

$$\mathbf{F} = \frac{Q_{m1} Q_{m2}}{4\pi\mu_0 r^2} \mathbf{e}_r \qquad (1.1)$$

Because single magnetic poles or magnetic charges have not been discovered yet (but they could have been generated at the beginning of the universe), one could imagine an approximation by very long, thin bar magnets in a distance r (unity vector $\mathbf{e}_r$; see Figure 1.1). The proportionality constant, in this case, $1/4\pi\mu_0$, depends on the actual system of units and is explained later.

The first attempt to explain the magnetic force was done by introducing a field **H** of magnetic lines of force, which is thought to originate at a magnetic pole, as shown in Figure 1.1. Consequently, the force acting on a magnetic pole in a field **H** is

$$\mathbf{F} = Q_m \mathbf{H} \qquad (1.2)$$

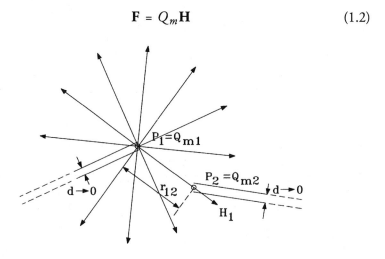

Figure 1.1 Long, thin magnets approximate magnetic charges of dipoles. (Only the field lines of the pole Q_{m1} are drawn.)

The magnetic field constant μ_0 was introduced in the Système International d'Unités (SI) to avoid a coefficient in (1.2).

Two magnetic poles (approximately a bar magnet with the length l) form a dipole with the Coulomb magnetic moment

$$\mathbf{j} = Q_m \mathbf{l} \tag{1.3}$$

experiencing a torque

$$\mathbf{T} = \mathbf{j} \times \mathbf{H} \tag{1.4}$$

in a homogeneous field and an additional force in an inhomogeneous field (e.g., $F_x = Q_m l \cdot \partial H/\partial x$ in x direction). Equations (1.3) and (1.4) describe the fundamentals of the first magnetic sensor. Some sources date the Chinese invention of the compass about 4,000 years in the past, and it remains the only technical application of magnetism until the nineteenth century.

1.1.1.2 Electrical Current and Field Strength

In 1820 Ørsted found a deviation of a compass needle near a current-carrying conductor. Ampère assumed from those results that a magnetic field H can also originate from moving electrical charges. He formulated the basic law of magnetomotive force:

$$NI = \oint_s \mathbf{H} \cdot \mathbf{ds} \tag{1.5}$$

which equals the ring integral of H over a closed path s to the ampere-windings (N conductors with current I). Therefore, the field H is measured in amperes per meter (A/m). With the newton (N) unit of force, kilogram meter per second-squared (kgm/s^2) = volt-ampere second per meter (VAs/m), (1.2) yields that Q_m is measured in Volt-seconds (Vs) in analogy to the electrical charge.

The law of Biot-Savart is used to calculate the field of general conductor arrangements:

$$d\mathbf{H} = \frac{I}{4\pi r^2} \mathbf{ds} \times \mathbf{e}_r \tag{1.6}$$

The current I flowing through a conductor part $d\mathbf{s}$ causes a field $d\mathbf{H}$ in a distance $\mathbf{r}$. Using (1.5) and (1.6), respectively, the circumferential field of a long conductor is $H = I/2\pi r$, the field along the axis of a long cylindrical coil (N turns, length $l \gg$ diameter d) is $H = NI/l$, and along the center x axis of a current loop with diameter d, it is

$$H = \frac{Id^2}{\sqrt{(d^2 + 4x^2)^3}} \tag{1.7}$$

The magnetic field can be visualized as lines of force. The tangent gives the direction, and their density (distance) determines the field strength. That is illustrated by Figure 1.2 for Ampère's law. Consider that (1.5) through (1.7) do not take into account the displacement current and thus are applicable only for slowly varying fields.

1.1.2 Magnetic Field and Matter

In general, the reaction of matter on a magnetic field $\mathbf{H}$ contributes to the induction

$$\mathbf{B} = \underline{\mu} \cdot \mathbf{H} \tag{1.8}$$

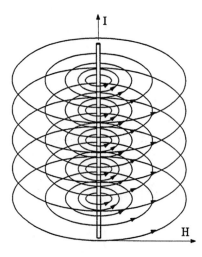

Figure 1.2 Magnetic lines of force H of a conductor with current I.

The permeability tensor $\underline{\mu}$ has to consider material properties like anisotropy, nonlinearity, inhomogeneity, and hysteresis. Only in very special cases is it a simple constant.

1.1.2.1 Flux Density

The basic law related to induction (see also Chapter 2),

$$V = -N\mathbf{A} \cdot \frac{\partial \mathbf{B}}{\partial t} \tag{1.9}$$

was found in principle by Faraday and Lenz: If the induction changes with time t in an area A enclosed by N turns of a conductor, a voltage V is generated. The magnetic flux

$$\phi = \int_A \mathbf{B} \cdot d\mathbf{A} \tag{1.10}$$

through this area is therefore measured in Volt-seconds (Vs), or weber (Wb) and the induction $\mathbf{B}$ is also called flux density with the unit Vs/m^2, or tesla (T). The definition of $\mathbf{B}$ is generally done by the Lorentz force:

$$\mathbf{F} = Q_e \mathbf{v} \times \mathbf{B} \tag{1.11}$$

An electrical charge Q_e moving with velocity $\mathbf{v}$ experiences a deflection force perpendicular to both $\mathbf{B}$ and $\mathbf{v}$ directions. Besides charge carrier scattering, this is the basic law for Hall sensors (see Chapter 5) and indirectly for every solid state magnetic sensor, because matter consists of moving charges. The torque

$$\mathbf{T} = I\mathbf{A} \times \mathbf{B} \tag{1.12}$$

on a conductor loop is proportional to the circuit area $\mathbf{A}$ and the current I; their product is the magnetic moment

$$\mathbf{m} = I\mathbf{A} \tag{1.13}$$

Consequently, the unit of μ is Vs/Am. In vacuum the effect of both fields $\mathbf{H}$ and $\mathbf{B}$ is in principle the same, but they have different units and values:

$$\mathbf{B} = \mu_0 \mathbf{H} \tag{1.14}$$

The magnetic field constant is $\mu_0 = 4\pi \cdot 10^{-7}$ Vs/Am or henry per meter (H/m) and is only a consequence of the units used. In the former centimeter-gram-second (cgs) unit system, both **B** and **H** had the same dimension.

In general, the energy per unit volume of a magnetic field is

$$W = \int \mathbf{H} \cdot d\mathbf{B} \tag{1.15}$$

In the case of linear dependence, for example, (1.14), the energy is

$$W = \frac{\mathbf{B} \cdot \mathbf{H}}{2} \tag{1.16}$$

The flux density is the only magnetic vector quantity that can be measured directly by using (1.11), and we could propose that all the other magnetic fields should be obsolete. But magnetism is a complex chapter of physics, and we need other magnetic quantities to find a practical approach to what we are trying to understand as "reality."

1.1.2.2 Polarization, Magnetization, Permeability, and Susceptibility

The flux density **B** consists of two contributions: the vacuum induction $\mu_0 \mathbf{H}$ and the polarization **J** of matter. With

$$\mathbf{B} = \mu_0 \mathbf{H} + \mathbf{J} \tag{1.17}$$

we find the same units for **J** and **B**. The basic idea is that there are magnetic dipoles with the average moment **j** in the volume V and that

$$\mathbf{J} = \frac{d\mathbf{j}}{dV} \tag{1.18}$$

is the density of those dipoles. As stated before, there are no real dipoles[1] because of the lack of isolated magnetic charges. A completely different

---

1. The difference between a fictitious dipole (producing only a dipole momentum) and a real dipole is that the latter can be divided into two monopoles. But we find more "dipoles" only if a magnetic compass needle is broken into arbitrary small parts. Therefore, (1.1) through (1.4) also are fictitious as long as the magnetic charge Q_m is not found. Magnetic field in the vicinity of the end of a long object has a typical "monopole" signature; the term monopole is used in geophysics (see Section 10.7).

approach to the magnetism of matter results from atomic physics: The moving unit charge ($e = -1.60 \cdot 10^{-19}$ As) of the electrons (mass $m_e = 9.11 \cdot 10^{-31}$ kg) are treated as currents—due to orbit and spin—producing a magnetic moment. Using Planck's constant ($h = 6.63 \cdot 10^{-34}$ Ws^2), the unit of the quantized orbit moment is the Bohr magneton,

$$\mu_B = \frac{eh}{4\pi m_e} \approx 9.27 \cdot 10^{-24} \text{Am}^2 \tag{1.19}$$

The quantum number of electron spin is 1/2, which leads to twice the gyromagnetic ratio (magnetic moment to mechanical impulse moment) compared to the orbit. When atoms condense to form a solid-state crystal, the orbits are fixed to a large extent with the atomic bond. Therefore, mainly the resulting spin moments—which are partially reduced by next-neighbor interactions—can be rotated by an applied field and contribute to the magnetization

$$\mathbf{M} = \frac{d\mathbf{m}}{dV} \tag{1.20}$$

It is the density of the average magnetic moments **m** that describes the macroscopic magnetic behavior equivalently to the polarization

$$\mathbf{J} = \mu_0 \mathbf{M} \tag{1.21}$$

and (1.17) yields

$$\mathbf{B} = \mu_0(\mathbf{H} + \mathbf{M}) \tag{1.22}$$

The material properties are also often characterized by the relative permeability tensor μ_r:

$$\mathbf{B} = \mu_0 \mu_r \mathbf{H} \tag{1.23}$$

Therefore, the relation between **M** and **H** can be expressed as

$$\mathbf{M} = (\mu_r - \underline{1}) \cdot \mathbf{H} = \kappa \cdot \mathbf{H} \tag{1.24}$$

where κ is the susceptibility tensor. Using (1.12), (1.13), (1.14), (1.20), and the general expression for energy $E = \int T d\varphi$ (φ is the angle between **A** and

B), we find the magnetostatic energy density (Zeemann energy) of a magnetized body to be

$$W_H = -\mu_0 \mathbf{M} \cdot \mathbf{H} \tag{1.25}$$

It should be stated that the atomic nucleus also has a magnetic moment (only about 1/2,000 of the magnetic electron spin moment). The dependence of its resonance frequency on an applied field is used for materials characterization and diagnostic techniques by nuclear magnetic resonance (NMR) (see Chapter 7).

Although the discussion about the aspects of magnetism is still in progress, especially about the nature of magnetic fields, the state-of-the-art electromagnetic theory is expressed by Maxwell's [13] equations:

$$\nabla \times \mathbf{H} = \mathbf{J}_c + \frac{\partial \mathbf{D}}{\partial t} \tag{1.26}$$

$$\nabla \times \mathbf{E} = -\frac{\partial \mathbf{B}}{\partial t} \tag{1.27}$$

$$\nabla \cdot \mathbf{D} = \rho_e \tag{1.28}$$

$$\nabla \cdot \mathbf{B} = 0 \tag{1.29}$$

Those equations [2] describe the relations between magnetic field **H**, current density $\mathbf{J}_c$, electrical field **E**, magnetic flux density **B**, dielectric displacement **D**, electrical charge density ρ_e, and time t. Applying the integration law of Stokes ($\int_A \nabla \times \mathbf{X} = \oint_s \mathbf{X} d\mathbf{s}$) for a vector **X** in an area A with the circumference s, (1.26) and (1.27) yield (1.5)—neglecting the displacement current density $\partial \mathbf{D}/\partial t$—and (1.9), respectively.

1.1.3 Units and Magnitudes of Magnetic Fields

Former cgs units are still found in magnetism literature. Table 1.1 lists the main conversions from cgs to SI.

There is a wide range of magnitudes of magnetic fields in the universe. To give an idea of the variety of measurement tasks, the following examples illustrate the full scale of flux density.

- Biomagnetic fields (brain, heart): $B \approx 10$ fT (1 pT)
- Galactic magnetic field: $B \approx 0.25$ nT

Table 1.1
Conversion of cgs Units to SI Units

Quantity	cgs-to-SI Conversion
Magnetic field intensity, H	1 Oe (Oersted) $\triangleq$ 1,000/4π A/m
Flux density, B	1 G (Gauss) $\triangleq$ 10^{-4} T
Magnetization, M	1 emu/cm^3 $\triangleq$ 1,000 A/m
Polarization, J	$J_{cgs} = J_{SI}/4\pi$
Permeability, μ	$\mu_{cgs} = \mu_{SI}/\mu_0$
Susceptibility, κ	$\kappa_{cgs} = \kappa_{SI}/4\pi$

- In the vicinity (distance about 1m) of electrical household appliances: $B \approx 600$ nT
- Earth's magnetic field (magnetic south is near geographic north): $B \approx 60 \ \mu$T
- Electrical power machines and cables (distance about 10m): $B \approx 0.1 \ldots 10$ mT
- Permanent magnets (surface): $B \approx 100$ mT $\ldots$ 1T
- Laboratory magnet: $B \approx 2.5$T
- NMR tomography (superconducting magnets with 1m diameter): $B \approx 4$T
- Nuclear fusion experiment: $B \approx 10 \ldots 20$T
- Short-pulse laboratory fields: $B \approx 60 \ldots 100$T
- White dwarf (small star with a density of 1,000 kg/cm^3): $B \approx 1$ kT
- Pulsar (tiny star with a density of 10^{10} kg/cm^3): $B \approx 100$ MT

The origin of the planetary magnetic fields is still under discussion. Current dynamo theories cannot explain both high efficiency and reversal of polarity. Recently, a new theory seems to describe successfully those open questions by rotating electric dipole domains [14].

1.2 Magnetic States of Matter

The effects of a magnetic field on matter are diverse. We touch on only diamagnetism and paramagnetism and emphasize the most important phenomena of magnetically ordered media: ferromagnetism and ferrimagnetism.

1.2.1 Weakly Magnetic Materials

If the magnetic field has almost no macroscopic effect on a material, one could believe that it is "nonmagnetic" ($|\kappa| \ll 1$). In fact, an applied field seems to "shine through" diamagnetic materials and is only weakly modified ("falsened") in paramagnetic matter.

1.2.1.1 Diamagnetism and Superconductivity

If a magnetic field is applied to an atom, an electrical field is induced following (1.27). The forces acting on the electrical charges accelerate the whole electron shell, and the result can be formally described by an additional rotational velocity (Larmor precession). The corresponding magnetic moment is antiparallel to the applied field. With Z electrons per atom and N atoms per cubic meter (effective equilibrium distance $2r$), the susceptibility is

$$\kappa = -\mu_0 N \frac{e^2 Z \langle r^2 \rangle}{6 m_e} \qquad (1.30)$$

The values of κ usually are very small ($\kappa \approx -10^{-5}$). Therefore, the atomic diamagnetic behavior is dominated by other forms of magnetism. A different form of diamagnetism can be observed in metals, where the conduction electrons are moving in helical trajectories according to (1.11). Important diamagnetic materials are the noble gases, semiconductors, water, and many metals (e.g., copper, zinc, silver, cadmium, gold, mercury, lead, and bismuth). Diamagnetic materials are used for the design of mechanical parts that must not disturb the magnetic field to be measured.

The magnetic behavior of superconductors (see Chapter 8) can be described as a strong form of diamagnetism. Because of the Meissner-Ochsenfeld effect, the material is completely shielded ($B \approx 0$) by nondissipative surface currents below a critical temperature and below a critical applied field; and $\kappa \approx -1$, in this case, according to (1.22) and (1.24). If the material consists of nonsuperconducting parts (e.g., grain boundaries, inclusions) or the applied field exceeds local critical values, B partially penetrates the material and $|\kappa|$ decreases.

1.2.1.2 Paramagnetism

In the case of a resulting magnetic moment **m** of the atoms that are not compensated, the magnetic field causes an alignment against the energy $skT/2$ of thermal motion [degree of freedom s ($s = 3$ for the monoatomic case) and Boltzmann constant $k = 1.38 \cdot 10^{-23}$ Ws/K]. In general, M is a nonlinear Brillouin function of H, but in the case of $mH/kT \ll 1$—high temperatures ($T \approx 300$ K) or low fields ($\mu_0 H \approx 1$ T), as is valid for most standard technical applications—the susceptibility is then

$$\kappa = \mu_0 \frac{Nm^2}{3kT} \qquad (1.31)$$

and the order of magnitude is $\kappa \approx +10^{-3}$. Another kind of paramagnetic behavior can be observed in metals, where an applied field leads to different energy levels for different orientations of electron spin. Establishing a common Fermi energy level[2] reduces those spin populations having a negative component with respect to the applied field at the expense of the other spin population (temperature-independent Pauli paramagnetism). Examples of paramagnetic materials are several gases, carbon, and several metals (e.g., magnesium, aluminum, titanium, vanadium, molybdenum, palladium, and platinum). One of the magnetic sensor applications of paramagnetic materials is the electron spin resonance (ESR) magnetometer (see Chapter 7).

1.2.2 Ferromagnetism and Ferrimagnetism

Solid state matter can exhibit a unique feature in special cases of atomic neighborhood order: spontaneous magnetization. Atomic moments (usually spin moments) can be aligned parallel (ferromagnetism) or antiparallel (antiferromagnetism) due to quantum-mechanical exchange coupling.[3] Ferrimagnetism is characterized by two or more sublattices with an antiparallel orientation of their magnetic moments that do not cancel each other. The possible spin alignments are shown in Figure 1.3. Other examples of magnetic behavior are metamagnetism, helimagnetism, or superparamagnetism.

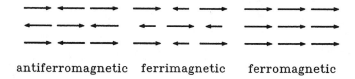

antiferromagnetic   ferrimagnetic   ferromagnetic

Figure 1.3 Spin alignment due to exchange coupling.

---

2. The Fermi level is the maximum energy of the conduction electrons (with probability).
3. Exchange coupling is based on electrostatic interaction and lowers the system's energy by aligning the uncompensated moments.

1.2.2.1 Spontaneous Magnetization

Only few elements and compounds are ferromagnetic at temperatures above 0°C (e.g., iron, nickel, cobalt, gadolinium, and chromium dioxide). Ferrites consist, in most cases, of iron oxide with oxides of manganese, nickel, copper, magnesium, yttrium, and so on (spinels and garnets). The main conditions for establishing a spontaneous magnetization

$$M_s = NmL\left(\mu_0 m \frac{H + \mu_0 w M_s}{kT}\right) \qquad (1.32)$$

below the Curie temperature $T_c = wN(\mu_0 m)^2/3k$ [the simplified presentation is a result of the classical ferromagnetism theory of Pierre Weiss (1907) by means of the Langevin function[4] $L(x) = \coth(x) - 1/x$ to demonstrate the principle] are summarized as follows:

- Resulting atomic magnetic moment m (transition elements: uncompensated spins in 3d or 4f shells);
- High atomic density N (condensed matter);
- Small range of the ratio of the interatomic distance to the diameter of the shell containing uncompensated spins (crystalline or amorphous solid state matter), leading to a large hypothetic exchange field proportional (Weiss constant w) to M_s: $H_E = \mu_0 w M_s$.

Many efforts have been made to find a quantum theory of magnetism, and there are two mutually exclusive models: The localized moment model works satisfactorily for rare earth metals, while the itinerant electron model describes quite well the magnetic properties of the 3d transition metals and their alloys. Figure 1.4 shows the Bethe-Slater curve, depicting the exchange coupling (represented by w) versus the ratio of the atomic distance to the uncompensated shell diameter, d_a/d_u. The larger w, the larger T_c is. The atomic spacing of the antiferromagnetic manganese can be enlarged to become ferromagnetic, for example, in manganese-aluminum-chromium (MnAlCr) or manganese-copper-tin (MnCuSn) (Heusler alloys).

At the Curie temperature, the energy kT_c of thermal motion equals the exchange energy, and M_s becomes zero. The susceptibility is then given by the Curie-Weiss law $\kappa = \mu_0 m N/3k(T - T_c)$ (paramagnetic behavior; see

---

4. The Langevin function is valid only in the classical limit: If spin quantization is taken into account, integration is replaced by summation, which leads to a more realistic but complicated Brillouin function.

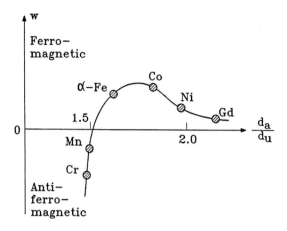

Figure 1.4 Exchange energy versus d_a/d_u.

Section 1.2.1.2, but with a temperature bias T_c). The solution of (1.32) is shown graphically in Figure 1.5 for cobalt, iron, and nickel.

1.2.2.2 Magnetic Anisotropy

Anisotropy is an important feature utilized in many sensor applications. This subsection describes only the most important kinds of anisotropy based on atomic or geometric aspects: magnetocrystalline, strain, and shape anisotropies. Other forms are surface anisotropy, exchange anisotropy, and diffusion anisotropy.

Magnetocrystalline Anisotropy

The electronic orbits and therefore the magnetic orbital moments are fixed in certain crystallographic directions. The magnetic coupling of spin and orbit moments is the reason that the spins are bound to certain directions in absence of an applied field H. To rotate the spins out of those easy directions of minimum energy, it is necessary to work against the magnetocrystalline anisotropy energy W_C. Cubic crystalline materials have a high degree of symmetry. Therefore, it is possible to describe the complex anisotropic behavior by a power expansion with only few constants (K_0 summarizes only direction-independent contributions), K_1 and K_2. For cubic crystals, we can write the energy density as

$$W_C = K_0 + K_1(\alpha_1^2\alpha_2^2 + \alpha_2^2\alpha_3^2 + \alpha_3^2\alpha_1^2) + K_2(\alpha_1^2\alpha_2^2\alpha_3^2) + \ldots \tag{1.33}$$

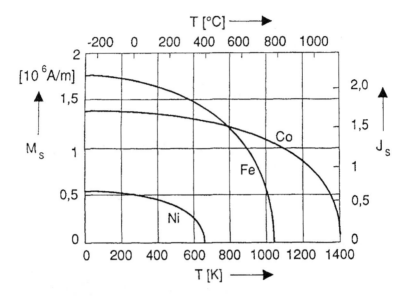

Figure 1.5 Dependence of M_s on temperature for the transition elements cobalt, iron, and nickel.

where α_i are the cosines of $\mathbf{M}_s$ with respect to the crystal axes [100], [010], and [001]. For hexagonal crystals, W_C depends only on the angle φ between $\mathbf{M}_s$ and the main axis [0001] of the crystal [in absence of anisotropy in the (0001) plane]:

$$W_C = K_0 + K_1 \sin^2\varphi + K_2 \sin^4\varphi + \ldots \quad (1.34)$$

It is also usual to define a fictitious anisotropy field H_k, which in the simplified case of uniaxial anisotropy and only one anisotropy constant can be written as

$$H_k = \frac{2K_1}{\mu_0 M_s} \quad (1.35)$$

The magnetocrystalline anisotropy constants can be determined from the torque $T = \partial W(\varphi)/\partial \varphi$ acting on an oriented single crystal sample (e.g., sphere, circular plate).

Figure 1.6 illustrates the energy surfaces (areas)—three-dimensional graphs of (1.33)—for different K_i. Iron has six easy directions (three easy axes) parallel to the edges, and nickel has eight easy directions (four easy

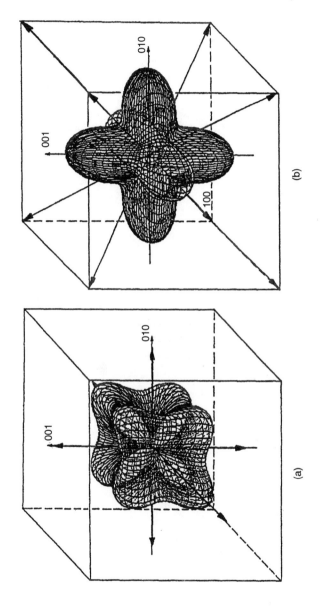

Figure 1.6 Energy areas of W_C for (a) iron ($K_1 = +48$ kJ/m^3, $K_2 = -9$ kJ/m^3) and (b) nickel ($K_1 = -4.5$ kJ/m^3, $K_2 = +2$ kJ/m^3).

axes) parallel to the space diagonals of the cubic crystals. A combination of materials with different signs of K_i can lead to a large number of easy directions or to an almost complete suppression of anisotropy, which is important for achieving high permeability (e.g., Permalloy).

Figure 1.7 shows the projection of $\mathbf{M}_s$ in the direction of the applied field $\mathbf{H}$ [energy W_H of (1.25)] for different materials and directions of $\mathbf{H}$. The curves can be calculated in principle by tracing the minima of the total energy $W_T = W_C + W_H$.

Distinct magnetocrystalline anisotropy can be macroscopically found only in single crystalline or grain-oriented materials. Otherwise, the grains are oriented randomly with respect to their easy directions, which leads to macroscopically isotropic behavior. But the magnetization process of polycrystals reflects the values of the anisotropy constants.

Magnetostriction and Strain Anisotropy

An applied field can induce a small anisotropy of the bonding forces and therefore a change of the lattice constants. The macroscopic result of that

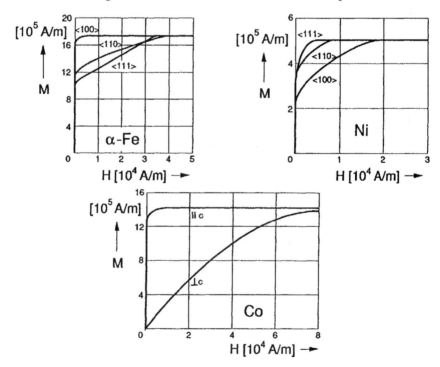

Figure 1.7 Projection of $\mathbf{M}_s$ in the $\mathbf{H}$ direction for different ferromagnetic crystals: α iron, nickel, and cobalt.

further manifestation of spin-orbit coupling is a relative length change, $\lambda_{\alpha,\beta} = \Delta l/l$, of the material at a changing state of magnetization (seldom is there volume change) in a measurement direction, defined by the direction cosines β_i with respect to the axes of a cubic crystal:

$$\lambda_{\alpha,\beta} = \frac{3}{2}\lambda_{100}\left(\alpha_1^2\beta_1^2 + \alpha_2^2\beta_2^2 + \alpha_3^2\beta_3^2 - \frac{1}{3}\right) \quad (1.36)$$

$$+ 3\lambda_{111}(\alpha_1\alpha_2\beta_1\beta_2 + \alpha_2\alpha_3\beta_2\beta_3 + \alpha_3\alpha_1\beta_3\beta_1)$$

where λ_{ijk} are the magnetostriction constants (maximum $\Delta l/l$ in the given crystallographic direction [ijk], occurring after a randomly demagnetized state). On the contrary, a mechanical stress σ can cause a change of the magnetic state that is established to support the strain by magnetostriction. There is an influence on the easy directions, described by the strain (magnetoelastic) anisotropy energy density

$$W_S = -\sigma(\lambda_{\alpha,\beta} + \lambda_0) \quad (1.37)$$

where λ_0 is a direction-independent constant and β_i are, in this case, the direction cosines of the applied stress σ with respect to the crystal axes. The effect of W_S on the easy directions (the total anisotropy energy is then $W_A = W_C + W_S$) is shown in Figure 1.8 for iron and nickel. If λ_{ijk} has a negative sign, the [ijk] direction becomes harder (higher anisotropy energy) at tensile stress, and vice versa. It is clear, therefore, that applied mechanical stress can change the magnetic state of a ferromagnetic medium. Inner stresses, for example, those caused by mechanical treatment, also will strongly influence the magnetic behavior of materials.

In the case of polycrystalline materials of isotropic orientation distribution of the crystal axes, the isotropic saturation magnetostriction is

$$\lambda_s = \frac{2}{5}\lambda_{100} + \frac{3}{5}\lambda_{111} \quad (1.38)$$

In the special case of "isotropic" magnetostriction ($\lambda_{100} = \lambda_{111} = \lambda_s$), the relative length change $\lambda_{\alpha,\beta} = \lambda_\varphi$ depends only on the angle φ between σ and $\mathbf{M}_s$:

$$\lambda_\varphi = \frac{3}{2}\lambda_s\left(\cos^2\varphi - \frac{1}{3}\right) \quad (1.39)$$

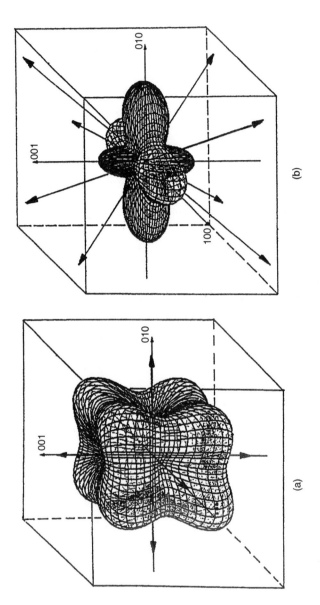

Figure 1.8 Magnetocrystalline and strain anisotropy energy areas (compressive stress in the [001] direction, for a given value of $\sigma < 0$) for (a) iron ($\lambda_{100} = +21 \cdot 10^{-6}$, $\lambda_{111} = -21 \cdot 10^{-6}$) and (b) nickel ($\lambda_{100} = -46 \cdot 10^{-6}$, $\lambda_{111} = -24 \cdot 10^{-6}$).

Shape Anisotropy

Shape effects play an important role in the design of magnetic sensors. A spontaneously magnetized body exhibits fictitious magnetic charges—north (+) and south (−) poles—at the surface, where the normal component of **M** is discontinous. As stated at the beginning of this chapter, those magnetic poles are considered the origin of a magnetic field[5] H_d, proportional to **M** (Figure 1.9) and oriented antiparallel to the magnetization within the magnetized body. The energy density of the so-called demagnetizing field of a general ellipsoid is

$$W_D = \frac{\mu_0 M_s^2}{2}(D_a \alpha_1^2 + D_b \alpha_2^2 + D_c \alpha_3^2) \tag{1.40}$$

where D_a, D_b, and D_c are the demagnetizing factors ($D_a + D_b + D_c = 1$) and α_i are the direction cosines of the magnetization with respect to the ellipsoid axes a, b, c. Formulas for demagnetizing factors of ellipsoids and cylinders are given in Chapter 2.

The energy areas of (1.40) show that W_D is minimum in the axis of a bar or in the plane of a disc (Figure 1.10). The magnetic fields are only homogeneous in an ellipsoid. In this case, the demagnetizing matrix

$$\underline{D_d} = (D_a, D_b, D_c) \tag{1.41}$$

defined by

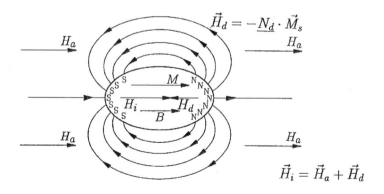

Figure 1.9 Demagnetizing field of a magnetized ellipsoid.

---

5. Because real magnetic charges do not exist, the field H_d originates from the moving electrical charges, producing the magnetic moments of the density M.

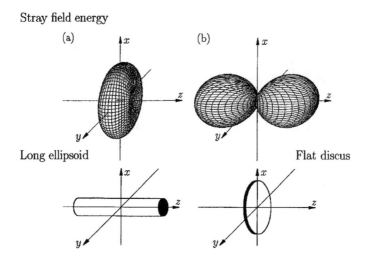

Figure 1.10 Demagnetizing stray field energy of ellipsoid approximations: (a) long bar and (b) flat disc.

$$\mathbf{H}_d = -\underline{D_d} \cdot \mathbf{M} \qquad (1.42)$$

can be calculated analytically [15]. Otherwise, it has a very complex structure and depends on the actual susceptibility. If the field **H** is applied, the inner, effective field

$$\mathbf{H}_i = \mathbf{H} + \mathbf{H}_d \qquad (1.43)$$

differs from **H** by $\mathbf{H}_d$ (lower if $\kappa > 0$ and greater if $\kappa < 0$).

If we visualize the effect of shape anisotropy in ferromagnetic materials, the total anisotropy energy $W_A = W_C + W_D$ of a disc of grain-oriented silicon iron (a material widely used for transformer cores and many other soft magnetic applications) is shown in Figure 1.11. Due to the large value of M_s, W_D overcomes W_C by about a hundred times if the magnetization is perpendicular to the disc plane.

1.2.2.3 Domain Structure

In Section 1.2.2.2, it was assumed that a ferromagnetic body exhibits spontaneous magnetization. But only some hard magnetic materials appear macroscopically magnetized in absence of an applied field. If the sample were magnetized with $\mathbf{M}_s$, the demagnetizing stray field energy W_D would be very high, as discussed previously. Therefore, $\mathbf{M}_s$ occupies the available easy

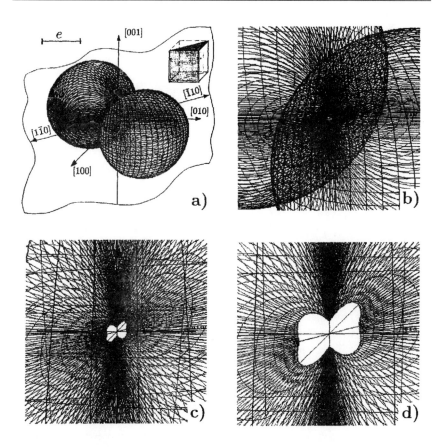

Figure 1.11 The spatial distribution of the total energy of ellipsoidal-shaped silicon-iron (FeSi) discs; a:b:c = 10:10:1, K_1 = 29 kJ/m^3, K_2 = –9 kJ/m^3: (a) energy scale e = 500 kJ/m^3; (b), (c), (d) with reduction of energy scale to 33% for each picture, the total energy reveals the magnetocrystalline energy in the sheet plane.

directions—given by the magnetocrystalline and strain anisotropy energy minima—to reduce the pole strength and herewith $\mathbf{H}_d$ and, consequently, W_D, as shown in Figure 1.12.

The volumes, magnetized uniformly with $\mathbf{M}_s$ in an easy direction, are the magnetic domains, divided by the domain walls. The walls also need energy to be established, which yields a lower domain size limit. The wall energy consists mainly of magnetocrystalline anisotropy and exchange energy (the contributions of stray fields are usually neglected). The wall thickness is a tradeoff to minimize the wall energy.

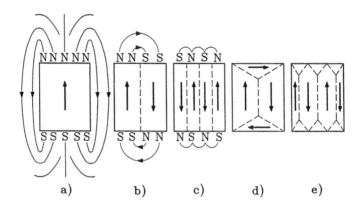

Figure 1.12 Reduction of stray fields (a, b, c) by orienting the spins in domains, magnetized uniformly in easy directions; (d) $\lambda_s \approx 0$; (e) compromise structure, $|\lambda_s| > 0$.

Domain structures and domain walls can have a very complex nature, caused by magnetostriction (both inner and applied stresses) and microscopic stray fields [16]. For instance, the domain structure of Figure 1.12(d) is favorable only if the magnetostriction is very low and no strain occurs. Otherwise, a compromise structure like that in Figure 1.12(e) is established.

Furthermore, the domain walls should provide a steady normal component of M_s, but in small volumes that law can be violated to reduce inner stray fields and stresses. The orientations of the easy directions may also differ for each crystal grain, or they can be randomly distributed in amorphous materials ($K_i \approx 0$). A typical width of a domain is 10–100 μm and a typical wall thickness is 10–100 nm, but much larger domains may exist in grain-oriented materials, thin films, or wires. Figure 1.13 shows a simplified 180-degree Bloch-type domain wall, where the magnetic moments are gradually rotated from one easy direction to the antiparallel one.

1.2.2.4 Magnetization Process

Figure 1.14 shows schematically the process of increasing the macroscopic magnetization by application of a field on a ferromagnetic material. In absence of **H**, the moments of domains are equally distributed along the four easy directions. At weak fields, the domain walls are moving, to increase the volume of the domains with a positive magnetization component with respect to **H**, at the expense of the other domains. That motion is caused by the torque acting on the magnetic moments within the tiny volume of the wall, where exchange energy and anisotropy energy are in a fine balance. These easy domain wall displacements are the reason for the high permeability of soft magnetic materials and determine their technical applications. Several

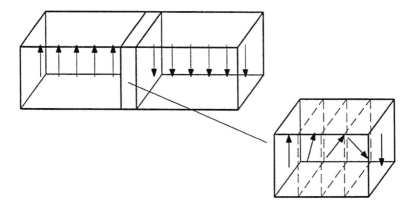

Figure 1.13 Charge of spin orientation (spin rotation) within a Bloch wall between antiparallel magnetized domains.

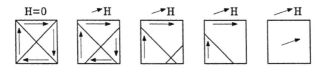

Figure 1.14 The effect of an applied field on a simplified domain structure: Rotation of the domain's magnetization starts before the end of the wall displacements only if a strong stray field can be avoided.

magnetic sensor concepts are based on the behavior of domain walls (see Chapter 9).

The increase of bulk magnetization by domain wall displacements in an applied field (energy W_H) is balanced by the shape anisotropy energy W_D and hindered by, for example, local stray fields and magnetostriction. If W_H is large enough to compete with the magnetocrystalline anisotropy energy, the domain's magnetization is rotated toward the field direction, tracing the minima of the total energy $W_T = W_C + W_S + W_D + W_H$. That simplified magnetization process for single crystals has been described by the phase rule of Néel, Lawton and Stewart [17], and is the basis for calculating magnetization curves of anisotropic materials [18].

In the case of nonideal material structures with grain boundaries, nonmagnetic inclusions, or cracks, for example, the domain walls are pinned, reducing the stray fields of those pinning sites and covering as many imperfections as possible. Figure 1.15 illustrates the mechanism of irreversible[6] domain wall displacements.

---

6. Rotation processes can also be irreversible in the case of materials with restrained domain structure (e.g., permanent magnets, fine particles, thin films).

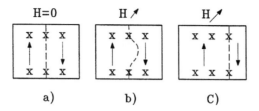

Figure 1.15 Reversible (a → b) and irreversible domain wall displacements (b → c).

Pinned at two imperfections in this case, the wall is moving reversibly at low H. Further field increase forces the wall to jump until it is pinned again by other impurities. The path of the domain wall differs for each cycle, and the so-called irreversible Barkhausen jumps are energy dissipative due to spin relaxation, magnetomechanical interaction, and microscopic eddy currents. Furthermore, the induction within the moving domain wall's volume because of spin rotation gives rise to the Barkhausen noise. Therefore, we find that the magnetization process depends on the magnetic history experienced by the material: The feature of hysteresis represents the widely known picture of magnetism.

1.2.2.5 Magnetization Curve

The initial magnetization curve is obtained when a field is applied on a previously completely demagnetized sample. Starting with reversible domain wall displacements at weak fields and proceeding with irreversible Barkhausen jumps, the saturation[7] M_s is finally reached by processes of magnetization rotation against the anisotropy energy. After H is reduced to zero, the sample remains magnetized at the remanence M_r. The magnetization becomes zero at the coercive field strength $H = -H_c$. The upper branch of the major hysteresis loop is completed at $M = -M_s$, and the lower branch is obtained by an analog procedure (Figure 1.16).

Demagnetization can be done by several methods, which yield different results. Heating the sample over T_c is a perfect erasure of the magnetic history. Applying a decaying ac field is the most usual way (Figure 1.17), but it yields no random distribution of the domain magnetization over the easy directions. An improvement is the demagnetization by a decaying rotating ac field, a technique used in geophysics. Demagnetization by mechanical

---

7. The saturation magnetization is almost identical to M_s. It differs from the spontaneous magnetization only by the effect of the field on aligning the magnetic moments ($H \to \infty$) against thermal disorder, as described by (1.32), which is very small at technical field values.

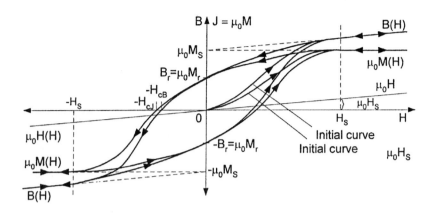

Figure 1.16 Hysteresis of $\mu_0 M(H)$ and $B(H)$.

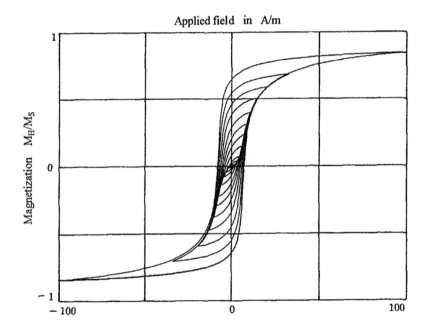

Figure 1.17 Demagnetization by application of a decaying ac field.

shock waves happens sometimes accidentially to permanent magnets (decaying stress oscillations changes the easy directions, which causes domain wall displacements of decreasing amplitude).

The dependence of B on H differs from $\mu_0 M(H)$ by $\mu_0 H$ (see Figure 1.16). The total permeability or amplitude permeability

$$\mu_t = \frac{B_t}{H_t} \tag{1.44}$$

depends on the operation point B_t, H_t. Important permeabilities are the initial permeability

$$\mu_i = \lim_{\substack{H \to 0 \\ B \to 0}} \frac{B}{H} \tag{1.45}$$

which is represented by the tangent line on $B(H)$ at the origin, the maximum permeability

$$\mu_{max} = \left. \frac{B}{H} \right|_{max} \tag{1.46}$$

which is the ascent of the tangent line from the origin to $B(H)$, and the reversible permeability

$$\mu_{rev} = \lim_{\Delta H \to 0} \left. \frac{\Delta B}{\Delta H} \right|_{B_t, H_t} \tag{1.47}$$

which is the limit of the superposition permeability for small ac fields at the operation point B_t, H_t

$$\mu_\Delta = \left. \frac{\Delta B}{\Delta H} \right|_{B_t, H_t} \tag{1.48}$$

The differential permeability

$$\mu_{diff} = \frac{dB}{dH} \tag{1.49}$$

is the derivative of the $B(H)$ curve.

Following (1.15), the energy loss W_L per unit volume and cycle that is finally converted to heat corresponds with the area of the hysteresis loop. To yield the power loss, we have to multiply W_L by the frequency f. Losses are distinguished between static hysteresis losses and frequency-dependent losses. The latter are distinguished between normal (due to bulk dB/dt) and anomal (due to additional dB/dt of large domain wall displacements) eddy-current losses.

To reduce the effects of hysteresis, like remanence and coercivity, the magnetization curve can be idealized by superposing an ac field of higher frequency (Figure 1.18), averaging the magnetization by tracing the centers of the minor loops. This technique, called magnetic shaking, is successfully used to improve the parameters of magnetic shielding (see Chapter 11) and in magnetic recording of analog signals. The total loss corresponds to the sum of the area of the major loop and the areas of the minor loops. The ideal magnetization curve[8] can be measured by performing a demagnetization at various bias fields (Figure 1.19).

Magnetization curves can be measured inductively at a ring core sample with different windings, valuating (1.5) and (1.9) for H and B, respectively.

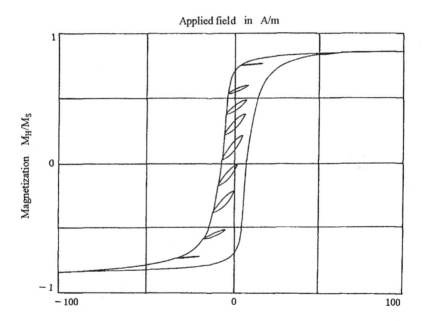

Figure 1.18 Minor loops and their permeability μ_Δ by application of a small ac field with a variable bias.

---

8. This can be experienced by the "magnetic screwdriver" effect. The long, hard, magnetic steel rod (easy axis of shape anisotropy) can be magnetized in the bias of the Earth's weak magnetic field by application of mechanical shock waves (hammer strokes), which affect decaying domain wall displacements. It can be demagnetized by performing that procedure perpendicular (hard axis) to the Earth's field.

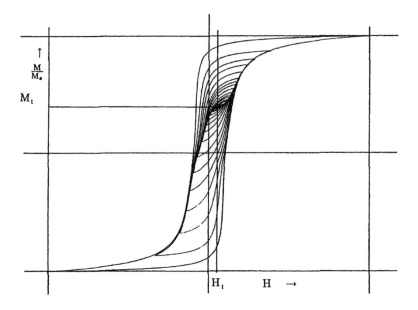

Figure 1.19 One point of the ideal magnetization curve by application of a decaying ac field with a variable bias.

If the sample is an open magnetic circuit, the demagnetizing field causes a flattening of the curve by decreasing the measured, apparent permeability

$$\mu_a \approx \frac{\mu_r}{1 + \mu_r D_d} \quad (1.50)$$

as a function of the relative permeability μ_r (approximation for $\mu_r \gg 1$) and following (1.23), (1.24), (1.42), and (1.43). The nonlinearity of magnetization curves is utilized in fluxgate sensors, for instance (see Chapter 3). Simulating the magnetic behavior of sensors, the calculation of hysteresis loops is a complex task [19], and several models have been developed, such as [20–22].

1.3 Magnetic Materials for Sensor Applications

Magnetic materials are classified by their coercivity H_c: soft ($H_c <$ 1000 A/m), medium ($H_c =$ 1 . . . 30 kA/m), and hard ($H_c >$ 30 kA/m). To imagine the wide range of the properties of modern materials, consider this example: If the plot of a soft magnetic hysteresis loop (H_c = 0.1 A/m) is 1 cm wide, the corresponding width for a typical hard magnetic material (H_c = 10^6 A/m) in the same scale would be 100 km! Magnetostrictive properties and thin films are also utilized for sensor applications. This section examines typical representatives.

1.3.1 Soft Magnetic Materials

Low coercivity and high permeability are achieved in various crystalline, amorphous, and nanocrystalline elements, alloys, and compounds. General rules for materials design are low magnetocrystalline and strain anisotropy or a large number of easy directions and negligible inner stresses. That is provided by amorphous materials without crystal structure ($K_i \approx 0$) and nanocrystalline materials, in which the small grain size averages out anisotropy. Anisotropic materials can be used advantageously by designing the flux path parallel to the easy directions [e.g., cores of grain-oriented silicon-iron (FeSi) steel]. Stress-relieve annealing is necessary after mechanical deformation in most cases to preserve the intrinsic material parameters. Finally, Barkhausen noise also must be considered for the suitability of a material for sensor applications. Table 1.2 presents basic properties of typical soft magnetic materials.

1.3.1.1 Crystalline Metals

Metallic crystals without imperfections are the classic soft magnetic materials. Typical examples are pure iron and nickel. Alloys widely used are nickel-iron (FeNi) [Permalloy, Supermalloy, Mumetal® with very high permeability], cobalt-iron (FeCo) [highest saturation], silicon-iron (FeSi) [increased resistivity, decreased eddy currents], and silicon-aluminum-iron (FeAlSi) [Sendust, mechanically hard].

1.3.1.2 Amorphous Metals

If an alloy is rapidly solidified from melt (typical cooling rate 10^6 K/s), it exhibits a topologic disorder without crystalline structure. A typical production process is melt spinning of thin wires and ribbons (5- to 50-μm thickness). Amorphous materials are alloys mainly based on iron and cobalt, with additions of boron and silicon. The main advantages are high permeability and low losses (maximum magnetization frequency up to 5 MHz). Disadvantages are the lower saturation magnetization ($\approx$ 1.6T) compared to crystalline iron alloys and the limited magnetic core design possibilities. Recently, amorphous materials have been successfully produced as tapes thicker than 50 μm and even in a bulk state [23].

Table 1.2
Maximum Permeability, Coercivity, and Saturation Flux Density (Quasistatic, $T = 300K$) of Selected Soft Magnetic Materials

Material	Composition	μ_{max}	H_c, A/m	B_s, T
Cobalt	$Co_{99.8}$	250	800	1.79
Permendur	$Fe_{50}Co_{50}$	5,000	160	2.45
Iron	$Fe_{99.8}$	5,000	80	2.15
Nickel	$Ni_{99.8}$	600	60	0.61
Silicon-Iron[a]	$Fe_{96}Si_4$	7,000	40	1.97
Hiperco	$Fe_{64}Co_{35}Cr_{0.5}$	10,000	80	2.42
Supermendur	$Fe_{49}Co_{49}V_2$	60,000	16	2.40
Ferroxcube 3F3[b]	Mn-Zn-Ferrite	1,800	15	0.50
Manifer 230[c]	Ni-Zn-Ferrite	150	8	0.35
Ferroxplana[d]	$Fe_{12}Ba_2Mg_2O_{22}$	7	6	0.15
Hipernik	$Fe_{50}Ni_{50}$	70,000	4	1.60
78 Permalloy	$Fe_{22}Ni_{78}$	100,000	4	1.08
Sendust	$Fe_{85}Si_{10}Al_5$	120,000	4	1.00
Amorphous[e]	$Fe_{80}Si_{20}$	300,000	3.2	1.52
Mumetal 3	$Fe_{17}Ni_{76}Cu_5Cr_2$	100,000	0.8	0.90
Amorphous[e]	$Fe_{4.7}Co_{70.3}Si_{15}B_{10}$	700,000	0.48	0.71
Amorphous[e]	$Fe_{62}Ni_{16}Si_8B_{14}$	2,000,000	0.48	0.55
Nanocrystalline	$Fe_{73.5}Si_{13.5}B_9Nb_3Cu$	100,000	0.40	1.30
Supermalloy	$Fe_{16}Ni_{79}Mo_5$	1,000,000	0.16	0.79

[a]Nonoriented; [b]at 100 kHz; [c]at 100 MHz; [d]at 1000 MHz; [e]annealed.

1.3.1.3 Nanocrystalline Metals

Annealing an amorphous material at the recrystallization temperature [≈ 500°C for boron-cobalt-iron (FeCoB)], nanocrystalline grains with about 10- to 15-nm diameter can be established. Although these materials exhibit excellent soft magnetic properties ($H_c \to 0$, $\mu_{max} > 10^5$ at relatively high saturation magnetization), the problems of high brittleness and poor mechanical engineering possibilities have prevented wide-scale sensor applications. Because of their large resistivity, they are currently used for high-frequency power transformer cores.

1.3.1.4 Soft Ferrites

In general, ferrites consist of about 70% iron oxide (Fe_2O_3) and 30% of other metal oxides [e.g., manganese oxide (MnO), magnesium oxide (MgO), nickel oxide (NiO), copper oxide (CuO), and iron oxide (FeO)]. They have become widely available because Fe_2O_3 is a recycling product at steel manufacturing (rust removal). Ferrites for microwave frequencies are almost perfect electrical insulators (resistivity $\rho > 10^7$ Ωm); they consist of oxides of iron, nickel, magnesium, manganese, and aluminum. Soft magnetic ferrites have a relatively small cubic magnetocrystalline anisotropy.

Because the moments in crystal sublattices have antiparallel orientations, the resulting M_s can be increased by substituting for manganese the almost nonmagnetic zinc [in the case of inverse spinels, manganese-zinc (MnZn) ferrites]. Besides the spinel structures, the garnets are another group of ferrimagnetic materials, for example, yttrium iron garnet (YIG). They are used mainly for magneto-optical and resonance applications. Yttrium orthoferrite ($YFeO_3$) has the highest domain wall velocity of magnetically ordered media and both high infrared transparency and Faraday rotation, and it is important for future magneto-optical sensors (see Chapter 6).

Advantages of ferrites are a high electrical resistivity (frequency limit up to 5 GHz), no oxidation, and the possibility of arbitrarily shaped cores. The main disadvantages are the lower saturation magnetization ($\mu_0 M_s \approx 0.5$T) compared to ferromagnets and only moderate permeability.

1.3.2 Hard Magnetic Materials

A permanent magnet is used to provide a magnetic field outside its boundary. In absence of both an additional magnetomotive force and an applied field, a permanent magnet is operated in the second quadrant of the hysteresis loop (demagnetization curve) as a consequence of (1.5), (1.42), and (1.43). A high stability is achieved if the knee of the $B(H)$ curve is far below the operation point; otherwise, a changing air gap or strong applied fields could partly demagnetize the magnet, especially at elevated temperatures. Besides high coercivity and high remanence magnetization, the maximum energy product $|BH|_{max}$ is important to choose the optimum operation point (smallest volume), as shown in Figure 1.20.

Basically, high remanence and high coercive field can be achieved in different ways by which the remanent magnetization process is accomplished.

- *Domain wall pinning.* Coercivity increases as the domain wall movements are impeded by intended, artificial imperfections. Pinning-type magnets have, in general, smaller coercivities and energy products compared to nucleation-type magnets.
- *Domain nucleation.* The grains are in a monodomain state due to a strong initial applied field. To reverse the magnetization, strong fields are necessary for the nucleation of oppositely magnetized

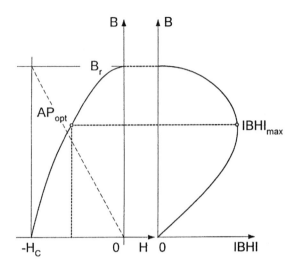

Figure 1.20 Demagnetization curve and energy product of permanent magnets.

domains, which results in high coercivity values (nucleation-type magnets).

- *Irreversible magnetization rotation.* Within small, isolated volumes (grains or particles) without domain walls but with high anisotropy, the magnetization reversal is possible only by rotation processes. When the spontaneous magnetization rotates beyond the direction of maximum anisotropy energy (see, for example, the energy area of the barlike ellipsoid in Figure 1.10), its position becomes unstable, and the magnetization jumps (switches) at a certain field strength irreversibly to the nearest easy direction. Corresponding to the switching field, the maximum coercivity values are achievable because of high shape and magnetocrystalline anisotropy energies.

1.3.2.1 Alloys

Domain wall pinning is achieved by adding chromium, cobalt, nickel, copper, vanadium, or carbon to steel (structural hardening). These permanent magnets have low H_c and $|BH|_{max}$, but they are very inexpensive, which provides several applications in low-cost sensor systems.

A first improvement of permanent magnetic properties was achieved with Alnico compounds, consisting of iron, cobalt, aluminum, nickel, copper, and titanium. Producing a controlled temperature gradient leads to the formation of preferentially oriented needlelike crystals (precipitation hardening). Additionally aligned in an applied field during cooling down from the melting temperature, the thin iron-cobalt needles of high anisotropy yield coercivities up to 180 kA/m and energy products of about 90 kJ/m^3.

1.3.2.2 Sintered Magnets

The state-of-the-art high-performance sintered magnets are samarium-cobalt (SmCo) ($H_c \approx$ 1.2 MA/m, $B_r \approx$ 1.1T, $|BH|_{max} \approx$ 250 kJ/m^3, $T_{max} \approx$ 300°C) and neodynium-iron-boron (NdFeB) ($H_c \approx$ 1.0 MA/m, $B_r \approx$ 1.3 T, $|BH|_{max} \approx$ 350 kJ/m^3, $T_{max} \approx$ 180°C). The single-phase SmCo$_5$ can be magnetized near to saturation by small fields (2 kA/m) because of easy domain wall displacements. But the high coercivity ($_MH_c$ up to 4 MA/m) is achieved only by saturating the grains (against inner stray fields) at strong fields (nucleation type). Oppositely, in the multiphase Sm$_2$Co$_{17}$ alloys, the domain walls are strongly pinned at the phase boundaries. The main disadvantages of the rare earth sintered magnets are poor mechanical properties (brittleness) and their high price.

1.3.2.3 Hard Ferrites

Magnetically hard barium ferrites and strontium ferrites have hexagonal crystal structures and possess high magnetocrystalline anisotropy. The coercivity is rather high ($H_c \approx$ 0.5 MA/m), but the low remanence of some 100 mT leads to a low-energy product ($|BH|_{max} \approx$ 35 kJ/m^3). On the other hand, ferrites are cheap, do not oxidize, and can be powdered and embedded in plastics to produce flexible magnets by injection die casting.

1.3.3 Magnetostrictive Materials

Magnetostrictive properties of materials are used in sensors for mechanical quantities and actuator systems. Applied stresses change the strain anisotropy energy and therefore influence the permeability (see Figure 1.8). Vice versa, applied fields change the magnetic state of the material and the elasticity via magnetostriction. Sensitive materials (large saturation magnetostriction λ_s at low saturation fields H_s) are amorphous wires and ribbons, for example, Co$_{68}$Ni$_{10}$(Si,B)$_{22}$ ($\lambda_s = -8 \cdot 10^{-6}$, $H_s =$ 200 A/m) and Fe$_{50}$Co$_{50}$ ($\lambda_s = +70 \cdot 10^{-6}$, $H_s =$ 10 kA/m). The rare earth–iron alloys exhibit highest magnetostriction at strong fields, for example, Terfenol (Tb$_9$Dy$_{24}$Fe$_{67}$; $\lambda_s = +2000 \cdot 10^{-6}$, $H_s =$ 200 kA/m).

1.3.4 Thin Films

Low-dimensional material shapes (thin films and fine particles) are of great importance in magnetic recording technology. But for several sensor applications (e.g., magnetoresistive sensors, miniature fluxgates, magneto-optical sensors), the magnetization processes in thin films and wires also have to be considered. In the latter structure, the magnetization process and the macroscopic behavior strongly depend on the various domain configurations (axial or radial domains, single domain state with nucleation, domain magnetization rotation in high frequency fields, and so on); it is described at the corresponding application.

Thin films of Permalloy (FeNi) are of interest for sensor applications (e.g., anisotropic magnetoresistance) because of their low magnetocrystalline anisotropy and low magnetostriction. They can be produced by sputtering, high-vacuum evaporation, and electroplating. Mechanically hard films are made of Sendust (FeAlSi) by sputtering only. Amorphous iron- and cobalt-based thin films are also used because of their excellent soft magnetic properties. Sandwich and multilayer structures of very thin films (1–10 nm) of iron, FeNi, FeCo, and chromium, copper, silver, or gold intermediate layers are made by molecular beam epitaxy or sputtering to form superlattices for spin valves and giant magnetoresistance sensors (see Chapter 4).

1.3.4.1 Magnetic Structure

The thickness of a thin film is usually of the order of magnitude of a typical domain wall in the bulk material, up to 100 nm. In principle, there would be three possibilities for establishing a domain structure (Figure 1.21): If the film is thick enough, the domain structure in Figure 1.21(a) is thinkable.

In the case of only two easy directions perpendicular to the film plane, the structure of Figure 1.21(b) could be established. That domain structure is stable only if

$$K_1 > \frac{\mu_0 M_s^2}{2} \qquad (1.51)$$

is valid. That is the case in garnets ($K_1 = 10^5$ J/m^3, $M_s = 1.8 \cdot 10^5$ A/m) and orthoferrites ($K_1 = 10^5$ J/m^3, $M_s = 1.0 \cdot 10^3$ A/m), which are important for magneto-optical applications.

In most cases, we have to deal with in-plane magnetization; see Figure 1.21(c). The spontaneous magnetization of metallic ferromagnets is so large that the demagnetizing stray field energy of the magnetic poles is much larger than the magnetocrystalline anisotropy energy.

If the thickness of the film is below the Bloch-type domain wall width, the rotation of the magnetic moments from one domain to the neighbor

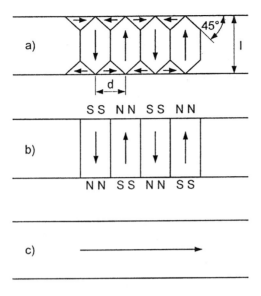

Figure 1.21 Possible domain structures in thin films.

domain direction will occur (in a simplified description) in plane[9] as a Néel-type domain wall. Because of the comparably high wall energy, thin films often appear uniformly magnetized in the film plane, and magnetization reversal can be achieved by coherent rotation (in real material geometries incoherent rotations also take place), depending on the applied field.

1.3.4.2 Coherent Rotation of Magnetization

Consider a spontaneously magnetized (M_s, angle φ with the easy direction) and approximately ellipsoidal ($a > b \gg c$) shaped film with uniaxial anisotropy, where the easy axis is parallel to the long ellipsoid axis a (Figure 1.22). The energy W_H of the applied field (angle ψ with the easy axis and components $H_x = H\cos\psi$, $H_y = H\sin\psi$)

$$W_H = -\mu_0 M_s H \cos(\psi - \varphi) \qquad (1.52)$$

was given by (1.25), and the direction-dependent terms of the magnetocrystalline anisotropy energy

$$W_C = K_1 \sin^2\varphi \qquad (1.53)$$

---

9. Depending on film thickness, exchange coupling, and anisotropy, mixed-type walls (e.g., cross-tie walls) also are known.

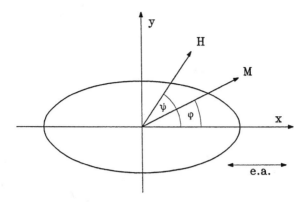

Figure 1.22 *H* and *M* in a thin-film ellipsoid and the coordinate system.

and of the shape anisotropy energy of demagnetizing stray fields

$$W_D = \mu_0 M_s \frac{D_b - D_a}{2} \sin^2 \varphi \qquad (1.54)$$

were given by (1.33) and (1.40), respectively. Using the simplifications $D_b - D_a = D_d$, $H_d = -D_d M_s$ (demagnetizing field), $H_k = 2K_1/\mu_0 M_s$ (anisotropy field), and the characteristic field $H_0 = H_k + H_d$, the zeros of the derivation of the total energy

$$W_T = W_H + W_C + W_D \qquad (1.55)$$

with respect to φ ($dW_T/d\varphi = 0$) give its extrema as a function of the parameter φ:

$$\sin \varphi = \frac{H \sin \psi \cos \varphi}{H_0 \cos \varphi + H \cos \psi} \qquad (1.56)$$

Depending on the applied field, the magnetization will make such an angle φ, where W_T has a minimum; the second derivative of W_T is negative in this case. The magnetization curves can be calculated for the projection of M_s in the H direction:

$$M_H = M_s \cos(\psi - \varphi) \qquad (1.57)$$

Stability considerations ($dW_T^2/d^2\varphi > 0$) lead to the well-known Stoner-Wohlfarth astroid [24] (for $-1 < H_y/H_0 < +1$)

$$|H_x|^{2/3} + |H_y|^{2/3} \leq |H_0|^{2/3} \qquad (1.58)$$

for the permitted solutions of φ. If H is within the astroid area, two solutions for M_H are possible. If the field strength is large enough to point out of the astroid area, only one solution is possible and switching occurs. That is shown by the calculated magnetization curves in Figure 1.23.

In the special case of $\psi = 90$ degrees, the magnetization curve is degenerated to a straight line $M_H = M_s \cdot H/H_0$. In the second special case, $\psi = 0$ degrees, the magnetization curve is a rectangle with the width H_0. In reality, the switching occurs often at fields lower than H_0 because of the nucleation of antiparallelly magnetized domains.

If there is an angle θ between the easy direction of uniaxial anisotropy and the long ellipsoid axis, the minima of W_T can be calculated by using a new coordinate system x', y' rotated by an angle δ from x, y. The modified characteristic field is

$$H_0 = \sqrt{H_d^2 + H_k^2 + 2H_d H_k \cos 2\theta} \qquad (1.59)$$

and the transcendent equation for δ is

$$\tan 2\delta = \frac{H_k \sin 2\theta}{H_d + H_k \cos 2\delta} \qquad (1.60)$$

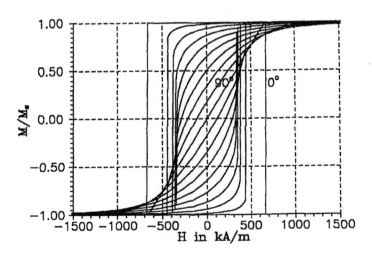

Figure 1.23 Magnetization curves $M_H(H)$ of a thin film depending on the angle between **H** and the easy axis.

In addition to the special cases of $\theta \approx 0$ degrees and $\theta \approx 90$ degrees, where the characteristic field is respectively the sum or the difference of H_d and H_k, the case of $\theta = 45$ degrees is of interest ($\tan 2\delta = H_k/H_d$ and $H_0 = \sqrt{H_d^2 + H_k^2}$). Much more complex calculations have to be done in the case of cubic anisotropy and three-dimensional field application. Solutions for the total energy minima can be found by numerically tracing the minima of the energy areas, shown in Figures 1.6, 1.8, 1.10, and 1.11.

Finally, the considerations of coherent magnetization rotation can be applied in analogy to thin, ellipsoidal-shaped wires in single domain state. The easy axis (a) is parallel to the longitudinal direction of the wire, and the hard axes ($b = c$) are perpendicular (in absence of magnetocrystalline anisotropy). The resulting magnetization curves correspond to those shown in Figure 1.23.

1.3.4.3 Exchange Anisotropy

If different magnetic materials (e.g., ferromagnetic and antiferromagnetic) are in intimate contact with each other or are separated by a layer thin enough to allow spin information to be communicated between the two materials, exchange coupling between the materials occurs and exhibits the phenomenon of exchange anisotropy (in contrast to magnetostatic coupling, which can arise from interfacial stray fields due to roughness of the boundary). This effect was observed first for fine cobalt particles with 20-nm diameter, which were covered by a thin layer of antiferromagnetic CoO, by a displaced M/H-characteristic of more than 80 kA/m [25].

That field-displaced loop was referred to as exhibiting exchange anisotropy as a result from an exchange coupling between the moments of cobalt and CoO. The vertical scale of the M/H loop of the sample reflects only the cobalt magnetization because the antiferromagnetic CoO is essentially unmagnetized at weak fields. The interfacial exchange coupling depends on the submicroscopic details of the interface. The coupling strength is usually described by an effective exchange field

$$H_E = \frac{J_E}{\mu_0 M_s d} \quad (1.61)$$

with respect to the interfacial exchange coupling energy J_E (typically 1 mJ/m^2), the film parameters M_s and K_1, and thickness d. Only if $J_E < K_1 d_a$, $M(H)$ curve is shifted unidirectionally (otherwise, only the coercivity is increased) because the interfacial coupling is too weak to reverse the spins throughout the entire thickness, d_a, of the antiferromagnetic layer.

That is the effect of exchange anisotropy, which is an important feature of magnetic multilayers, as described in Chapter 4.

1.4 Sensor Specifications

Besides the well-known parameters for other sensors and instruments, magnetic sensors have specific peculiarities given by the nonlinear materials and complex effects used. We also mention the basic problems associated with the sensor calibration. The calibration instruments, especially the shieldings, are described in Chapter 11.

Manufacturers' datasheets sometimes give only the sensitivity or "resolution" but do not mention temperature stability of the sensitivity and offset, perming, hysteresis, and other tricky parameters. In general, all the absolute specifications should be given in the units of the measured field; the sensitivity should be referred to the sensor output; and the measuring conditions should be specified.

1.4.1 Full-Scale Range, Linearity, Hysteresis, and Temperature Coefficient of Sensitivity

These characteristics can be measured in calibration coils (e.g., the Helmholtz coil pair). The coils should be far from ferromagnetic objects, which could cause nonlinearity, and should be much larger than the tested sensor to ensure that the field is homogeneous in the sensor location. The disturbing fields should be small; the ideal calibration coil is located in the thermostated nonmagnetic house far from the sources of the magnetic pollution. Nighttime is the ideal time for precise measurements because of low traffic. The worst disturbing fields are caused by dc currents from electric trains, streetcars, and subways. Simultaneous monitoring of the Earth's field variations is recommended; it may prevent invalid calibration during magnetic storms. Averaging may reduce the uncertainty.

If absolute calibration of the sensor sensitivity is required, the testing coil should be calibrated (usually with the resonant magnetometer) and special care taken to align the sensor sensing axis to the direction of the testing field.

Calibration of dc magnetic sensors is usually done by field steps of both polarities to correct eventual changes of the sensor offset and the background field. The field steps should have sufficient duration because the time constant of the sensor as well as the time constant of the calibration coil may be long. The current into the coil should be measured synchronously with the sensor output. The calibration current may heat the testing coil, an effect that may change its constant: The coil temperature should be monitored by measurement of the voltage drop.

1.4.2 Offset, Offset Temperature Coefficient, and Long-Term Stability

Offset and offset stability are tested in magnetic shieldings (see Chapter 11). It is important to check the shielding remanence by moving the sensor inside. If necessary, ferromagnetic shieldings can be demagnetized. Good cylindrical shielding has an axial rest field below 10 nT and a radial rest field below 2 nT. If the remanent field is stable and homogeneous, it can be corrected by averaging the sensor reading from two opposite sensing directions (or more directions equally distributed in direction). Offset can be measured with an abolute precision of 0.1 nT by this technique. It should be mentioned that some sensors cannot measure zero field; a typical example is resonant sensors (see Chapter 7).

Temperature changes of offset are tested by the thermostat inside the magnetic shielding. The thermostat should be made of nonmagnetic material and electrically insulating material; otherwise, thermoelectric currents generated by temperature gradients in the testing device may cause disturbing magnetic fields. There should be good temperature insulation between the thermostat and the shielding, so that the shielding temperature is constant. The temperature changes should be slow enough: The magnetic sensor offset is often more sensitive to temperature gradients than to slow changes, because temperature gradients cause large mechanical stresses due to different dilatations of the sensor structure.

It is also necessary to check whether the offset changes with temperature are reversible and reproducible. Irreversible changes often occur at higher temperatures. The offset temperature coefficient (tempco) is usually given in parts per million per degree Celsius (ppm/°C). In many cases, offset tempco changes with temperature and also piece to piece, so typical or maximum values are often given. Offset interval in a given temperature range is a more convenient value than tempco, especially for large temperature intervals.

1.4.3 Perming

Perming is offset change after the magnetic shock. It is similar to hysteresis, but the applied field may be much higher than the full-scale range. All the sensors containing ferromagnetic material are susceptible to perming. The only solution is periodical remagnetization of the sensor, which ensures the defined magnetic state. In the case of fluxgate sensors (see Chapter 3), the remagnetization is performed by the excitation current; if the excitation amplitude is high enough, no significant perming is observed. Ferromagnetic magnetoresistors can be flipped to reduce the perming and crossfield responses. Perming may also appear at semiconductor sensors with ferromagnetic flux concentrators.

1.4.4 Noise

Noise is a random variation of the sensor output when the measured value is zero. Ultralow frequency noise is similar to long-term stability: The zero fluctuations are usually referred to as *noise,* while the trend is referred to as *time drift.* Noise properties are characterized by its power spectrum. Power spectrum density $P(f)$ at a given frequency f can be used to compare the noise properties of sensors. If the peak-to-peak (p-p) value of the noise or root mean square (rms) value is given, the frequency range from f_L to f_H should be specified.

The magnetic sensor noise is usually something between the ideal white noise and the ideal $1/f$ noise. White noise has a power spectrum density P independent of frequency; for noise of $1/f$ character $P(f) = P(1)/f$ [nT²/Hz], where $P(1)$ is the power spectrum density at 1 Hz. In the case of $1/f$ noise, the precise specification of f_L is much more important than that of f_H. Sensor data like "noise p-p (or rms) value from dc to 10 Hz" have sense only for white noise, which is rare.

Although the peak-to-peak noise value may be tricky and usually is not precisely defined, it is convenient and practical to show the sensor performance. The low-frequency limit f_L is often simply taken as the inverse value of the observation period; f_H is often taken as the corner frequency of the low-pass filter in the signal path. The rms value can also be calculated by integration of the power spectrum density (PSD) over the required frequency range (from f_L to f_H); in digital processing, the integration is replaced by summation.

$$N_{rms} = \sqrt{\int_{f_L}^{f_H} P(f)\,dt} \qquad (1.62)$$

In the case of the white noise,

$$N_{\text{rms}} = \sqrt{P(f_H - f_L)} \quad (1.63)$$

In the case of the $1/f$ noise,

$$N_{\text{rms}} = \sqrt{P(1)\ln(f_H/f_L)} \quad (1.64)$$

The noise spectrum density is often given, in units of nanotesla per square root Hertz (nT/$\sqrt{\text{Hz}}$) or picotesla per square root Hertz (pT/$\sqrt{\text{Hz}}$). The conversion is straightforward: For example, 9 nT^2/Hz corresponds to 3 nT/$\sqrt{\text{Hz}}$, often written as 3nT/sqrtHz. For $1/f$ noise, the $\sqrt{P}$ value decreases with $\sqrt{f}$: In our example, if the noise is 3 at 1Hz, it is $3/\sqrt{10} = 0.95$ at 10 Hz, and $3 \cdot \sqrt{10} = 9.5$ nT/$\sqrt{\text{Hz}}$ at 0.1 Hz.

If we sum the noise contributions N_i from several (independent) sources or several (nonoverlapped) bandwidths, we should calculate the total noise N

$$N^2 = \sum_i N_i^2 \quad (1.65)$$

1.4.5 Resistance Against Environment (Temperature, Humidity, Vibrations)

Temperature range is often limited by package technology. Table 1.3 lists the minimum and maximum values.

Permanent magnets have a limited operation temperature: NdFeB works up to 100°C, while SmCo works up to 300°C; ferrites and AlNiCo magnets can be used over 400°C.

Table 1.3
Temperature Range

Sensor Type	Minimum Temperature	Maximum Temperature
Silicon Hall	1.5 K < –40°C	150°C
GaAs Hall		175°C
InSb magnetoresistor	–60°C	200°C
Anisotropic magnetoresistance (AMR)		85 (potentially < 200)°C
Giant magnetoresistance (GMR)		150°C

Humidity destroys high-temperature superconducting quantum interference device (SQUID) sensors (Chapter 8). Self-capacitance of the coils increases with moisture; sensor coils may be impregnated, but most of the impregnation processes worsen the coil temperature stability. A number of sensors fail if they are wet from condensed moisture. One solution is to keep the sensor and electronics in a sealed package with desiccant.

1.4.6 Resistance Against Perpendicular Field and Field Gradient

Vector magnetic sensors, which contain ferromagnetic material such as ferromagnetic magnetoresistors and fluxgates, are also sensitive to fields perpendicular to their sensing direction. The "crossfield effect" is nonlinear so it cannot be easily corrected. The errors of this origin are small for the Earth's field: 20 μT perpendicular field in horizontal direction may cause 1nT to 10 nT error of fluxgate sensor (see Chapter 3) and similar error for GMR and AMR sensors. Large perpendicular fields may cause gross errors—the sensor characteristics may be distorted or even reversed, and in the case of magnetoresistors the error may remain after the perpendicular field disappears. As with the perming effect, the sensor should be remagnetized.

Resonant magnetometers fail in large field gradients: The broadening of the resonant curve causes a decrease of the signal, which may cause decrease in the accuracy or even a false reading. Thus, resonant sensors cannot be used for measurements inside buildings in the close vicinity of ferromagnetic objects. On the other hand, precisely defined field gradients are used in NMR tomography.

1.4.7 Bandwidth

The sensor bandwidth is usually given for a specified decrease in sensitivity (e.g., −3 dB, −10%), and phase characteristics usually are not specified. Increasing the sensor bandwidth usually causes an increase in the noise level.

When testing the bandwidth, care should be taken to check the frequency characteristics of the testing coils and monitor the testing current. Also, eddy currents in conducting structures may cause false calibration.

1.4.8 Other Parameters

Sensor power is critical for battery-operated devices. Lowering power usually means compromises with sensor noise and stability. In the case of ac-excited sensors, the power consumption can be decreased by lowering the operation frequency, but the bandwidth would be reduced.

The sensors most susceptible to radiation are semiconductor sensors, but they withstand a much higher dose than typical computer memory.

The sensor cost is often an important parameter. Simple Hall sensors cost less than $1, but the price is growing quickly with performance: Highly linear calibrated Hall sensors may cost $1,000. High-performance sensors often are not sold alone, but as part of the magnetometer. That is usually the case of resonance sensors and high-performance fluxgates.

References

[1] Becker, R., and W. Döring, *Ferromagnetismus*, Berlin, Germany: Springer, 1939.

[2] Stratton, J. A., *Electromagnetic Theory*, New York, London: McGraw-Hill, 1941.

[3] Bozorth, R. M., *Ferromagnetism*, New York: van Nostrand, 1951.

[4] Kneller, E., *Ferromagnetismus*, Berlin, Germany: Springer, 1962.

[5] Brown, W. F., *Magnetostatic Principles in Ferromagnetism*, Amsterdam, North Holland: Interscience, 1962.

[6] Chikazumi, S., *Physics of Magnetism*, New York: Wiley, 1964.

[7] Hofmann, H., *Das elektromagnetische Feld*, Wien, New York: Springer, 1974.

[8] Heck, C., *Magnetic Materials and Their Technical Applications*, New York: Crane and Russak, 1974.

[9] Wohlfarth, E. P. (ed.), *Ferromagnetic Materials*, Amsterdam, North Holland: Elsevier, 1980.

[10] Boll, R., *Weichmagnetische Werkstoffe*, Berlin, München: Siemens AG, 1990.

[11] Jiles, D. C., *Introduction to Magnetism and Magnetic Materials*, New York: Chapman and Hall, 1991.

[12] Fasching, G. M., *Werkstoffe für die Elektrotechnik*, Wien, New York: Springer, 1994.

[13] Maxwell, J. C., *A Treatise on Electricity and Magnetism*, Oxford, England: Clarendon Press, 1873.

[14] Paschke, F., "Rotating Electric Dipole Domains as a Loss-Free Model for the Earth's Magnetic Field," *Session Reports of the Austrian Academy of Sciences*, Vol. 207, 1998, pp. 213–228.

[15] Osborn, J. A., "Demagnetizing Factors of the General Ellipsoid," *Physical Review*, Vol. 67, 1945, pp. 351–357.

[16] Hubert, A., and R. Schäfer, *Magnetic Domains*, Berlin, Heidelberg, New York: Springer, 1998.

[17] Lawton, H., and K. H. Stewart, "Magnetization Curves for Ferromagnetic Single Crystals," *Proc. Royal Society London*, Vol. A193, 1948, pp. 72–88.

[18] Hauser, H., "Energetic Model of Ferromagnetic Hysteresis 2: Calculation of the Magnetization Curves of (110)[001] FeSi-Sheets by Statistic Domain Behaviour," *J. Applied Physics*, Vol. 77, No. 6, 1995, pp. 2625–2633.

[19] Bertotti, G., *Hysteresis in Magnetism*, San Diego, CA: Academic Press, 1998.

[20] Preisach, F., "Über die magnetische Nachwirkung," *Zeitschrift für Physik,* Vol. 94, 1935, pp. 277–298.

[21] Jiles, D. C., and D. L. Atherton, "Model of Ferromagnetic Hysteresis," *J. Magn. Magn. Mater.,* Vol. 61, 1986, pp. 48–58.

[22] Hauser, H., "Energetic Model of Ferromagnetic Hysteresis," *J. Applied Physics,* Vol. 75, No. 5, 1994, pp. 2584–2597.

[23] Inoue, A., and A. Makino, "Preparation and Soft Magnetic Properties of Fe-Based Bulk Amorphous Alloys," *J. Physique IV,* Vol. 8, 1998, pp. 3–10.

[24] Stoner, E. C., and E. P. Wohlfarth, "A Mechanism of Magnetic Hysteresis in Heterogenous Alloys," *Philosophical Trans. of the Royal Society London,* Vol. A240, 1948, pp. 599–642.

[25] Meiklejohn, W. H., and C. P. Bean, "New Magnetic Anisotropy," *Physical Review,* Vol. 102, No. 5, 1956, pp. 1413–1414.

2

Induction Sensors
Pavel Ripka

The basic description of induction sensors starts from the Faraday law:

$$V_i = d\Phi/dt = d(NA\mu_0\mu_r(t)H(t))/dt \qquad (2.1)$$

where V_i is the voltage induced in a coil having N turns (instantaneous value); Φ is the magnetic flux in the coil, $\Phi = BA$; A is the core cross-sectional area; H is the magnetic field in the sensor core; and $\mu_r(t)$ is the sensor core relative permeability (core may be ferromagnetic or air).

Thus, we can write the general equation for induction sensors:

$$V_i = NA\mu_0\mu_r dH(t)/dt + N\mu_0\mu_r H dA(t)/dt + NA\mu_0 H d\mu_r(t)/dt \qquad (2.2)$$

Basic induction or "search" coils are based on the first term of (2.2). The middle term describes rotating coil sensors (Section 2.5), where $A(t)$ is the effective area in the plane perpendicular to the measured field. The last term is the basic fluxgate equation (fluxgate sensors are covered in Chapter 3).

A typical induction coil magnetometer consists of a multilayer solenoid. We start our description with air coils (Section 2.1). They are very stable and linear, but their sensitivity is limited. Typical optimized air coils have a large diameter and are relatively short. Coils with a ferromagnetic core (Section 2.2) have larger sensitivity, but they are less stable. They should be long and thin to have low demagnetization. Modern low-noise induction coils are usually working in the current-output mode. A low-noise voltage or current preamplifier is usually mounted in close proximity to the induction coil, and it should be considered part of the sensor. Induction sensors are passive, and they should be distinguished from inductance sensors (Section 12.1), which are based on the change of the sensor inductance and which need excitation.

Here we briefly mention some applications of induction magnetometers (more about applications in Chapter 10): In geophysics, they serve to measure micropulsations of the Earth's magnetic field (1 mHz–1 Hz frequency range); in audiofrequency applications, they are also used in magnetic recording techniques. Magnetotelluric exploration is the measurement of secondary magnetic fields caused by the Earth's currents after artificial excitation at frequencies up to the audio range. Geophysical exploration may also use natural electromagnetic field variations in the 1 Hz-to-20 kHz band [1]. In space research, induction coils are used in plasma experiments (Section 10.9). Magnetic antennas are used for navigation and communication (submarines, trains, etc.) (Section 10.3). Important are velocity sensors and position detectors (examined in Chapter 12). One of the most important applications is the electromagnetic compatibility (EMC) measurement: Extremely low frequency (ELF) magnetometers measure the magnetic field at the power frequency (50 or 60 Hz). They are designed to measure the fields from electric appliances and power lines. ELF magnetometers often also have "wideband" (WB) rms value output, which typically covers the 20 Hz–50 kHz frequency range, and rms very low frequency (VLF) output (2–50 kHz). Three-axial magnetometers of this type usually calculate the total field from three rms values, measured by three orthogonal coils: That can cause large errors for rotating fields [2].

Air-cored coils together with an integrator (fluxmeter) are used to map the dc induction B by an extraction method (Section 2.5.2) and also to measure field intensity H (Section 2.5.4). It is important to properly calibrate each individual induction coil. The coil constant can be reliably calculated only for very simple air coils with well-defined geometry. The sensitivity calibration techniques are explained in Chapter 11.

2.1 Air Coils

Air coils (or loop antennas) contain no nonlinear magnetic materials, so they have linear amplitude characteristics and their parameters are very stable in time. Their frequency characteristics are linear in a certain frequency range: at low frequencies for voltage output and at middle frequencies for current output. The temperature coefficient of the sensitivity depends mainly on the thermal expansion of the used materials, so it can be made very small and predictable.

Any air coil that serves for the generation of magnetic fields (such as Helmholtz coils) can also be used as a pickup coil. Pickup coils not only can measure the outside field, they also are used to measure the magnetic moment of objects located inside them, and that is the principle of rotating sample magnetometers.

The distributed-coil capacitances of multiturn coils together with their high inductance cause resonances at relatively low frequencies; the output voltage response at higher frequencies is thus nonlinear. The parasitic capacitances may be temperature sensitive, so that the frequency characteristics change with temperature. Current output is more convenient: Short-circuiting the coil inductance L with a small serial resistance R using a current-to-voltage converter suppresses the effect of the parasitic capacitance C. While the ideal induced voltage V_i is proportional to the frequency, the output current response is flat for $f >> R/2\pi L$.

Increasing the coil diameter increases the sensitivity, so the R/L factor depends on the amount of copper; the number of turns and the wire thickness are selected to match the input noise characteristics of the amplifier. A disadvantage of the large area induction coils is their susceptibility to vibrations, which change the effective coil area and thus, in the presence of the dc field, cause noise. It is therefore necessary to construct them mechanically stable and mount them securely [3].

An example of an induction magnetometer with an air coil is ACM-1 (Meda). The coil diameter is 45 cm, and it has a rectangular cross-section of 2.5 cm by 2.5 cm. The sensor frequency range is 20 Hz to 10 kHz (−3 dB), the dynamic range of the internal I/V converter is 800 nT peak-to-peak (p-p). The sensor noise is 0.3 pT/$\sqrt{Hz}$ at 20 Hz, accuracy is 1% in the whole frequency range, and the output scale value is 25 mV/nT. The geometry of the air coil is shown in Figure 2.1.

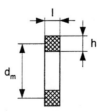

Figure 2.1 Geometry of the air coil.

While the cylindrical coils measure the average field over their volume, the ideal spherical coil measures the field value at its center [4]. Inversely, a current into the spherical coil generates an internal homogeneous field. A good approximation of the spherical coil is used for the compact spherical coil (CSC) fluxgate magnetometer (Chapter 3).

If the measured field is not homogeneous, the designer should consider the sensitivity to field gradients. Axially symmetric coils are insensitive to all even gradients. In the case of cylindrical coils, the sensitivity to odd gradients (spatial harmonics) is minimal for a length-to-outside diameter of 0.67. It is possible to construct coils insensitive to third- and fifth-order gradients and having very small sensitivity to seventh-order gradients [5]. If we want to measure ac field gradients, it is possible to construct gradiometric induction coil systems that are insensitive to a homogeneous field (so-called harmonic coils); the simplest type is a Helmholtz coil pair connected antiserially. (Helmholtz coils are a popular source of magnetic field; see Chapter 11.)

Proper shielding and grounding of the induction coils and their cables are important to prevent capacitive couplings (i.e., sensitivity to electrical fields). Sensor shielding must not create any short-circuited turns. Copper is a common material for coils, but where weight is critical, aluminum coils can be used. As shown in Table 2.1, the density of aluminum is three times lower than that of copper. Although aluminum has higher resistivity, an aluminum coil has only 45% the weight of a copper coil having the same voltage sensitivity and noise [6].

High-frequency coils are sometimes wound from *litze* (woven) wire to reduce the effect of eddy currents. Some search coils are made by printed circuit board (PCB) technology. PCB-fabricated coils are flexible, and they have a well-defined geometry. However, the achievable number of turns is limited. Coils made by thin film or CMOS technology have the disadvantage of higher resistance. Sputtered 1μm-thick search coils are used for the measurement of flux distribution in individual sheets in laminated cores [7].

Table 2.1
Basic Parameters of Wire Materials

Parameter	Copper	Aluminum
Density	8.9 g/cm^3	2.7 g/cm^3
Specific electrical resistivity	0.0178*10^{-6} Ωm	0.027*10^{-6} Ωm
Tempco of resistance	0.39%/°C	0.4%/°C
Coefficient of linear expansion	16.6 ppm/°C	25 ppm/°C

2.1.1 Voltage Sensitivity at Low Frequencies

If the parasitic capacitances are negligible and the average turn area is A, we can write this equation for the induced voltage:

$$V_i(t) = NAdB(t)/dt \qquad (2.3)$$

Suppose the measured field B is periodical with period $T = 1/f$. After integration between two zero crossings t_1 and t_2,

$$\int_{t_1}^{t_2} V_i(t)dt = NA \cdot \Delta B = NA(B_{max} - B_{min}) \qquad (2.4)$$

because B reaches its maximum and minimum values at the time when $V_i(t)$ goes through zero. The arithmetic mean value of the induced voltage will be

$$V_{mean} = \frac{2}{T}\int_{t_1}^{t_2} V_i(t)dt = 2fNA(B_{max} - B_{min}) \qquad (2.5)$$

(because we know that V_i has no dc component). Usually $B_{max} = -B_{min}$ and if the coil is cylindrical and has a mean diameter of d_m, then

$$V_{mean} = 4fNAB_{max} = fN\pi d_m^2 B_{max} \qquad (2.6)$$

Here we notice that at low frequencies, where we can neglect the parasitic capacitances, the voltage sensitivity is proportional to field frequency. Equation (2.6) is valid for any arbitrary antiperiodical waveform of B. (*Antiperiodical* means that positive and negative parts of the waveform are symmetrical, that is, there are no even harmonic components.)

To find the peak value of a nonsinusoidal field, the induced voltage must be measured by a voltmeter measuring the arithmetic mean value (rectified average value), not by a true rms voltmeter. A suitable instrument is, for example, a Keithley 2001. Knowledge of the signal frequency is necessary for the calculation. If the waveform of the magnetic field is requested, the induced voltage must be integrated by an analog integrator or, after the sampling and ADC conversion, numerically integrated.

If B is a sinewave, we can write the peak (maximum) value of the induced voltage:

$$V_p = 2\pi f N A B_{max} = N d_m^2 f B_{max} \pi^2 / 2 \qquad (2.7)$$

or for its rms value the well-known formula

$$V = 4.44 N A f B_{max} = 1.11 N \pi d_m^2 f B_{max} \qquad (2.8)$$

The coil dc resistance can be calculated as

$$R_{DC} = 4\rho N d_m / d_w^2 \qquad (2.9)$$

where ρ is the coil wire specific resistivity (see Table 2.1), d_w is the wire diameter, and d_m is the mean coil diameter.

We can also easily derive the coil winding mass:

$$m_w = \pi^2 \gamma N d_m d_w^2 / 4 \qquad (2.10)$$

where γ is the density of the winding wire (usually copper, rarely aluminum; see Table 2.1).

2.1.2 Thermal Noise

The coil produces thermal noise, which has (in the frequency band of Δf) the rms value

$$V_{Nrms} = \sqrt{(4 k_B T R_{DC} \Delta f)} \qquad (2.11)$$

and the noise density is

$$V_N = \sqrt{(4 k_B T R_{DC})} \ [V/\sqrt{Hz}] \qquad (2.12)$$

where T is the absolute coil temperature (K) and k_B is Boltzmann's constant.

Let us recalculate the coil thermal noise to the equivalent noise density of the measured field. Using (2.7) and after some calculation, we obtain, in agreement with [6],

$$B_{N\tau} = \frac{8\sqrt{k_B T\rho}}{\pi^2 f} \frac{1}{d_w\sqrt{Nd_m^3}} \qquad (2.13)$$

and for an equivalent field thermal noise $B_{N\tau}$ as a function of the coil mass we can derive:

$$B_{N\tau} = \frac{4\sqrt{k_B T\rho\gamma}}{\pi f} \frac{1}{d_m\sqrt{m_w}} \qquad (2.14)$$

An important design rule is this: The thermal noise depends only on the coil mass and diameter, not on the wire diameter. The air coil diameter should be as large as possible. The noise voltage is of the white type, but because the voltage sensitivity is proportional to f, the field noise is proportional to $1/f$. Another important factor is the input noise of the preamplifier. Because the amplifier noise depends on the coil resistance and also on frequency, design rules become complex and lead to an optimum wire diameter (Section 2.3). We should also keep in mind that in very "thick" coils, the turn diameter is not constant.

2.1.3 The Influence of the Parasitic Capacitances

Distributed capacitance, inductance, and resistance of the coil cause several resonant frequencies. The main parasitic capacitance (parallel to the coil terminals) is responsible for the first (bottom) resonant frequency f_{r1}. Because f_{r1} is the basic limiting frequency, a simple equivalent circuit is sufficient for the description (Figure 2.2) [8].

The coil inductance can be estimated from the following two popular empirical formulas for short coils with a large mean diameter d_m (the optimum shape for an air coil) [10]:

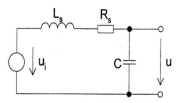

Figure 2.2 Equivalent circuit of the voltage-output induction coil (from [9]).

$$L = \frac{\mu_0 d_m N^2}{2} \ln\left(\frac{4 d_m}{\sqrt{l^2 + h^2}} - 0.83\right) \quad (2.15)$$

or the formula from [11]

$$L = \frac{78.7 d_m^2 N^2}{3 d_m + 9l + 10h} \quad (2.16)$$

where h is the coil height and l is the length (see Figure 2.1).

At high frequencies, L_s decreases because of the eddy currents, which produce a magnetic field decreasing the main field [6, 11]. But for most cases, L can be considered to be constant. R_s consists of the dc resistance and ac losses caused at higher frequencies by skin effects, proximity effects and eddy currents in the shielding and other parts of the sensor. The skin effect is stronger for thick wires: The R_{ac} is 2% of R_{dc} if the penetration depth δ is equal to the wire radius. For 0.1-mm copper wire, that happens at approximately 200 kHz; for 1-mm wire, at 18 kHz. In the case of densely wound multiturn coils, the proximity effect may be stronger than the basic skin effect.

The parasitic capacitances of coils are discussed and derived in [11]. In solenoids, the self-capacitance is proportional to

$$C \sim l(d_m/h) \log(c/d_w) \quad (2.17)$$

where c is the wire pitch, d_m is the main diameter, and h is the coil height (Figure 2.1).

From (2.17) it follows that short disk coils have a lower capacitance than long solenoids for the same number of turns and mean diameter. The parasitic capacitances can then be reduced by using thicker wire insulation, layer spacing, or special winding techniques (wild winding, cross winding). A good technique is to divide the solenoid into n sections, connected in series (split coil): For a sufficiently large section spacing, the capacitance is reduced by $1/n^2$.

The actual value of the coil capacitance greatly depends on the winding geometry, so the deviation from calculated results may be large. The actual value can be calculated from the resonant frequency. The resonant frequency f_r can easily be measured if the coil is voltage driven from an audio generator: With increasing frequency, the coil current decreases, as the ωL component of the impedance dominates. At f_r the current reaches its minimum; if the frequency is further increased, the current increases because the parasitic impedance has a capacitive character. The general rule is that the coil can be used without special care for frequencies up to $f_r/5$.

The typical frequency characteristics of the sensitivity for the voltage-output induction coil of the thick solenoid type is shown in Figure 2.3. The Q factor of the resonance is determined by the coil resistance and losses in the parasitic capacitances. Sometimes a parallel damping resistor is added to decrease Q and linearize the characteristics so the coil can be used closer to f_r.

2.1.4 Current Output (Short-Circuited Mode)

If the induction coil equivalent circuit (see Figure 2.2) is connected to the current-to-voltage converter (Figure 2.4), the capacitance of the coil is virtually short-circuited and thus effectively eliminated. The useful frequency range of the induction coil is then considerably expanded [8, 9].

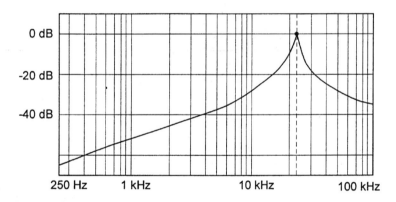

Figure 2.3 Frequency characteristics of the thick air solenoid with voltage output (from [9]).

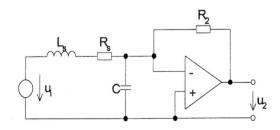

Figure 2.4 Short-circuited induction coil (from [9]).

If, for a sinusoidal waveform, we express the maximum value of the induced voltage using the formula $V_p = 2\pi f N A B_m$, from (2.7), the output voltage of the current-to-voltage (I/V) converter with the feedback resistor R_2 can be found as

$$V_2 = \frac{R_2}{\sqrt{(2\pi f L_s)^2 + R_s^2}} NA2\pi fB \qquad (2.18)$$

The frequency characteristics are shown in Figure 2.5. At higher frequencies, where $2\pi f L_s \gg R_s$, the output voltage is independent of frequency:

$$V_2 = \frac{R_2}{L_s} NAB \qquad I = \frac{NA}{L_s} B \qquad (2.19)$$

Because we suppose a sinewave shape of B, then (2.18) and (2.19) are correct for instantaneous, peak, rms, and mean values; it is just necessary to use the same value on both sides. If B is expressed as a maximum value, V will also be a maximum value.

If we suppose that $d_m \gg l, h$, then $L \sim N^2 d_m$ and the current sensitivity depends on d_m/N; here, the sensitivity is not dependent on the wire diameter d_w. The short-circuited induction coil can be used for observing nonsinusoidal fields, because (2.19) can be rewritten in the time domain, so that I and V_2 follow the waveform of B for $f \gg f_1 = R_s/L_s$. Furthermore, $R_s \sim N d_m/d_w^2$, so that $f_1 \sim 1/N d_w^2$, and the only way to decrease f_1 without decreasing the sensitivity is to increase the wire diameter d_w (with already mentioned limitations from skin effects and increase of coil weight).

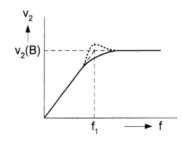

Figure 2.5 Frequency characteristics of the current-output induction sensor. The dotted line shows the (over)compensation by a serial capacitor.

Some frequency compensation may be necessary to guarantee the stability at higher frequencies and also possibly linearize the characteristics at low frequencies. The most simple method is to use a serial capacitor at the input. The approximate low frequency limit of the frequency-independent current-output induction magnetometer is the coil corner frequency f_1. The upper frequency limit is usually determined by the I/V converter.

The air-core induction coil should be designed according to the required frequency range and noise properties. In general, current-output mode and a large coil diameter are preferred. The number of turns should be fitted to match the amplifier input (Chapter 4). The design of a three-axis search-coil magnetometer is well documented in [10], where the coil parameters are (largest coil): d_m = 0.33m, l = 17 mm, h = 21 mm, N = 4,100 turns, R = 1,560Ω, L = 11.9H, f_c = 20 Hz. The frequency range is 20 Hz to 2.5 kHz (−1dB, after compensation at a low-frequency corner), and the current output sensitivity is 34 mA/mT.

The voltage output is suitable for measurements at ultralow frequencies, such as slow variations of the geomagnetic fields. For such applications, induction coils with ferromagnetic cores (which are often working in the short-circuited mode) or fluxgate magnetometers may be preferable.

2.2 Search Coils With a Ferromagnetic Core

In the design of an induction sensor with a ferromagnetic core, the demagnetization effects cannot be neglected. Because of that, H in the sensor core is substantially lower than the measured field H_0 outside the sensor core. Therefore, we must write for the flux density within the core:

$$B = \mu_0 \mu_r H_0 / [1 + D(\mu_r - 1)] = \mu_0 \mu_a H_0 \qquad (2.20)$$

where D is the effective demagnetizing factor and μ_a is the apparent relative permeability.

In general we suppose that the measured fields are small and thus μ_r is constant. Figure 2.6 illustrates the geometry of an induction sensor with ferromagnetic core.

2.2.1 Voltage Output Sensitivity

After inserting the ferromagnetic core, the voltage sensitivity is increased by μ_a:

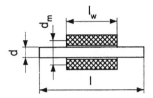

Figure 2.6 Geometry of an induction sensor with ferromagnetic core.

$$V_i = NA\mu_a dB(t)/dt \qquad (2.21)$$

The apparent permeability can be measured by comparing the inductance L of the coil with the core and L_0 with the core removed. The first estimate is $\mu_a = L/L_0$, but a more precise analysis has to take into consideration the different cross-sectional areas of the core and the coil. The analysis of the demagnetization effect D and of the inductance L can be found in [12–14] and references listed therein. For rotational ellipsoids having the length-to-diameter ratio m, the demagnetization factor D does not depend on μ_r, and it can be exactly calculated using Stoner's formula:

$$D = \frac{1}{m^2 - 1}\left[\frac{m}{\sqrt{m^2 - 1}}\ln(m + \sqrt{m^2 - 1}) - 1\right] \qquad (2.22)$$

For long ellipsoids, that can be simplified to

$$D = (\ln 2m - 1)/m^2 \qquad (2.23)$$

The effective demagnetization factor for long rods ($m = l/d > 10$) was approximated for $\mu_r = \infty$ by the Neumann-Warmuth empirical formula, which gives similar numerical results:

$$D \approx [2.01 \log(m) - 0.46]/m^2 \qquad (2.24)$$

For finite permeability, D is decreased. Demagnetization factors for prolate ellipsoids and cylinders are plotted versus m in Figure 2.7.

Ferromagnetic cores in sensors generally should be long and thin. The use of ferromagnetic cores has serious drawbacks. The core permeability changes with time and temperature and after the sensor is subjected to vibrations. The permeability also changes if the core is dc magnetized—it

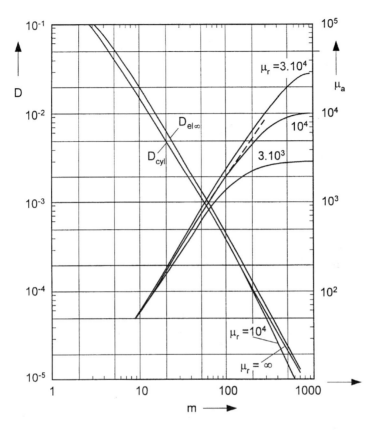

Figure 2.7 Demagnetization factors of the prolate ellipsoids and cylindrical rods; in the case of rods, μ_r is a parameter (after [6, 14]).

may happen if it is exposed to the Earth's field. The permeability is dependent on amplitude and frequency of the measured field; that dependence may cause sensor nonlinearity and distortion. Thus, it is important to design the sensor so that its constant is not very sensitive to changes of the core material permeability μ_r. If the material permeability is high, the apparent permeability μ_a is mainly given by the core geometry. That approach is illustrated by Figure 2.8. If m is extremely high, such as $m = 1,000$, μ_a is very high for high value of relative permeability (e.g., $\mu_r = 100,000$), but in this case μ_a is very sensitive to a variation of μ_r. However, if we use a high-permeability core (e.g., having μ_r larger than 50,000 for the whole temperature range) and moderate m (e.g., 100), we may reach a moderate value of $\mu_a = 2,500$, which is relatively insensitive to variations of μ_r. Moderate values of μ_a also keep the core dc magnetization caused by the

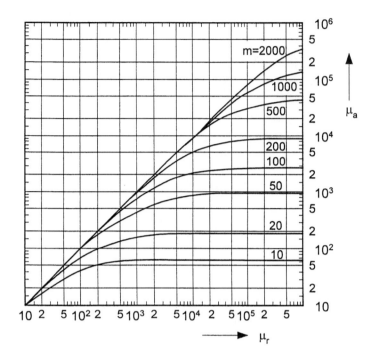

Figure 2.8 The apparent permeability μ_a of the cylindrical rods as a function of the length-to-diameter ratio m and the material relative permeability μ_r (after [14]).

Earth's field low. A Permalloy core with $\mu_a = 20{,}000$ would be saturated even by the Earth's field of 50 μT.

In other words, good design should satisfy the condition $\mu_r D \gg 1$, so that (2.20) is simplified and $\mu_a \cong 1/D$ becomes practically independent of μ_r. If that is satisfied, and we estimate D from (2.23) and substitute $A = \pi d^2/4$ and $m = l/d$ into (2.21), we obtain the following for the output voltage:

$$V_i \cong N \frac{\pi}{4} \frac{l^2}{\ln(2l/d) - 1} dB/dt \qquad (2.25)$$

If the measured field is sinewave shaped with a frequency $\omega = 2\pi f$, then we can write for the ideal frequency-normalized voltage sensitivity

$$S_{lf} = \frac{V}{Bf} = \frac{\pi^2 l^2 N}{2\ln(2l/d) - 1} \qquad (2.26)$$

That formula fits well the measured sensitivities of sensors having m between 20 and 100 [15].

Now let us discuss the optimum core diameter if the sensor maximum length is given. Figure 2.9 shows the situation for a 2m-long core with a 1m-long winding of 100,000 turns. Increasing the core diameter to more than 20 mm brings an increase of the core and winding weight but almost no increase in sensitivity (due to the demagnetization).

The coil length should be less than the core length to avoid the edge effects. The voltage sensitivity at the core ends may drop down to 10% of the maximum sensitivity in the middle part. On the other hand, a short coil with the same number of turns has a longer wire and thus a higher dc resistance; other parameters such as parasitic capacitance and ac resistance may also be worse. The reported coil lengths are between 50% to 90% of the core length; values between 75% to 90% may be recommended. Coil shapes other than solenoids have shown no significant advantages.

In any case, the induction sensors with ferromagnetic cores should be field calibrated. That is usually done in a calibration system, which has to be either much larger than the calibrated coil, or corrections for the calibration field nonhomogeneity should be done. For very large stationary coils, calibration is usually performed by comparing them with calibrated air coils using the Earth's natural field variations. An alternative calibration method is to

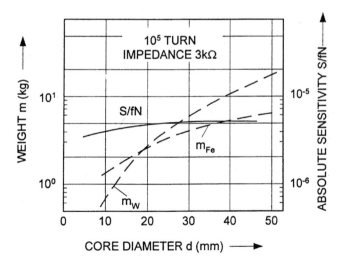

Figure 2.9 The absolute sensitivity S/fN (normalized to one turn, solid line), core weight m_c and winding weight m_w (dashed lines) of 2m sensor as a function of the core diameter. The 100,000 turns winding is 1m long (after [15]).

use a small field source (calibrated coil or rotating permanent magnet) at a large and known distance.

2.2.2 Thermal Noise of the Cored Induction Sensor (Voltage Output)

We should take into consideration that the mean coil diameter d_m is larger than the core diameter d and also that the voltage sensitivity of a cored solenoid is increased by a factor of μ_a according to (2.21). By similar calculations to those that we did for the air coil, we obtain for the equivalent field thermal noise B_{NT} as a function of the coil mass:

$$B_{NT} = \frac{4\sqrt{k_B T \rho \gamma}}{\pi f} \frac{d_m}{\mu_a d^2 \sqrt{m_w}} \quad (2.27)$$

Again, in this simplified model the noise does not depend on the wire diameter or the number of turns but only on the weight of the winding. It is also clear that the coil mean diameter d_m should be kept low by tight winding. The dependence on the core diameter is not simple, because it is related to μ_a: If we increase d, then we should substantially increase l and thus the core weight to keep μ_a constant.

Keep in mind that the coil thermal noise voltage is only one component of the magnetometer noise, which represents the theoretical low limit (for noiseless amplifier): The calculations show that a sensor containing 2.5 kg of copper and 2.5 kg of magnetic material having μ_r = 10,000 and a length-to-diameter ratio of 100 has a theoretical noise limit of 30 fT/$\sqrt{Hz}$ at 1Hz [6], which is 15 times lower than realistic values achievable with up-to-date real sensors having the same weight and 1m length (MEDA MGC-3 [2]). The influence of the amplifier noise is discussed in Section 2.3. Magnetic noise of the core is usually considered to be negligible.

2.2.3 The Equivalent Circuit for Cored Coils

The inductance of induction coils having short windings ($l_w \ll l$) in the middle of the long high-permeability cores was evaluated in [13] to be

$$L = \frac{N^2}{2} \frac{\mu_0 \pi l}{\ln 2m - 1} \quad (2.28)$$

If we suppose that $\mu_a = 1/D = m^2/(\ln 2m - 1)$ according to (2.23), we get

$$L = N^2 \frac{2\mu_0 \mu_a A_{\text{core}}}{l} \qquad (2.29)$$

Notice that this is twice the value derived in [16] for long winding on a core with effective length l.

If the winding is not short, the inductance is lower [6]. For a winding having 50% of the core (physical) length, the inductance is 90% of the ideal value; for 100% length, the inductance drops to 30% of the ideal value calculated from (2.28). Lukoschus [17] determined and experimentally verified by the empirical formula for l_w/l in the range of 0.1 to 0.8:

$$L = N^2 \frac{\mu_0 \mu_a A_{\text{core}}}{l} \left(\frac{l_w}{l}\right)^{-3/5} \qquad (2.30)$$

The self-capacitance of coils with an electrically conducting core cannot be easily calculated, especially in the case of shielded coils as capacitances between the shielding, core, and winding become dominant. The resistance R_s is increased by losses in the core: eddy-current losses ($R_e \sim f^2$) and hysteresis losses ($R_h \sim f$). Eddy-current losses are suppressed by using laminated cores. Ferrite cores have very small eddy-current losses, but they are less effective for low- and middle-frequency coil cores, because they have only a small permeability (initial relative permeability μ_i max. 10,000) compared to that of Permalloy (μ_i may be 100,000). Cobalt-based amorphous materials have some advantages over Permalloys: They do not need to be annealed, they are not so sensitive to vibrations, and their eddy currents are lower because they have higher resistivity; the disadvantage is their low saturation magnetization B_s. Good candidates for the future are nanocrystalline alloys: They have a large B_s, and they can be annealed to have a large constant permeability so that high linearity can be obtained.

The cored induction coils optimized for voltage output usually have a large L, a small C (because of the split winding), and often a high Q factor, which results in a large step-up near the resonance frequency f_r. A parallel damping resistor R_d is sometimes used to lower Q and linearize the frequency characteristics.

2.2.4 Cored Coils With Current Output

We can substitute L from (2.29) into the equation for the output voltage of the I/V converter (for sinewave B, $f \gg f_1 = R_s/L_s$) and we obtain

$$V_2 = \frac{R_2}{L} NA_{\text{core}} \mu_a B = \frac{R_2 l}{2N\mu_0} B \qquad (2.31)$$

and for the short-circuited current:

$$I = \frac{NA_{\text{core}} \mu_a}{L} B = \frac{l}{2N\mu_0} B = \frac{l}{2N} H \qquad (2.32)$$

Notice that this is half the short-circuited current of the very long, zero resistance solenoid used as a search coil. It appears as if the output current does not depend on the core parameters, but remember that this is true only for long, high-permeability cores.

2.3 Noise Matching to an Amplifier

The amplifier noise usually plays an important role in the induction magnetometer design. Because both the input current and the voltage noise are frequency dependent, it is not possible to globally optimize the magnetometer performance over the whole frequency range. The usual procedure is to fit the required noise level at two or three fixed frequencies. A good start for the design is to fix the sensor length and weight. Several iterations using numerical simulation tools are always necessary. Also, the selection of the amplifiers is not trivial, because they differ in the values and frequency dependencies of the input current noise I_{na} and voltage noise V_{na}.

The best modern low-noise operational amplifiers suitable for voltage amplification have a very low V_{na} but a rather high I_{na}. Those are two reasons why the coil resistance should be kept low—$R_{dc} \cong 100\Omega$ is recommended in [6], which contrasts with the traditional induction coils having huge N and R. Figure 2.10 shows noise characteristics of EMI induction coils [18].

It should be clearly stated that the general trend is the current-output mode except for the geophysical induction magnetometers working at ultralow frequencies (from 100 μHz). Some of those instruments use magnetic feedback—the voltage amplifier output is fed back through the feedback resistor into a compensation winding [6]. The resulting characteristic is similar to that of the current-output coil, that is, the influence of the parasitic capacitance is suppressed, and the sensitivity is constant in a wide frequency range. The use of a feedback coil separate from the sensing coil and having different parameters gives another degree of freedom in the magnetometer

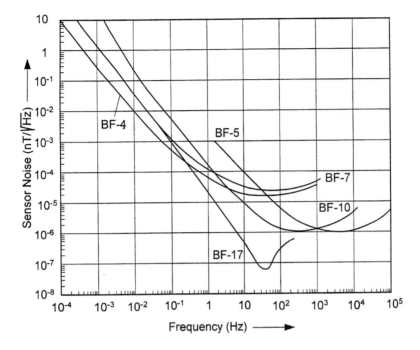

Figure 2.10 Noise of the EMI induction sensors (after [18]).

design. If the size and weight of the low-frequency magnetometer are strictly limited, as in the case of satellite instruments, fluxgate magnetometers (Chapter 3), which also measure the dc component, may be preferable.

2.4 Design Examples

Table 2.2 shows examples of induction magnetometers, and Table 2.3 lists examples of induction coil designs.

2.5 Other Measuring Coils

2.5.1 Rotating Coil Magnetometers

Rotation coil magnetometers have a flat frequency response starting from dc. If the effective coil area a (i.e., perpendicular to the direction of B) is changing by rotation, $a(t) = A_0 \cos\omega t$, then the induced voltage according to (2.1) will be

Table 2.2
Induction-Coil Magnetometers

Type	Frequency Band (Hz)	Noise@Frequency ($pT/\sqrt{Hz}$@Hz)	Length (m)	Core Diameter (mm)	N (Turns)	Weight (kg)	Producer or Reference
Air cored							
MEDA ACM-1	20–10k	0.32 pT@20 Hz 0.1 pT@100 Hz	0.4				MEDA
With ferromagnetic core							
Low frequency	0.3m–200	0.1 pT@1 Hz	1			15	LEMI [19]
Middle frequency	1–20k	10 fT@1 kHz	0.35			0.45	
High frequency	10–600k	2 fT@50 kHz	0.55			0.07	
MGC-3	0.2–1k	2.5 pT@0.2 Hz 0.5 pT@1 Hz 10 fT@1 kHz	1			5.5	MEDA
MGC-1	5–10k	0.5 pT@10 Hz 0.1 pT@100Hz 20 fT@1 kHz	0.25			1	
MGC-1A	30–10k	0.4 pT@100 Hz 60 pT@1 kHz	0.16			1	
MGCH-2	10–100k	2.5 pT@10 Hz 0.1 pT@100 Hz 10 fT@1 kHz 4 fT@10 kHz	0.325			0.71	

Table 2.2 (continued)
Induction-Coil Magnetometers

Type	Frequency Band (Hz)	Noise@Frequency (pT/√Hz@Hz)	Length (m)	Core Diameter (mm)	N (Turns)	Weight (kg)	Producer or Reference
IRM HF antenna	10k–2M	10 fT@10 kHz	0.26	1 cm^2		5	Ferrite core [20]
MFS 05 (broadband)	0.25m–8k	1 pT@0.1 Hz 0.1 pT@1 Hz	1			14	Metronix (flux feedback)
KIM 879 (audio)	1–20k	6 pT@1 Hz 0.1 pT@10 Hz	0.9 ferrite	22	10k	8	
L'Aquila	1m–10	0.44 pT/√Hz@1 Hz	2	15	200k		Feedback [21]
BF-17	0.3m–30	1 pT@0.1 Hz 20 fT@1 Hz 0.5 fT@10 Hz					EMI [18]
BF-4	0.1m–100	0.2 nT@1 MHz 0.5 pT@0.1 Hz 80 fT@1 Hz					
BF-6	2–10k	0.1 pT@10 Hz 2 fT@100 Hz					
Stationary		10 pT@0.1 Hz	2	20	100k	80	[15]
Portable		—	1	20	100k	—	
Satellite		—	0.4	5	20k	—	

Table 2.3
Examples of Induction Coil Designs

Type	Frequency Band (Hz)	Sensitivity V/(THz)	L (H)	R (Ω)	Noise or Minimum Working Frequency, f_1	f_r (Hz)	N (Turns)	l (m)	d (mm)	Weight (kg)	Producer or Reference
Ferromagnetic-core coils											
			10k				190k +2k	0.26		10	U. Newcastle [22]
Example	1–10k	3,200	70	3k	5 pT/√Hz@1 Hz	4,200	10k	0.352	6	—	[6]
UC	1–10k	50 mV/nT	500	2k	0.8 pT/√Hz@1 Hz	—	45k	0.6	16	2.5	Current output [1]
FC-1	5–1k	270	7.5	850	—	4,400	—	0.26	25	0.273	MEDA
Air coils											
Type 1	–200 Hz	Voltage out	5,600	151k	$f_1 = 39$ Hz	35k	168k	0.26	10	1	[23]
Type 2	1m–10k	Current out	10.4	404	resolution 100 pT @ 10 mHz	150	10k	0.3	2m	0.25	[3]
Campbell	4m–10	—	—	—			16k				
Spherical			70m	480	$f_1 = 1,100$ Hz	30k	NA = 3.6 m^2				[9]
Cylindrical			94	48	$f_1 = 80$ Hz	23k	NA = 2.3 m^2				

$$V_i = -BNA_0 \omega \sin \omega t \qquad (2.33)$$

Rawson-Lush produced a range of "spinning coil" magnetometers with a range of 50 nT to 20 mT and a subnanotesla resolution [24]. A sophisticated pneumatically powered magnetometer with air bearings and mercury contacts is described in [25]. A 20-mm-diameter coil spinning at 20,000 revolutions per minute gives a resolution of 10 pT and an offset below 100 pT. An instrument called a Rocoma is now produced by RS Dynamics.

2.5.2 Moving-Coils, Extraction Method

Small search coils connected to an electronic integrator can be used to measure B: The induced voltage is integrated while the coil is extracted from the measuring location M with the field B_M to place with "zero" field B_Z [26]:

$$\int V_i dt = -NA(B_Z - B_M) \qquad (2.34)$$

This technique for precise mapping of the magnetic field is described by Green [5]. The coil movement is computer controlled; after scanning each row, the coil is returned to the "null" position to check the integrator drift. Another coil of the same geometry may be fixed inside the magnet and connected antiserially to cancel the signal deviations due to the source fluctuations. Linear drift, caused, for example, by constant thermoelectric voltages, can be corrected. The integrator drift is linear within several minutes after proper compensation; therefore, the integrator should be periodically zeroed and analog drift compensated.

The dc-magnetic induction in the specimen or part of the magnetic circuit can be measured by wrapping around it several turns of wire connected to the input of the integrator (fluxmeter). Another possibility is to use a sliding search coil. The magnetic moment of a permanent magnet can be measured using a Helmholtz coil connected to a fluxmeter by the flux change produced after the magnet is inserted into the coil [27].

2.5.3 Vibrating Coils

Vibrating coils are used to measure field gradients. Their main application is in the measurement of magnetization of samples. The coil is in the vicinity of the sample and measures the field difference between two positions.

Periodic excitation allows the extraction of weak signals by a synchronous detector. In general, vibrating sample or rotating sample magnetometers are the preferred solutions, but some types of samples (such as liquids, dusts, or brittle solids) are sensitive to vibrations and acceleration, so vibrating or rotating coils are used instead.

2.5.4 Coils for Measurement of *H*

Coils in connection with an analog integrator (or in the ac case, a voltmeter measuring the mean value) can also be used to measure $H \cdot l$ or the magnetic potential drop (voltage) U in amperes. *H*-coils are flat coils that serve to measure *H* near the surface of a ferromagnetic material. They are based on the fact that the tangential component of *H* does not change across the material boundary. *H*-coils are used in single-sheet testers (devices for the measurement of hysteresis loops of electrical steels and other soft magnetic materials). *H*-coils are sometimes doubled to compensate for the linear decrease of *H* with distance from the sample surface: One coil is close to the surface and the second coil is slightly displaced; the value of *H* inside the sample is extrapolated. *H*-coils are also sometimes used for measuring the parameters of magnetically hard materials in electromagnets.

2.5.4.1 Rogowski-Chattock Potentiometer

The Rogowski-Chattock potentiometer measures the magnetic potential drop (voltage) *U* (in amperes) between two points. It consists of a long slim coil wound on a nonmagnetic and nonconducting core. The typical shape is a semicircle with the ends of the coil lying in one plane [28]. If the potentiometer lies flat on the surface of a ferromagnetic material (Figure 2.11) and there is no electric current flowing inside the semicircle, we can write

$$\oint H dl = 0 \qquad (2.35)$$

and thus

$$\int_X H_{\text{Fe}} dl = \int_Y H_{\text{air}} dl = U_{AB} \qquad (2.36)$$

where H_{Fe} and H_{air} are the field intensities inside the specimen and in the air, respectively.

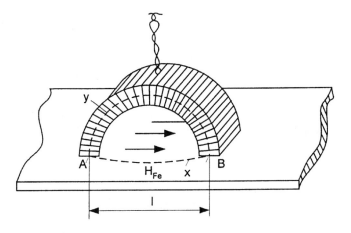

Figure 2.11 The Rogowski-Chattock potentiometer.

For a short path, we can write $U_{AB} = H_{Fe} l_{AB}$. If the winding density is high and constant, we can write the following equation for the coil flux of the potentiometer:

$$\sum_N \Phi_i = \mu_0 A \sum_N H_i \cong \frac{\mu_0 A}{l} \int_A^B H dl \qquad (2.37)$$

Semicircular potentiometers are used to measure H in ferromagnetic samples. In the case of ac magnetization, the induced voltage is measured (mean value, if nonsinewave) and eventually integrated (if the waveform is also required). The dc value of H can be measured, if we integrate the induced voltage while the coil is removed from the sample surface to a sufficiently large distance, where the stray field is negligible.

2.5.4.2 Straight Potentiometer

A straight potentiometer (potential coil) together with an integrating fluxmeter is used to measure the magnetic potential drop between two points. The induced coil output voltage is integrated while one coil end is moved between two points; the other coil end is supposed to be at zero or constant field [27]. The influence of the Earth's field can be compensated for by another coil having the same coil constant but a smaller size, attached to the "far end" of the potentiometer and connected antiserially. This technique is used for mapping H in permanent magnet structures.

Potentiometers are made as single-layer solenoids with a very small area. A large number of turns are necessary to obtain sufficient sensitivity. The winding uniformity is very important. Potentiometers should be calibrated because their sensitivity constant cannot be calculated from the dimensions with sufficient precision.

Circular Rogowski coils of the ring shape are used to measure large ac and transient currents [29]. The device contains no magnetic material, so it is extremely linear; again, the uniformity of the winding is critical.

References

[1] Labson, V. F., et al., "Geophysical Exploration With Audiofrequency Natural Magnetic Fields," *Geophysics*, Vol. 50, 1985, pp. 656–664.

[2] Macintyre, S. A., *ELF Magnetic Field Measurement Methods and Accuracy Issues*, brochure, MEDA Inc., 458 Spring Park Place, Herndon, VA 22070.

[3] Stuart, W. F., "Earth's Field Magnetometry," *Reports on Progress in Physics*, Vol. 35, 1972, pp. 803–881.

[4] Brown, F. W., and J. H. Sweer, "The Flux Ball: A Test Coil for Point Measurements of Inhomogeneous Magnetic Fields," *Rev. Sci. Instr.*, Vol. 16, 1945, pp. 276–279.

[5] Green, M. I., "Search Coils," in S. Turner (ed.), *Measurement and Alignment of Accelerator and Detector Magnets*, Geneva, Switzerland: CERN, 1998.

[6] Dehmel, G., "Induction Sensors," in R. Boll and K. J. Overshott (eds.), *Magnetic Sensors*, Weinheim, Germany: VCH, 1989, pp. 205–255.

[7] Ilo, A., et al., "Sputtered Search Coils for Flux Distribution Analyses in Laminated Magnetic Cores," *J. Phys. IV France*, Vol. 8, 1998, pp. 733–736.

[8] Macintyre, S. A., "Magnetic Field Sensor Design," *Sensor Review*, Vol. 11, 1991, pp. 7–11.

[9] Kašpar, P., and P. Ripka, "Induction Coils: Voltage Versus Current Output," *Proc. Imeko World Congress*, Vienna, 2000, pp. 55–60.

[10] Macintyre, S. A., "A Portable Low Noise Low Frequency Three-Axis Search Coil Magnetometer," *IEEE Trans. Magn.*, Vol. 16, 1980, pp. 761–763.

[11] Welsby, W. G., *The Theory and Design of Inductance Coils*, London: MacDonald, 1960.

[12] Primdahl, F., et al., "Demagnetising Factor and Noise in the Fluxgate Ring-Core Sensor," *J. Phys. E: Sci. Instrum.*, Vol. 22, 1989, pp. 1004–1008.

[13] Kaplan, B. Z., "A New Interpretation of the Relationship Existing Between Demagnetizing Factor and Inductance," *IEEE Trans. Magn.*, Vol. 30, 1994, pp. 2788–2794.

[14] Bozorth, R. M., and D. M. Chapin, "Demagnetizing Factors of Rods," *J. Applied Physics*, Vol. 13, 1942, pp. 320–326.

[15] Saito, T., et al., "Development of New Time-Derivative Magnetometers To Be Installed on Spacecrafts" (in Japanese), *Bull. Inst. Space Aeronaut. Sci.*, Univ. Tokyo, 1980, pp. 1419–1429.

[16] Primdahl, F., et al., "The Short-Circuited Fluxgate Output Current," *J. Phys. E: Sci. Instrum.*, Vol. 22, 1989, pp. 349–353.

[17] Lukoschus, D. G., "Optimization Theory for Induction-Coil Magnetometers at Higher Frequencies," *IEEE Trans. Geosci. Electr.*, Vol. 17, 1979, pp. 56–63.

[18] www.emi.com.

[19] Berkman, R. J., B. L. Bondaruk, and V. Korepanov, "Advanced Flux-Gate Magnetometers With Low Drift," *Proc. Imeko World Congress*, Tampere, Finland, 1997, pp. 121–126.

[20] Hausler, B., et al., "The Plasma Wave Instrument on Board the AMPTE IRM Satellite," *IEEE Trans. Geosci. and Rem. Sens.*, Vol. 23, 1985, pp. 267–273.

[21] Cantarano, S., and M. Vellante, "A Facility for Measuring Geomagnetic Micropulsations at L'Aquila, Italy," *Il Nuovo Cimento*, Vol. 6C, 1983, pp. 40–48.

[22] University of Newcastle, Ground-Based Magnetometer Stations, http://plasma.newcastle.edu.

[23] Prance, R. J., T. D. Clark, and H. Prance, "Compact Broadband Gradiometric Induction Magnetometer System," *Sensors and Actuators A*, Vol. 76, 1999, pp. 599–601.

[24] Lush, M. J., "Rotating Coil Gaussmeters," *Instruments and Control Systems*, May 1964, pp. 111–113.

[25] Prihoda, K., et al., "MAVACS—A New System for Creating a Non-Magnetic Environment for Paleomagnetic Studies," *Geologia Iberica* 12/1988-89, p. 223. Technical data at www.rsdynamics.com.

[26] Jiles, D., *Introduction to Magnetism and Magnetic Materials*, London: Chapman & Hall, 1998.

[27] Steingroever, E., and G. Ross, *Magnetic Measurements With the Helmholtz Coil and the Potential Coil*, brochure, Dr. Steingroever GmbH, Emil Hoffmann Strasse 3, Köln, D-50996, Germany.

[28] Crangle, J., *Solid State Magnetism*, New York: Van Nostrand Reinhold, 1991.

[29] Ramboz, J. D., "Machinable Rogowski Coil, Design and Calibration," *IEEE Trans. Instrum. Meas.*, Vol. 45, pp. 511–515.

3

Fluxgate Sensors
Pavel Ripka

Fluxgate sensors measure the magnitude and direction of the dc or low-frequency ac magnetic field in the range of approximately 10^{-10} to 10^{-4} T. The basic sensor principle is illustrated in Figure 3.1. The soft magnetic material of the sensor core is periodically saturated in both polarities by the ac excitation field, which is produced by the excitation current I_{exc} through the excitation coil. Because of that, the core permeability changes, and the dc flux associated with the measured dc magnetic field $\mathbf{B}_0$ is modulated; the "gating" of the flux that occurs when the core is saturated gave the device its name. Figure 3.2 shows the simplified corresponding waveforms. The device output is usually the voltage V_I induced into the sensing (pickup) coil at the second (and also higher even) harmonics of the excitation frequency. This voltage is proportional to the measured field.

A similar principle was used earlier in magnetic modulators and magnetic amplifiers, but in those cases the measured variable was a dc electrical current flowing through the primary coil. A comprehensive bibliography of early fluxgate papers was collected by Primdahl [1–3]. The first patent on the fluxgate sensor (in 1931) was credited to H. P. Thomas [4]. Aschenbrenner and Goubau worked on fluxgate sensors from the late 1920s; by 1936 they reported 0.3-nT resolution on a ring-core sensor [5]. Sensitive and stable sensors for submarine detection were developed during World War II. Fluxgate magnetometers were used for geophysical prospecting, airborne field mapping, and later for space applications. Since Sputnik 3 in 1958, hundreds of fluxgate magnetometers (most of them three-axis) have been launched. Fluxgate magnetometers worked on the Moon [6] and in deep space

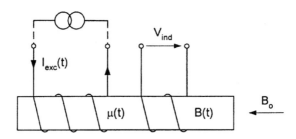

Figure 3.1 The basic fluxgate principle. The ferromagnetic core is excited by the ac current I_{exc} of frequency f into the excitation winding. The core permeability $\mu(t)$ is therefore changing with $2f$ frequency. If the measured dc field $\mathbf{B}_0$ is present, the associated core flux $\Phi(t)$ is also changing with $2f$, and voltage V_{ind} is induced in the pickup (measuring) coil having N turns (after [19]).

[7]. Since the 1980s, magnetic variation stations with fluxgates supported by a proton magnetometer [8] have been used for observing changes in the Earth's magnetic field [9]. Fluxgate compasses are extensively used for aircraft and vehicle navigation (Chapter 10). Förster [10] started to use the fluxgate principle for the nondestructive testing of ferromagnetic materials. The fluxgate principle is also used in current sensors and current comparators [11]. Compact fluxgate magnetometers are used for navigation, detection and search operations, remote measurement of dc currents, and reading magnetic labels and marks. Magnetic sensor applications are discussed in Chapter 10.

Fluxgate sensors are solid-state devices without any moving parts and they work in a wide temperature range. They are rugged and reliable and may have low energy consumption. They can reach 10-pT resolution and 1-nT long-term stability; 100-pT resolution and 10-nT absolute precision is standard in commercially produced devices. Many dc fluxgate magnetometers have a cutoff frequency of several hertz, but when necessary, fluxgates can work up to kilohertz frequencies. Fluxgates are temperature stable: The offset drift may be 0.1 nT/°C, and sensitivity tempco is usually around 30 ppm/°C, but some fluxgate magnetometers are compensated up to 1 ppm/°C. Most of the fluxgate sensors work in the feedback mode; the resulting magnetometer linearity typically is 30 ppm.

Fluxgates are the best selection if resolution in the nanotesla range is required. They may have a noise level comparable to that of a high-temperature superconducting quantum interference device (SQUID), but a much larger dynamic range. If pT or even smaller fields are measured, a low-temperature SQUID should be used (SQUIDs are covered in Chapter 8). For higher fields, the fluxgate competitors are the magnetoresistors,

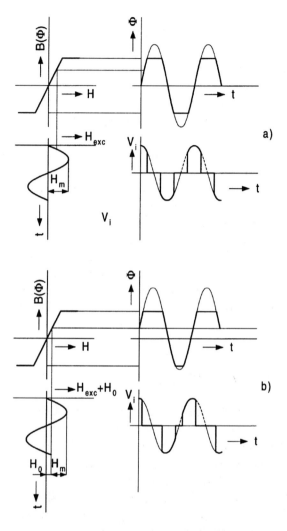

Figure 3.2 Simplified fluxgate waveforms: (a) in the zero field; (b) with measured field H_0.

especially anisotropic magnetoresistance (AMR) sensors (Chapter 4). Commercially available AMR magnetoresistors have a resolution worse than 10 nT, but they are smaller and cheaper and consume less energy than fluxgates.

Gordon and Brown [12] and Primdahl [13] have written excellent review articles on fluxgate sensors. The most important source of information about the development of the fluxgate sensors in Russia and the former USSR are the books written by Kolachevski [14] and Afanasiev [15]. Japanese fluxgate designs are reviewed in [16]. A lot of information can be found in the VCH monograph on magnetic sensors: A chapter on fluxgate sensors is interesting, but it focuses on the time-domain signal processing principles, which were once promising but later turned out to be not as advantageous as traditional second harmonic detection [17]. The chapter covering applications of the same book written by Gibbs and Squire [18] still contains valuable information about fluxgate sensors. The last review on fluxgates was published in 1992 [19]. A remarkable advanced fluxgate magnetometer was designed for the Oersted satellite launched in 1999 to map the Earth's magnetic field; construction details about that instrument can be found in [20]. The detailed instructions on how to build and calibrate a simple fluxgate magnetometer can be found in [21].

The typical modern low-noise fluxgate magnetometer is the parallel type with ring-core sensor, but double-rod sensors also have a lot of advantages. A phase-sensitive detector extracts the second harmonic in the induced voltage, and the pickup coil also serves for the feedback. Other designs are used for specialized purposes, such as miniature rod-type sensors for nondestructive testing or position sensing.

3.1 Orthogonal-Type Fluxgates

Figure 3.1 showed the configuration of the most widely used parallel type of fluxgate, for which both the measured and the excitation fields have the same direction. Another type of fluxgate, called the orthogonal sensor, has an excitation field perpendicular to the sensitive axis of the sensor, which is identical to the ideal axis of the sensing coil.

Probably the first orthogonal fluxgate sensors were patented by Alldredge [22]. Two suggested sensor configurations are shown in Figure 3.3: They have a core in the form of a ferromagnetic wire (a) or tube (b). A current through the core excites the first type; such configuration has the principal disadvantage that the excitation field in the center of the core is zero, which causes problems with sensor remanence. The second type is excited by a single wire in the tube or by a toroidal coil. The gating mechanism of the orthogonal type fluxgate is explained in [23]. Primdahl constructed a magnetometer using an orthogonal-type fluxgate sensor with a ferrite tube core and published the design details and the results of calibration and testing [13].

The miniature planar sensor described by Seitz [24] is also of the orthogonal type. Another orthogonal sensor made from Permalloy film elec-

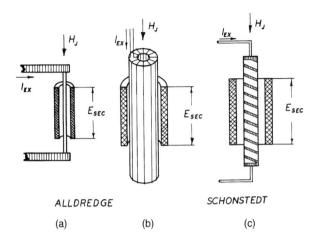

Figure 3.3 Orthogonal-type fluxgates: (a) with wire core; (b) with tube core; and (c) helical core (from [23]).

trodeposited on a solid copper rod was reported by Gise [25]. A mixed orthogonal-parallel sensor was constructed from the helical core formed by a tape wound on a tube [26], as shown in Figure 3.3(c). The hairpin sensor developed by Nielsen et al. [27] uses the same principle, but the anisotropy of the sensor core is not caused by its shape or stress; the helical anisotropy is induced by annealing the tape under torsion. Some sensors, like the one described by Perlov et al. [28], use a rotating excitation field, in which they mix both parallel and orthogonal modes of operation.

This book concentrates on parallel type fluxgates, which generally have better parameters. Thus, if the fluxgate type is not explicitly specified, we always have in mind the common parallel-field type.

3.2 Core Shapes of Parallel-Type Fluxgates

The basic single-core design is being used in simple devices like the one described by Rabinovici and Kaplan [29]. The main problem is the large signal on the excitation frequency at the sensor output, because the device acts as a transformer. Thus, double cores (either double-rod or ring-core) are normally used for precise fluxgates. The basic parallel-type fluxgate configurations are shown in Figure 3.4.

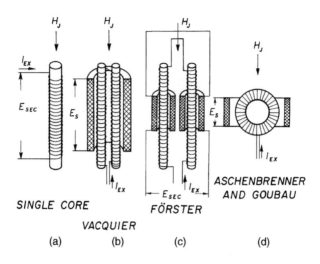

Figure 3.4 Parallel-type fluxgates: (a) single-rod sensor; (b) double-rod sensor of Vacquier type; (c) double-rod sensor of Förster type; and (d) ring-core sensor (from [23]).

3.2.1 Single-Rod Sensors

Sensors with a single open core are used for magnetometers using time-domain detection, the devices described by Sonoda and Ueda [30] and Heinecke [31] being typical examples, and they are often used for auto-oscillation or magnetic multivibrator sensors, such as in [32]. Principles of these devices are described in Section 3.5. Single strips of amorphous alloy were also used in simple magnetometers by Ghatak and Mitra [33] and Zhang et al. [34].

3.2.2 Double-Rod Sensors

Most magnetometers use the conventional method of evaluating the second harmonics of the output signal. In such cases, great difficulties may arise from the presence of a large signal at the excitation frequency and odd harmonics, caused by the transformer effect between the excitation and sensing windings. The large part of that spurious signal is eliminated in the two-core sensor consisting of symmetrical halves excited in opposite directions, so that the mutual inductance between the excitation and measuring coils is near zero. Figure 3.4(b) shows the double-rod sensor with a common pickup coil (Vacquier type). Such a sensor was developed by Moldovanu et al. for the INTERBALL satellite instrument [35] and later tested with different kinds of core materials [36, 37]. Förster [10] used two identical pickup coils connected serially; such a configuration, as shown in Figure 3.4 (c), allows easier matching of the cores and adjustment of the sensor balance by moving the cores with respect to their coils. A 50-cm-long fluxgate of this type with 10-pT resolution was constructed for a geophysical observatory [38].

The large geometrical anisotropy gives the main advantages of the rod-type sensors: great sensitivity and resistance against perpendicular fields. There are also important disadvantages: Sensor cores with open ends usually are noisier and their offset less stable with temperature and time than the closed-core sensors. The open rods are more difficult to saturate, so those sensors are also more energy consuming and more susceptible to perming effects (i.e., offset change after a shock of a large field).

Weyand and Bosse proposed the H-shaped core as a compromise between the double-rod and closed-core designs [39]. The excitation coils are positioned on the horizontal parts of H, while the detection coils are on the inner vertical parts. The outer legs serve as flux concentrators for increasing the sensitivity. The principal weakness of this type of design is that the flux concentrators are not saturated by the excitation field, so their remanent field is not removed and the sensor would have large perming effect errors.

3.2.3 Ring-Core Sensors

The widely used ring-core sensor is shown in Figure 3.4(d). The excitation coil is in the form of an anuloid, while the pickup coil is a solenoid. Ring-core sensors can be regarded as a form of balanced double sensor: The two half-cores are the parts of the closed magnetic circuit. The core usually consists of several turns of thin tape of soft magnetic material. The ring-core sensor design was used as early as 1928 by Aschenbrenner and Goubau [5]. Sensors made from sheets in the shape of flat rings or race-tracks have been also reported [40, 41]. Although the ring-core sensors have, in principle, low sensitivity due to the demagnetization, ring-core geometry was found to be advantageous for the low-noise sensors for several possible reasons:

- It allows fine balancing of the core symmetry by rotating the core with respect to the sensing coil.
- The possible mechanical stress in the core is uniformly distributed.
- The open ends usually accompanied by regions of increased noise are absent. Possible tape ends play only a minor role.

The size of the core affects the sensor sensitivity, as analyzed by Gordon et al. in [42]. Although the problem is complex due to demagnetization effect and nonlinearity, the sensitivity generally increases with the sensor diameter. With the given diameter, there is always an optimum of other dimensions for the best performance; usually it must be found experimentally. The core dimension and number of excitation and measuring turns also play a crucial role in matching the excitation and interfacing electronic circuits. A typical low-noise sensor size is a 17- to 25-mm diameter ring-core wound from 4 to 16 turns of tape 1- to 2-mm wide and 25-μm thick.

An original mechanical design used for the sensors was developed by Nielsen et al. for the Oersted satellite [20]. The amorphous tape is wound in the groove in the *inner* surface of the ceramic support ring (bobbin). The tape is kept in position by its own spring force, but to increase the pressure and make it uniform, Nielsen later used an additional nonmagnetic stainless-steel spring of complex shape.

To reach high sensor symmetry, the excitation coil should be wound with high uniformity. That is a difficult task even if numerically controlled winding machines are used. A good technique is to keep the wire turns dense at the inner diameter of the bobbin.

3.2.4 Race-Track Sensors

A race-track (oval) sensor is shown in Figure 3.5. This core geometry has a lower demagnetization factor than that of the ring-core, so the sensor sensitivity is higher and the sensor is less sensitive to perpendicular fields. Race-tracks, on the other hand, still have the advantages of closed-type sensors. Race-track sensors wound from tape were reported by Gordon and Brown [12]. Such sensors have a potential problem from the higher tape stress in the corners.

Sensors made of etched sheets of race-track shape have been reported by Ripka in [43]. The core is made of eight layers of 35-μm-thick amorphous ($Co_{67}Fe_4Cr_7Si_8B_{14}$) material. Some of the sensors have 8-pT rms noise in the 50 mHz to 10 Hz band. The core length was 70 mm, core width 12 mm, and the track width 2 mm.

The main problem of the race-track sensors are the large spurious signals; they cannot be simply balanced as the ring-cores can. A modified race-track core that allows for a final adjustment is suggested in [44, 45]. The track width varies slightly, changing along the core length. By sliding the pickup coil out of its symmetrical position, the effective cross section

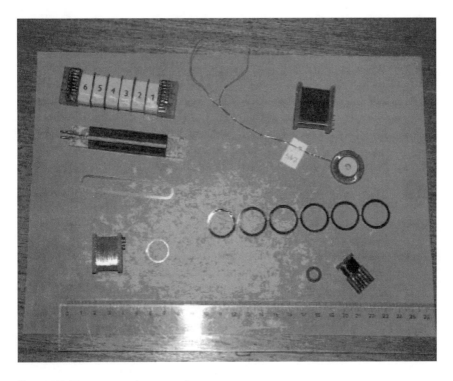

Figure 3.5 Ring-core and race-track sensors.

of the core halves can be changed, which allows adjusting for the unbalance caused by core nonhomogeneity or excitation winding imperfections.

3.3 Theory of Fluxgate Operation

The classic description of the fluxgate principle is given in [42]. Many other papers are based on idealized magnetization characteristics $B(H)$ of the sensor core and excitation field waveform $H(t)$. Primdahl [23] verified the validity of such a semigraphical description by observing the B-H curves on a two-core sensor. A similar description for the ring-core sensor from [46] is shown in Figure 3.6. An idealized hysteresis loop of one-half of the sensor (Φ_1 vs. H_{exc}) is identical to the characteristics of the magnetic material as long as the magnetic circuit is closed, as shown in Figure 3.6(a). When the external measured dc magnetic field is present, the characteristic curve is distorted, as shown in Figure 3.6(b). For some critical value of H_{exc}, that half of the core, in which the excitation and the measured fields have the same direction,

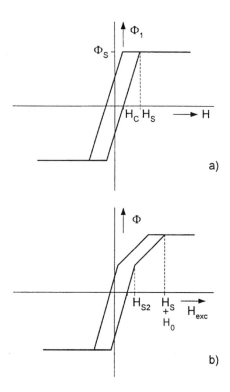

Figure 3.6 Ring-core fluxgate model: magnetization characteristics of the half-core (a) without external field; (b) with external dc field H_0 (after [19]).

becomes saturated. At that moment, the reluctance of the magnetic circuit rapidly increases (because the flux is "gated"), and the effective permeability of the other half-core is decreased. In addition, the characteristic is shifted by the external field along the H axis.

The characteristic for the second half-core is symmetrical with respect to the Φ axis. By summing up those two loops, we obtain the transfer function, that is, the Φ versus H_{exc} characteristic, which is shown in Figure 3.7. The height of the transfer function (which corresponds to the peak-to-peak change of pickup coil flux) increases with the measured field. The mentioned dependence is linear up to high field intensities, for which the whole sensor becomes saturated. That principle has also been used for evaluation of the sensor output by integrating the induced voltage and measuring the peak-to-peak (p-p) value of the waveform obtained that way. Figure 3.8 shows the actual shape of the dynamical hysteresis loop and the transfer function measured at 1 kHz with an oval-shaped core from amorphous

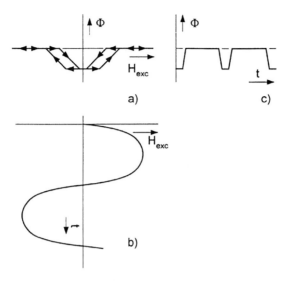

Figure 3.7 Ring-core fluxgate model, derivation of the pickup coil flux: (a) transfer function Φ vs. H_{exc}; (b) excitation field; and (c) pickup coil flux (after [19]).

material. Thus, if we know the excitation field waveform and transfer function, we can construct the flux waveform, and by taking its derivative, we obtain the waveform of the induced voltage.

Kaplan analyzed the fluxgate sensors from the perspective of a field theory and showed its duality with electric field sensors in [47–49].

3.3.1 The Effect of Demagnetization

If we assume a constant pickup coil area in the general induction sensor equation (2.2), we have:

$$V_i = NA\mu_0\mu \, dH(t)/dt + NA\mu_0 H \, d\mu(t)/dt \tag{3.1}$$

We see that the basic induction effect (first term) is still present in fluxgate sensors: In some cases, it can cause interference, but sometimes it can be used simultaneously with the fluxgate effect to measure the ac component of the external field (Section 3.13). But here we concentrate on the fluxgate effect, the second term in (3.1). The time dependence of the core permeability is caused by the periodical excitation field. The given formula can be used for long rod-type sensors. For the more often used ring cores, we should consider the demagnetization effect, that is, recognize that H in

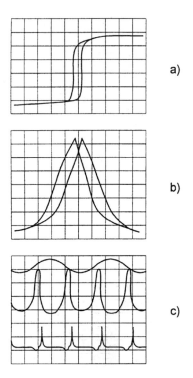

Figure 3.8 Measured ring-core fluxgate waveforms: (a) dynamic hysteresis loop; (b) transfer function; and (c) induced voltage (after [19]).

the core material is substantially lower than the measured field H_0 outside the sensor core. Thus, we must write for the flux density within the core:

$$B = \mu_0 \mu H_0 / [1 + D(\mu - 1)] = \mu_0 \mu_a H_0 \qquad (3.2)$$

where D is the effective demagnetizing factor (as seen from the pickup coil) and μ_a is the apparent permeability, $\mu_a = \mu/[1 + D(\mu - 1)]$, for very high μ, $\mu_a \to 1/D$.

If we consider the demagnetization, the equation for the fluxgate output voltage becomes more complex:

$$V_i = NA \frac{dB}{dt} = NA\mu_0 H_0 \frac{(1 - D)}{\{1 + D[\mu(t) - 1]\}^2} \frac{d\mu(t)}{dt} \qquad (3.3)$$

The detailed analysis of the demagnetization effect can be found in [50] and the references listed therein. A good estimate for the effective demagnetization factor of tape-wound ring cores is

$$D = kT/d \qquad (3.4)$$

where T and d are the core thickness and the diameter and k is constant depending on the pickup coil geometry and other factors, but not on core permeability and excitation parameters.

As we see, D approximately does not depend on the tape width w; that holds for $w \gg T$.

Primdahl measured $k = 0.22$ for 17-mm-diameter ring-cores wound of 1-mm-thick tape. The pickup coil was wound on the 23-by-23-by-5.7-mm bobbin. The typical values he measured on a 10-wrap core were $D = 0.0032$, $\mu = 33,000$, and $\mu_a = 300$ for no external dc field and zero excitation.

If we substitute those values into (3.3), we obtain for zero excitation:

$$V_i = 10^{-4} NA\mu_0 H_0 d\mu(t)/dt = 10^{-4} NA\mu_0 H_0 (d\mu(t)/dH_{exc})(dH_{exc}/dt) \qquad (3.5)$$

where A is the core cross-sectional area and N is the number of turns of the pickup coil, which would suggest that the sensitivity of the ring-core sensor is 10,000 times lower than that of a very long rod core. In fact, the real situation is much better: For zero excitation current, the core permeability is constant so that $d\mu(t)/dH_{exc} = 0$ and therefore V_i is zero regardless of the core shape. The fluxgate effect appears at the moment the core becomes saturated; then the permeability decreases and the effect of demagnetization is lower. For $\mu(t) = 1,000$ the ring-core sensor sensitivity is only 0.6 of the ideal rod sensor. The local distribution of the demagnetizing field in the ring-core is theoretically analyzed in [51].

The given formulas show that the analytical description of the voltage-output fluxgate mechanism is complex. Further analysis by Fourier transformation may follow to calculate the sensor sensitivity, as was done by Primdahl [23], Burger [52], and later Nielsen et al. [53]. A computer simulation of the fluxgate magnetometer using PSPICE® and the Jiles-Atherton model of the hysteresis loop was performed in [54]. However, such descriptions have limited practical use. The general practical rules for achieving high sensitivity can be deducted from (3.3) and also from practical experience rather than complex theoretical models:

1. Voltage sensitivity increases with N (for high N, sensitivity is limited by other factors, such as coil parasitic capacitance).
2. Sensitivity increases monotonously with D.
3. For given D, sensitivity increases with T for small T, but this dependence saturates for big T.
4. The core material should have a steep change of permeability when coming into saturation.
5. The preferred excitation current waveform is squarewave.
6. Voltage sensitivity increases with excitation frequency (because $dH_{exc}/dt \sim f$), until parasitic effects (which change the shape of the hysteresis loop) become important.

In real applications, the voltage output is tuned, either intentionally to utilize parametric amplification or unintentionally by parasitic pickup coil capacitance (Section 3.7). Also, the real excitation current waveforms are very different from the ideal shapes (such as sinewave, squarewave, and triangle), which further limits the practical applicability of the voltage-output fluxgate theory. In contrast, the current output fluxgate description (Section 3.8) is quite simple and matches the reality much better.

3.4 Core Materials

It is difficult to discuss the selection of the core material generally, because it depends on the type and the geometry of the sensor, on the type of processing of the output signal, and also on the excitation frequency and required temperature range. However, there are general requirements for the material properties.

- High permeability (permeability may be further reduced intentionally by thermomagnetic treatment to reduce the noise);
- Low coercivity;
- Nonrectangular shape of the magnetization curve (points 2 and 3 are equivalent to the smallest possible area of the B-H loop);
- Low magnetostriction;
- Low Barkhausen noise;
- Low number of structural imperfections, low internal stresses;
- Smooth surface;

- Uniform cross section and large homogeneity of the parameters;
- Low saturation magnetization;
- High electrical resistivity.

All known studies of core material composition and processing parameters have shown that the minimum noise is achieved for near-zero magnetostriction alloys. References [13, 42, 55] discuss the sensor core material properties. The material traditionally used for the sensor cores is high-permeability, low-magnetostriction Permalloy (or Mumetal) in the form of a thin tape. The Permalloy 81.6 Ni 6 Mo developed by the Naval Ordnance Laboratory [56] for a low-noise NASA magnetometer is still the superior material with the lowest noise. Electrodeposited Permalloy [24, 25] and ferrite [13] were also used for the sensor core but generally with worse results. Thin-film sensors made by vacuum deposition were reported by Hoffmann [57]. Orthogonal fluxgates with a ferrofluid core also show no advantages over the traditional design [58]. Fluxgate effects were also reported in high-temperature superconductors [59, 60].

Amorphous magnetic materials are magnetic glasses produced by rapid quenching. They started to be used for fluxgate cores from the early 1980s. Properties of those alloys as sensor materials are discussed by Mohri [61]. A study concerning the noise of the amorphous magnetic materials was performed by Shirae [62], who found that

- Low-magnetostriction cobalt-based alloys are suitable for fluxgate applications.
- Room temperature noise decreases with Curie temperature.
- Annealing of the tape may further decrease the noise level.

Narod at al. [63] tested a number of alloys and observed the $1/f$ characteristics of the noise spectrum, which was already known for crystalline materials. The minimum noise level for annealed cobalt-based alloy was 10^{-4} nT^2/Hz (= 10 pT/$\sqrt{\text{Hz}}$) at 1 Hz. The measured sensor was a 23-mm ring-core. Nielsen et al. have used stress annealing of the tape to produce a creep-induced anisotropy with the ribbon axis being a hard direction. In materials having that type of anisotropy, the rotational magnetization process dominates over the domain wall movement, which reduces Barkhausen noise [64]. The magnetization processes of those materials were studied by Tejedor [65]. For a 17-mm-diameter ring-core sensor, the lowest noise level was

11 pT rms (64 mHz to 12 Hz), noise density was $1.8 \cdot 10^{-5}$ nT^2/Hz (or 4.2 pT/$\sqrt{\text{Hz}}$) at 1 Hz [66].

The amorphous core properties can be further improved by chemical etching and polishing [67, 68].

Nanocrystalline alloys were also tested for fluxgate cores, but until now they exhibited no specific advantage over amorphous materials [69–71].

3.5 Principles of Fluxgate Magnetometers

Second harmonic detection of the output voltage is still the most frequently used; the classic fluxgate magnetometer working on this principle will be described as an introduction. Besides that, a number of other principles for the processing of the sensor output voltage appeared, but in general they brought no significant advantage except of simplification of the circuitry. The fluxgate output is often tuned to enhance the sensitivity; problems with output tuning are discussed in Section 3.7. Fluxgate may also work in the short-circuited mode; current output is analyzed in Section 3.8.

3.5.1 Second-Harmonic Analog Magnetometer

This section describes a block diagram of a typical magnetometer working on second harmonics principle and discusses crucial parts of the magnetometer electronics. The sensor output is amplitude modulated by the measured field and the phase-sensitive detector (PSD) demodulates it back to dc or near-zero frequency. Because the sensor itself has a linear range limited to typically 1 μT, fields larger than that usually have to be compensated.

Figure 3.9 shows the block diagram of a feedback magnetometer. The analog feedback loop has a large gain, so that the sensor works only as a zero indicator. The output variable is then the current into the compensation coil. The sensor nonlinearity and the nonstability of its sensitivity are suppressed by the feedback gain. The sensor is excited by the generator (GEN) working on the frequency f. The power amplifier (PA) is often just a totempole connected pair of low on-resistance (e.g., hexfets) transistors. The generator circuits also produce the $2f$-squarewave signal as a reference for the PSD. Proper phase shift of that signal must also be performed. The use of a phase-locked loop (PLL) to generate reference signal [72] is not recommended because it decreases the circuit stability. The pickup (detection) coil for voltage output typically has 2,000 turns. The first harmonic and other spurious signals at the sensor output are sometimes too high for the PSD,

Fluxgate Sensors

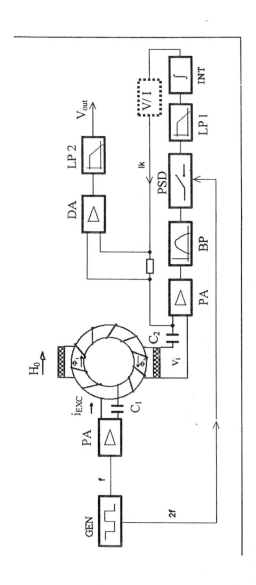

Figure 3.9 Block diagram of the second harmonic fluxgate magnetometer. The generator *GEN* produces the excitation frequency *f* and the reference frequency 2*f* for the PSD.

so preamplification and filtration have to be performed in a bandpass (BP) filter. Classic active RC filters are preferred, because the synchronous filters were shown to be noisy [73]. To obtain sufficient amplification, an integrator (INT) is introduced into the feedback loop. The loop signal is fed back to the feedback (compensation) coil.

Additional feedback current generated by the digitally controlled low-noise and stable current source can be added to increase the range of the instrument. The dynamic range of the basic analog feedback magnetometer may be 120 dB, which is sufficient for most applications: It allows construction of the instrument with 100 μT range, making it possible to measure the Earth's field with 0.1-nT resolution. The compensation current usually is measured as a voltage across the highly stable sensing resistor; two principles of analog-to-digital converters (ADCs) are used for that application: (a) sigma-delta ($\Sigma\Delta$) ADCs may have resolution of 16 to 24 bits, and their speed/resolution may be software configured; (b) integrating (e.g., dual slope) converters are slower, but they better suppress the power-line frequency.

The excitation generator produces a sinewave or a squarewave frequency between 400 Hz and 100 kHz, 5 kHz being typical for crystalline core materials. An increase in the frequency increases the sensitivity to some point at which eddy currents in the core material become important. Thus, sensors made from thin tape with large electrical resistance can be operated at higher frequencies. Increasing the excitation frequency accelerates the dynamical performance of the sensor (Section 3.13). The excitation current should have a large amplitude and a low second harmonic distortion, which may cause a false output signal. The excitation circuit can also be tuned (Section 3.6).

An increasing number of turns of the pickup coil increases the sensor sensitivity, but there are limitations caused by parasitic self-capacitances, which create a resonant circuit. The voltage output is often tuned to increase the sensitivity (Section 3.7). The input amplifier has to be decoupled by the serial capacitor C_2 to prevent any dc current from flowing through the pickup coil and causing sensor offset. The feedback current source should be of large output impedance to prevent short-circuiting of the sensor output, which would lower the sensitivity.

In simple magnetometers, the pickup coil also serves for the feedback, but that requires some tradeoffs. The pickup coil should be close to the sensor core to keep the air flux low. The feedback field should be homogeneous; therefore, a large feedback coil is required. Thus, in precise magnetometers, the two coils are separated. Even in that case, the impedance of all connected circuits must be kept high to prevent decrease of the sensitivity, because the interaction between the two coils is high. To the contrary, the mutual inductance between the excitation and the pickup coil is very low, so the output impedance of the excitation generator has no direct influence on the sensitivity.

Classic analog magnetometers measure only the second harmonic. If we use a gated integrator instead of classical PSD, we also use the higher even harmonics, which may increase the sensitivity and lower the noise [74].

3.5.2 Digital Magnetometers

The feasibility of digital signal processing of the fluxgate output was shown in [75]. The first real-time fluxgate magnetometer based on programmable gate arrays (FPGA) was built at the Max-Planck Institute in Berlin [76].

A fully digital fluxgate magnetometer performs the analog-to-digital conversion of the sensor output signal right after the preamplification and possibly analog prefiltering to suppress the unwanted signals:

- High-frequency components, which may cause aliasing;
- Feedthrough voltage at the excitation frequency.

The harmonic distortion in the ADC would cause a false signal output. As the feedthrough changes with temperature, that effect could degrade the offset stability.

The phase-sensitive detection and additional filtration are performed numerically in a digital signal processor (DSP). The digital magnetometer based on an ADSP21020 CPU is described in [77]. The instrument has an rms noise of 71 pT in the band of 0.25 Hz to 10 Hz (400 pT p-p), while the noise of the fluxgate sensor of the same type used in an analog magnetometer was significantly lower (15 pT rms in the 0.03 Hz to 12 Hz frequency band, 80 pT p-p).

A digital fluxgate magnetometer was flown on board the Swedish Astrid-2 satellite [78]. The advantage of using the digital detection is that the reference signal can have an arbitrary shape so that it can perfectly match the measured signal, better even than the variable-width detector of the switching type. Thus, the information about the measured field can be fully used in an optimum way.

Another approach was suggested by Kawahito et al. [79, 80]. They used an analog switching-type synchronous detector followed by an analog integrator, and a second-order $\Delta\Sigma$ modulator. The magnetic feedback loop is closed via a 1-bit digital/analog converter (DAC) (which guarantees 1-bit linearity), followed by an analog lowpass filter; the problem was the excessive noise of the device, which had the magnetic circuit integrated on the same chip as the processing electronics.

Theoretically, it would be possible to combine the two mentioned principles in a magnetometer employing digital detection and $\Delta\Sigma$ modulation in the feedback. To achieve the high dynamic range, however, large oversampling certainly would be necessary.

In any case, the feedback of the digital magnetometers should include a voltage-to-current converter to eliminate the influence of the changing resistance of the feedback winding.

It should be noted that the increased noise level of the mentioned instrument is not a property of digital processing but is due to the design compromises to lower the power consumption of the magnetometer electronics. DSP-based laboratory lock-in amplifiers such as SR 830 have very low noise, so they can be used for testing high-performance fluxgate sensors.

3.5.3 Nonselective Detection Methods

Although the PSD detection of the second harmonic component of the sensor output voltage is the most usual method, several other detection methods have appeared to process the output signal in the time domain.

The peak detection method is based on the fact that, with an increasing measured field, the voltage peaks of the sensor output are increasing in one polarity and decreasing simultaneously in the opposite polarity. The difference between positive and negative peak value is zero for the null field and may be linearly dependent on the measured field within a narrow interval. Probably the best magnetometer based on this principle was constructed by Marshall in 1971 [81]. He used a 4-79 Mo Permalloy ring core and reached 0.1-nT resolution with a portable instrument.

Robertson presented a 1-mm-long single-core sensor. Using differential peak detection, a similar sensor excited at 40 MHz had 250 pT/$\sqrt{Hz}$@10Hz noise [82, 83].

Figure 3.10 illustrates the principle of the phase-delay method. We assume the simple magnetization characteristics without hysteresis and a triangle waveform of the excitation current. For the time intervals t_1, t_2 between two succeeding output voltage pulses, we can write

$$t_1 = T/2 - 2\Delta t = T/2 - TH_0/H_m \quad (3.6)$$
$$t_2 = T/2 + 2\Delta t = T/2 + TH_0/H_m$$

where T is the period of the excitation current, H_m is the excitation field maximum value, and H_0 is the measured field.

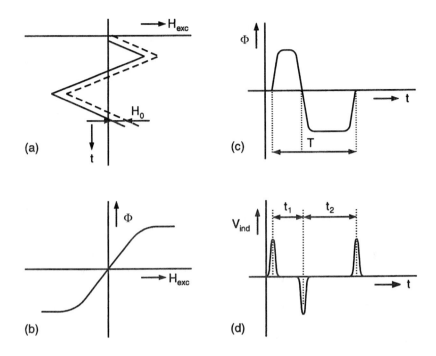

Figure 3.10 Phase-delay method: (a) excitation field with dc shift by measured field H_0; (b) magnetization characteristics; (c) flux; and (d) induced voltage (after [19]).

Those relations were used by Heinecke in a digital magnetometer described in [31]. The time intervals t_1, t_2 were measured by a counter with the reference frequency n times higher than the excitation oscillator frequency. If the time $(t_2 - t_1)$ is equal to N periods of the reference oscillator, then we can write

$$t_2 - t_1 = 2TH_0/H_m = NT/n \quad (3.7)$$

$$H_0 = H_m N/2n \quad (3.8)$$

The resolution of this method is limited by the maximum counter frequency. Heinecke reached 2.5-nT resolution with a 10 MHz basic oscillator. The resolution was improved to 0.1 nT by summing 100 time intervals, but that caused a significant limitation of the sensor's dynamic response, because the excitation frequency was only 400 Hz. The complications with noise from fast digital signals of the counter and other drawbacks limit the performance of the magnetometers based on this method. A magnetometer working on a similar principle but with analog output is described by Rhodes [84]. A Heinecke-type magnetometer with an H-core sensor design had a resolution of 10 nT, but the stability of such a sensor is questionable due to an expected large perming effect [39].

The relaxing-type magnetometer uses a single core saturated by unipolar pulses and measures the length of the relaxation pulse after the excitation field is switched off. The instrument has ±200 μT range, 5% linearity error, and about 0.5 nT p-p noise [85].

3.5.3.1 Sampling Methods

In the sampling method used by Son [86], the instantaneous value of the excitation current at the time of zero-crossing of the core induction depends on the measured field. In an ideal case, the sensitivity is not dependent on the excitation frequency, amplitude, or waveform. Son reached 0.1-nT resolution and 5-μV/nT sensitivity using two open 4-mm long cores made from amorphous water-quenched Vitrovac 6030. The magnetometer works in an open loop, but the linearity error is only 0.02% in the 400-μT range; stability data were not given. A similar principle was used by Sonoda and Ueda in their feedback magnetometer employing a single-rod sensor [30].

Other exotic nonselective methods, such as calculating the power difference between the two half-periods of the induced voltage [87], have been tried but with no significant advantage.

3.5.4 Auto-Oscillation Magnetometers

Auto-oscillation magnetometers are considered a separate group, although most of them are similar to the previously described magnetometers. Mohri constructed a magnetometer based on a two-core multivibrator. He experimented with various configurations of 2-mm-long to 100-mm-long amorphous cores and reached 1-nT resolution [88]. A magnetic multivibrator constructed by Takeuchi and Harada [32] consists of a single-core sensor, capacitor, and operation amplifier forming the oscillating circuit. The multivibrator duty cycle depends on the amplitude of the measured field. Resolution of 0.1 nT in this very simple device was reported, although no information was given on the device stability (often the missing information of many fluxgate papers; poor stability is a weak point of simple fluxgate magnetometers). Kozak et al. improved the circuit by adding another op amp as a current source [89]. Other auto-oscillation magnetometer designs have the oscillator frequency as the output variable. Such a principle is used in PMI's fluxgate compass sensors (although the manufacturer calls them

"Magneto-inductive sensors" and claims they are not based on fluxgate technology).

Auto-oscillation devices should not to be confused with another class of magnetometers, which uses the sensor core as part of an excitation multivibrator, where the oscillation parameters are not affected by the measured field. An example is the open-loop high-field magnetometer for the Pioneer XI satellite with 1-mT range [90].

Neither the nonselective method nor the auto-oscillation method has reached the parameters of the best low-noise and long-term stable magnetometers based on the conventional second-harmonics principle. Nevertheless, they may find application in simple low-cost and low-power instruments used for various indication and search purposes.

3.6 Excitation

The excitation current has to be free from any distortion at the second harmonic, because that can leak into the sensor output through the inductive coupling caused by a nonideal balance of the sensor or capacitance coupling and cause a false signal. Berkman has shown that the higher even components in the excitation are dangerous [91]. The amplitude of the excitation current must be large enough to deeply saturate the sensor core in each cycle to remove any remanent effect. Because the excitation field is attenuated in the central part of the core by the eddy currents and because some magnetically harder regions in the material may exist, the excitation current peak value has to be 10 to 100 times higher than required for the "technical" saturation. High narrow peaks of the excitation current can be achieved by using a tuning capacitor either parallel to the excitation winding (for current-mode or high source impedance) or connected serially (for voltage-mode excitation), as shown in Figure 3.11. The typical tuned current waveform is shown in Figure 3.12 (upper trace). Tuning also may decrease the second harmonic distortion of the excitation current [92].

3.7 Tuning the Output Voltage

In voltage-output sensors, the parasitic self-capacitance and the inductance of the pickup coil form a parallel resonant circuit. We can find the multiple resonant peaks by changing the excitation frequency. Because of the nonlinear character of the circuit, the resonant frequency also depends on the excitation

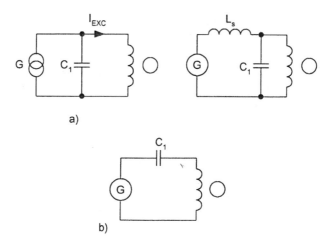

Figure 3.11 Tuned excitation circuit: (a) parallel; (b) serial.

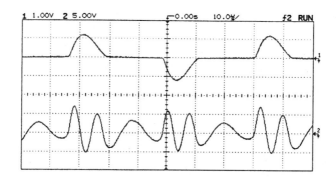

Figure 3.12 Excitation current (upper trace, 1A/div) and the untuned voltage output (lower trace) of the ring-core sensor in 5μT-uncompensated measured field.

amplitude. Sometimes the circuit is unstable, that is, it oscillates at some harmonics even without any external field. In general, this kind of resonance is unwanted because the coil self-capacitance is unstable and temperature dependent. Splitting the pickup coil into separate sections lowers the self-capacitance. Many studies on fluxgate sensitivity are influenced by such resonance, so they are valid only for the particular pickup coil, and their results should be generalized only with great care.

Some of the fluxgate magnetometers intentionally use parametric amplification by tuning the sensor voltage output by a parallel capacitor. Gains of as high as 50,000 are achievable, but the circuits are extremely sensitive to parameter variations and have a narrow bandwidth [87]. An increase in the sensitivity by a factor of 10 to 100 is reasonable; for a high quality factor (low pickup coil resistance), the circuit may again become unstable. This situation arises more often for very sensitive fluxgates, for example, race-track core sensors.

Figure 3.12 [93] shows the unloaded output voltage of a race-track fluxgate sensor for the measured field of 5μT. Higher even harmonics are dominant due to parasitic self-capacitance resonance of the pickup coil. Figure 3.13(a) shows the same sensor tuned into an unstable mode by a parallel capacitor with C_2 = 11 nF. The large amplitude oscillations are present even for the zero measured field. The sensor may be stabilized by damping resistor either in series or parallel to the tuning capacitor. Figure 3.13(b) shows the same sensor stabilized by a serial resistor of 10Ω. It should be noted that the value of the damping resistor necessary to stabilize the sensor is small compared to the dc resistance of the pickup coil, which was 45Ω.

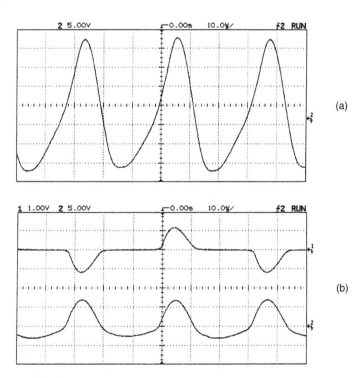

Figure 3.13 (a) Oscillations at the output of the unstable tuned sensor; (b) stabilized sensor tuned to second harmonic, output voltage for 5μT-measured field.

Ring-core sensors are less sensitive than racetrack sensors due to the higher demagnetization factor. If their pickup coil is wound of thin wire, they usually are unconditionally stable (for each value of the tuning capacitor and measured field and any excitation parameter).

Although the resonating circuit is simple, the precise analytical description is complex, because the effect is again strongly nonlinear. Since the first study by Serson and Hannaford [94], parametric amplification and its stability have been analyzed by many authors [95–98]. Primdahl and Jensen [99] point to some disadvantages of the parametric amplification: The sensor degradation they observed is most pronounced close to the stability limit. Some authors observe that proper moderate tuning of the sensor output increases the sensor sensitivity (typically 10 times) but also suppresses the magnetometer noise to about one-half. They explain that this is because the tuning concentrates the broadband output power to the second harmonic frequency [100]. They also conclude that moderate tuning does not degrade the temperature stability of the sensor offset. But this subject should be further studied, because the mentioned results are not in accordance with Primdahl and Jensen [99] and Pedersen [101].

3.8 Current-Output (or Short-Circuited) Fluxgate

In the conventional fluxgate magnetometer, the output of the pickup coil is connected to an amplifier with a large input impedance, so that the voltage induced into this coil forms the output of the sensor. Primdahl et al. [102] short-circuited the pickup coil by the current-to-voltage converter with a very low input impedance and used the current-output mode of operation. The amplitude of the current pulses was shown to depend linearly on the measured field. Low input impedance of the electronics eliminates problems with the stray capacitance of the coil and cable, and the design of the low-noise input amplifier is simplified. A fluxgate magnetometer working on this principle is described in [53]. Rather than direct evaluation of the peak-to-peak value by means of a peak detector (which is, in general, a noisy device), they use the sampling circuit. Primdahl et al. [66] have shown that by using the gated integrator (controlled rectifier of the switching type with an adjustable gate width), the maximum sensitivity is achieved for a specific phase delay and the width of the reference voltage when the shape of the pulse is best fitted. Because that principle uses the information from all even harmonic components (the weight of each given by the spectrum of the reference), it is not possible to use a classical bandpass input filter. Switching filters with the comb characteristics were shown not to be useful for fluxgate applications because of their distortion and limited dynamic range. The demands on detector circuits are very high in the case of current output, because very small signals have to be processed in the presence of large overcoupled disturbing signals. There are two main sources of feedthrough: (1) air mutual inductance between the excitation and pickup coils and (2) flux leakage from the core. Those two contributions cannot be simultaneously nulled, as shown in [103].

The sensitivity of the short-circuited fluxgate increases with increasing sensor length and cross section (the latter dependence saturates for thick sensors because of the demagnetization) as usual for voltage-output sensors but decreases with the increasing number of turns with certain practical limitations. The sensitivity of the 17-mm toroidal core sensor was about 40 nA/nT compared with about 20 μV/nT sensitivity that can be reached for an untuned voltage-output sensor of the same dimension. Although the short-circuited fluxgates have shown practical advantages, in general they have very similar parameters as the traditional voltage-output design.

3.8.1 Broadband Current Output

The circuit diagram of the current-output fluxgate is shown in Figure 3.14. The pickup coil is short-circuited by the current-to-voltage converter with the feedback resistor R. The capacitor C is used to prevent any dc input current of the op amp from flowing through the pickup coil (such current would cause sensor offset). In the traditional broadband case, C is very large, but it can also be used for tuning (Section 3.8.2).

The current-output fluxgate model based on circuit analysis was presented in [102, 66, 104].

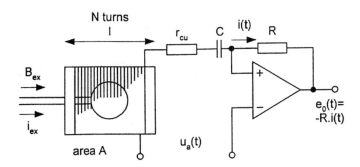

Figure 3.14 Circuit diagram of the current-output fluxgate (after [93]).

The basic equation for the circuit in Figure 3.14 is

$$\frac{d\Phi}{dt} + i(t)r_{Cu} + \frac{1}{C}\int i(t)dt = 0 \qquad (3.9)$$

To incorporate the measured dc field B_0 into the circuit equations, we replace its effect by the equivalent coil current i_{EQ},

$$i_{EQ} = \frac{l}{\mu_0 N} B_0 \qquad (3.10)$$

The total pickup coil flux is

$$\Phi = [i_{EQ} + i(t)]L(t) \qquad (3.11)$$

The effective length of the coil is defined here as $l = \mu_0 NI/B_0$, where B_0 is an external field, which is canceled by a dc compensation current I into the pickup coil. We suppose that l is dependent only on the coil geometry, not on the sensor core properties nor on the mode of the excitation. The value of l is higher than the physical length of the coil and can also be determined from the inductance of the pickup coil with removed (or completely saturated) core [50].

The apparent permeability μ_a is defined as $\mu_a = \mu_0 B/B_0$, where B is the magnetic field inside the core. Apparent permeability is lower than the core material permeability because of the core demagnetization. It is dependent not only on the core size and properties and the mode of excitation but also on the geometry of the pickup coil.

After substitution of (3.10) and (3.11) into (3.9), we obtain

$$i_{EQ}\frac{dL(t)}{dt} + \frac{d}{dt}[i(t)L(t)] + i(t)r_{Cu} + \frac{1}{C}\int i(t)dt = 0 \qquad (3.12)$$

Here $L(t)$ is a periodic function of frequency $2f$, for which

$$L(t) = \mu_0 \frac{N^2}{l} A\mu_a(t) \qquad (3.13)$$

where l is the effective length of the coil, A is its cross-sectional area, N is the number of turns, and $\mu_a(t)$ is the (modulated) apparent permeability of the sensor core.

In the untuned case, where $C \to \infty$, we have

$$d\Phi/dt + i(t)r_{Cu} = 0 \tag{3.14}$$

If the coil losses are small, then the copper resistance r_{Cu} can be neglected, as discussed in [102]. Then, (3.14) is simplified to $d\Phi/dt = 0$, or

$$\Phi = (i_{EQ} + i(t)) \cdot L(t) = \text{constant} \tag{3.15a}$$

and from that

$$i(t) = -i_{EQ} + \Phi/L(t) \tag{3.15b}$$

The output current cannot have any dc component, that is, it has a zero time-average, $<i(t)> = 0$

$$0 = -i_{EQ} + \Phi<1/L(t)> \Rightarrow \Phi = i_{EQ} \cdot L_{G0} \tag{3.16}$$

where L_{G0} is the geometric mean value of the pickup coil inductance

$$1/L_{G0} = <1/L(t)> \tag{3.17}$$

From that, we have the basic short-circuited fluxgate equation:

$$i(t) = i_{EQ}\left[\frac{L_{G0}}{L(t)} - 1\right] \tag{3.18}$$

It was shown that in practical cases the core permeability, and thus the pickup coil inductance, rapidly changes between two values: L_{max} for the linear part of the hysteresis loop and L_{min} for the saturated sensor core. Thus, the ideal short-circuited current is a squarewave. The peak-to-peak value of the output current was derived in [66] as

$$i_{p-p} = i_{EQ}\frac{L_{G0}}{L_{max}}\left(\frac{L_{max}}{L_{min}} - 1\right) \tag{3.19}$$

The actual waveforms of the current output sensor are shown in Figure 3.15: The current impulses decay because of ohmic losses, and the waveform is also distorted by the feedthrough from the excitation.

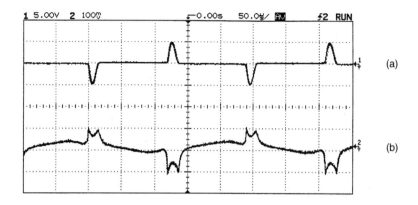
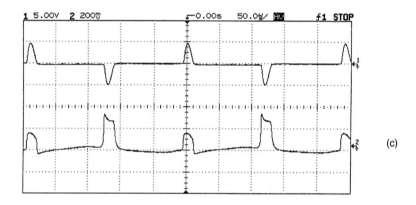

Figure 3.15 Waveforms in the broadband current-output fluxgate: (a) excitation current (0.5 A/div); (b) output current for $B_0 = 0$ (20 μA/div); and (c) output current for $B_0 = 5$ μT (40 μA/div) (from [93]).

3.8.2 Tuning the Short-Circuited Fluxgate

It was recently shown that the current fluxgate can be tuned by a serial capacitor [104]. The sensitivity was increased five times while the level of the spurious feedthrough remained the same. Serial tuning is easily made only by decreasing the value of the input decoupling capacitor. The simplified analytical description was shown to fit the measured parameters well. If the pickup coil is well tuned, we can limit the solution only to the second harmonic output current:

$$i(t) \cong i_a \cos(2\omega t) + i_b \sin(2\omega t) \tag{3.20}$$

The circuit analysis performed in [104] leads to the matrix equation:

$$\begin{bmatrix} \dfrac{r_{Cu}}{2\omega L_0} & \dfrac{-L_4}{2L_0} \\ \dfrac{-L_4}{2L_0} & \dfrac{r_{Cu}}{2\omega L_0} \end{bmatrix} \cdot \begin{bmatrix} i_a \\ i_b \end{bmatrix} = \begin{bmatrix} 0 \\ \dfrac{L_2}{L_0} \cdot i_{ex} \end{bmatrix} \quad (3.21)$$

and the stability condition is $r_{Cu} > \omega L_4$, where L_2 and L_4 are second-order and fourth-order Fourier components of $L(t)$ and L_0 is a zeroth-order Fourier component, which is the arithmetic mean value $L_0 = <L(t)>$.

The current-output fluxgate resonance condition is

$$C = \frac{1}{(2\omega)^2 \cdot L_0} \quad (3.22)$$

Notice that the arithmetic mean value L_0 differs from the geometric mean value L_{G0}, which is important for the untuned current-output equations.

An example of tuned current-output fluxgate waveforms is shown in Figure 3.16 for the same sensor that was shown in Figure 3.15. Here, the ideal sinewave also is distorted by feedthrough from excitation.

3.9 Noise and Offset Stability

An important factor that affects the precision of the fluxgate magnetometer is the stability of the sensor zero. The changes of the offset usually are divided into two components (although they may be caused by similar effects): (1) noise as a relatively fast variation and (2) long-term offset stability.

Sensor noise has been investigated by many authors, starting from the classic study by Scouten [55, 105]. It was demonstrated that the noise level can be decreased using high peak value of the excitation current [106], a rule that has been confirmed by many others. This was interpreted as if there existed small volumes inside the material that are more difficult to magnetize than the rest, and so they are not necessarily saturated during each period of the excitation field. The uncertainty of magnetization of these regions is one of the sources of the sensor noise and offset. These regions often are associated with structural (such as dislocations or inclusions) or surface imperfections of the core: It has been shown that improving the

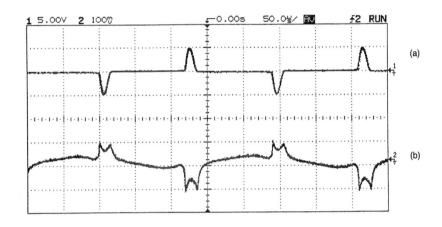

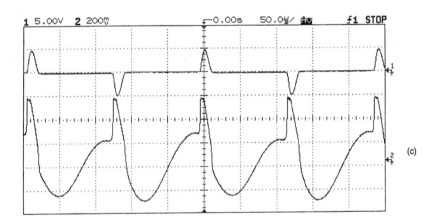

Figure 3.16 Waveforms in the tuned current-output fluxgate: (a) excitation current (0.5 A/div); (b) output current (20 μA/div) for $B_0 = 0$; and (c) output current (40 μA/div) for $B_0 = 5$ μT (from [93]).

structural and surface quality by annealing and etching or polishing reduces noise.

Sensor noise depends mostly on the core material, but the geometric aspect is also important; the demagnetization factor determines how the material noise is calculated to the effective noise value on the sensor input. Primdahl [50] observed that the internal noise in the ring core is practically independent of the core cross section, that is, the magnetic noise within the core is well correlated. Thus, it makes no sense to increase the core volume and expect "averaging-off" of the noise. The dependence of the noise level on the excitation frequency for Permalloy cores was discussed by Afanasenko and Berkman [107]. They showed that the effective noise level decreases up to a certain limit for increasing frequency. That effect can be explained by a change in the sensor sensitivity.

Decrease in the sensor noise with increasing temperature was observed first for low Curie point alloys. With increasing temperature, the core permeability increased and the saturation magnetization decreased simultaneously. A similar effect was observed for low-noise Permalloys [14]. The effect of decreasing the room-temperature noise with Curie temperature was also observed for amorphous alloys [63]. Shirae [62] reported 2.5 pT rms (100 mHz to 16 Hz) for cobalt-based amorphous alloy having Curie point $T_C = 50°C$.

The techniques used to measure the spectrum of the sensor noise are described in Chapter 11. Fluxgates have a noise of $1/f$ nature from millihertz up to kilohertz frequencies, that is, $P(f) = P(1)/f$ nT^2/Hz, where $P(1)$ is the power spectrum density at 1 Hz. The rms level of the noise, N_{rms}, in the frequency band from f_L to f_H is then given by the expression

$$N_{rms} = \sqrt{\int_{f_L}^{f_H} P(f)\,df} = \sqrt{P(1)\ln(f_H/f_L)} \qquad (3.23)$$

The noise spectrum density is often given in nT/$\sqrt{Hz}$ or pT/$\sqrt{Hz}$. The conversion is straightforward, for example, 9 pT^2/Hz corresponds to 3 pT/$\sqrt{Hz}$, as was explained in Section 1.4.4.

Primdahl et al. [66] have found that the fluxgate noise peak-to-peak level typically is approximately six times larger than the rms value in the same frequency band.

Figure 3.17(a) shows the noise spectrum for the sensor with Permalloy low-noise core produced by Infinetics for NASA (type S1000-C31-JC-2239C). The measurement was performed by an HP 3566A digital spectrum analyzer using overlapping averaging from 1,000 samples to smooth the plot and remove the distortion caused by the time window. The noise spectral density was 3.8 pT/$\sqrt{Hz}$ at 1 Hz, and the rms value calculated from the measured spectrum was 8.76 pT (64 mHz to 10 Hz). The time plot of the last one of the 1,000 (overlapping) 32-second time intervals is in Figure 3.17(b). The rms value calculated from the estimated $P(1)$ using (3.23) is 8.81 pT, which is close to the measured value.

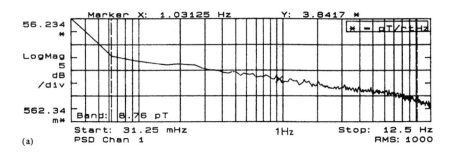

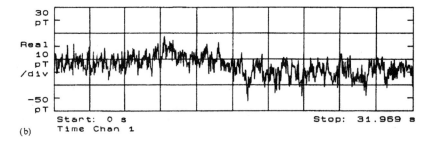

Figure 3.17 (a) Noise spectrum for the superior sensor; (b) time plot (from [19]).

For ring-core and race-track sensors, the noise level reaches its minimum at a certain low value of the cross-sectional area. Low-noise Vacquier-type sensors should have the excitation coils longer than the cores and the detector coil shorter than the cores. Moldovanu has shown that a 65-mm-long sensor with cores of 1-mm wide, 25-μm thick, stress-annealed Vitrovac 6025 has the same 11 pT rms (64 mHz to 10 Hz) level of noise as a 17-mm-diameter ring-core sensor from the same material [108].

A noise study on amorphous tape performed by Nielsen [109] has shown how weak our knowledge is on this subject. His results may indicate that the loop parts with constant permeability do not contribute to the sensor noise as much as the near-saturation region, which is where the fluxgate effect occurs. Another surprising observation was that the eddy currents through changing contact resistance between the tape layers do not contribute to the sensor noise; tape insulation brought no improvement.

3.9.1 Zero Offset

Zero offset of the sensor and its changes may be partially caused by some of these factors:

- Magnetically hard regions already mentioned;
- Thermal and mechanical stresses;
- Inhomogeneities of the core and winding together with changes in the magnetic properties of the core material, the parameters of the excitation field, and temperature.

To increase the temperature and long-term stability of both the sensor offset and sensitivity, it is useful to match the thermal expansion coefficient of all the sensor parts to reduce internal stresses. Nonmagnetic metal Inconel 625 or ceramics (corundum or machinable Macor) are used for the bobbin. For Permalloy cores, which need to be annealed at higher temperatures, bobbins are made from corundum ceramics. Gordon and Brown [12] have measured zero offset changes less than ±50 pT during 24 hours and temperature shift less than 100 pT from −40°C to +70°C using 81 Mo Permalloy for the sensor core. Those results probably are the best ever reported. More recently, the long-term stability of magnetometers on magnetic observatories was compared during the geomagnetic observatory workshops [110, 111]. Drift less than 1 nT/year can be achieved using currently available instruments in a temperature-controlled room. The temperature coefficient of the compensation field may be as low as 2 ppm/°C with quartz coil frames. That corresponds to approximately 0.1 nT/°C temperature dependence during measurement of the Earth's field.

The best temperature offset stability reported for amorphous sensors was 1 nT in the −20°C to +60°C range [20]. Diaconu reported stability of 3.7 nT and 2.2 nT in the −70°C to +25°C range for Vacquier-type sensors having cores of Vacoperm 100 and cobalt-based amorphous wires, respectively [112].

3.9.2 Offset From the Magnetometer Electronics

The main source of the magnetometer offset is usually its electronics. A useful tool for distinguishing between the offset contributions is to incorporate switches into critical magnetometer points [53]. For example, reversing the excitation coil will reverse only the offset from excitation. Simultaneous flipping of the pickup coil and feedback coil will change the polarity of all the offsets except those generated in the processing circuits. The offset of the processing circuits is given not only by the offset of the dc-coupled op amps (starting the PSD) but also by the second harmonic distortion of the ac amplifiers (before PSD), and the dynamic distortion caused by a finite slew rate of the op amp and finite switching time of the switches in PSD (see also Section 3.5.1).

3.9.3 Other Magnetometer Offset Sources

Temperature gradients in conducting materials close to the sensor induce dc currents, which may generate magnetic fields. Some materials in the supporting mechanical structures may be magnetic: Remanent field looks like magnetometer offset. Also, some electronic components are magnetic (tantalum capacitors, some ceramic chip carriers, etc.). Some relays and electric motors contain strong permanent magnets. Also, dc currents may cause offsets; a good technique is to use twisted conductors to minimize the current loops and keep the current far from the sensor. While the magnetic field at the distance d from a permanent magnet is decreasing with $1/d^3$, the field created by a current conductor is dropping only with $1/d$. The magnetometer on the Oersted satellite was mounted on a boom 6m off the satellite body to suppress interference, especially from the solar cell currents [20]. A practical guide for magnetic cleanliness is [113].

3.10 Crossfield Effect

Many vector magnetic field sensors have a nonlinear response to magnetic fields perpendicular to their sensing direction ("crossfields"). The crossfield effect is dramatic in anisotropic magnetoresistors [114], but it also may be found in fluxgate magnetometers. In general, the crossfield effect can be suppressed by core shape (in rod-type or race-track sensors) or by total compensation of the measured field, not only the component in the sensing direction (as in compact spherical coil design described in Section 3.14).

Crossfield-effect errors of more then 20 nT in the Earth's field were observed in ring-core fluxgates [115, 116]. Brauer showed that a variation in magnetic susceptibility along the core might result in a nonlinear perpendicular field response [117]. It was also shown that the crossfield response is temperature dependent: Errors as high as 2 nT/°C in the Earth's field were observed on low-cost sensors, which should be compared with their 0.1 nT/°C offset drift [118]. The crossfield effect may be the dominant source of error when three-axial measurements are made in the presence of the Earth field. It can be reduced by making the core more homogeneous, decreasing the ring-core diameter, or selecting another core geometry.

3.11 Designs of Fluxgate Magnetometers

Fluxgate magnetometers can be made very resistant against environmental factors such as vibrations, radiation, and temperature: Sensors developed for deep drilled wells work up to 200°C. There are many fluxgate designs for specific purposes. The two extremes are simple low-power, low-cost instruments and precise, stable, low-noise, expensive station magnetometers.

3.11.1 Portable and Low-Power Instruments

An ultralow-power two-axis magnetometer has been developed for vehicle detection [119]. The instrument power consumption was 0.72 mW, linearity 5% in the $80\mu T$ range, noise 1.5 nT p-p, but huge temperature offset drift of $6\mu T$ existed in the $-40°C$ to $+70°C$ range. The offset drift was reduced to the order of nT/°C when the power consumption was increased to 5 mW.

A three-axis vector portable analog fluxgate magnetometer is described in [120]. The instrument has 300 mW consumption from ±6V source, a range of ±100 μT with 0.01% linearity. The effective resolution is 1 nT, and response time to a large field step is 4 ms. Innovated sealed ring-core fluxgate sensors made of etched rings ensure high resistance against vibrations and mechanical shocks.

3.11.2 Station Magnetometers

Together with a proton magnetometer, a single-axis fluxgate magnetometer mounted on top of a nonmagnetic theodolite telescope is the standard instrument for absolute measurements at magnetic observatories. The so-called fluxgate theodolite is routinely used for the measurement of the magnetic declination D and inclination I. The procedure is described in detail in [121]. An early development of such an instrument with many still useful details is described in [94].

Three-axis fluxgate magnetometers are also used for the recording of magnetic variations. Long-term and temperature stability of these instruments is critical. The coils are often wound on quartz tubes, which may improve their temperature coefficient to 2 ppm/°C. The sensor head is mounted on a pillar in a thermostated nonmagnetic chamber. Marble is often used for supporting structures because it is a dimensionally stable, nonmagnetic, and electrically insulating material that is also easily machinable and cheap. The possible pillar tilting can be corrected for by the use of suspended sensors [122]. The best fluxgate magnetometers for observatories have long-term offset stability of 1 nT/year. Periodical calibration by absolute measurements further reduces the absolute error of the observatory records to 1 nT. Comparisons and testing results of station magnetometers were performed at observatory workshops [110, 111]. Observatory fluxgate magnetometers are made by GEM, EDA, Narod Geophysics, Dowty, Scintrex, Thomson-Sintra, Danish Meteorological Institute, Shimadzu [123], and other companies.

Battery powered sea floor triaxial gimbaled fluxgate magnetometers are described in [124, 125].

3.12 Miniature Fluxgates

Many applications, including magnetic ink reading, safety and security sensors, and sensor arrays, require a very small sensor size. The process of the fluxgate sensor miniaturization is complicated because the magnetic noise dramatically increases with decreasing sensor length. Small-size fluxgates are made of open or closed cores from amorphous or Permalloy wires, tapes, and deposited structures. Up to now, the quality of sputtered or electrodeposited Permalloy has not been sufficient for low-noise fluxgate applications, so patterns etched of amorphous tape are often used for the sensor core.

The compensation sensor manufactured by Siemens-VAC Hanau (Germany), which has a Permalloy wire core, may also work as a single-core fluxgate sensor [126]. A number of simple multivibrator-type fluxgate magnetometers have been reported from Japan (see Section 3.5). A 15-mm-long hairpin sensor was made up of a strip with helical anisotropy [27]. The sensor has 5 nT/$\sqrt{Hz}$ noise (averaged in the 64 mHz to 10 Hz band) and 4 nT/°C temperature drift in the 25°C to 50°C interval.

The simple PCB construction of the 15-mm-long fluxgates is described in [127]. The annealed core made of amorphous foil is sandwiched between two layers of PCB, which have outer metal layers forming the halves of the winding. The layers are then connected by electroplating.

A fluxgate sensor with all the coils made by planar technology was developed by Vincueria [128]. The sensor has two parallel cores of 24-mm-long amorphous strips, and the excitation and sensing flat coils are 10-μm thick. Figure 3.18 shows how the magnetic field of the sensor core strip can be sensed by a pair of flat coils. A simpler and much smaller planar fluxgate sensor with flat coils is described in [129, 130]. The sensor core is in the form of two serially configured 1.4-mm-long strips of sputtered Permalloy 2-μm film. The flat excitation coil saturates the strips in opposite directions, and the differential flux is sensed by two antiserially connected

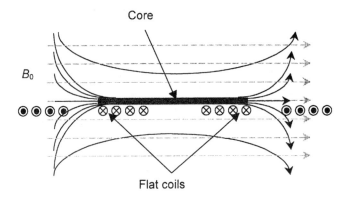

Figure 3.18 Planar fluxgate with a pair of flat pickup coils (from [132]).

flat pickup coils. The maximum sensitivity of 73 V/T was reached for a 1 MHz/150 mA p-p excitation current. A similar sensor having three flat excitation coils is described in [131]. The sensor response covers fields up to 250 μT. In the ±60μT range, the linearity and hysteresis error is below ±1.2%. The error of angular response to a 50-μT field is ±1.6%

The orthogonal fluxgate with flat excitation and pickup coil is described in [133]. The sensor 10-mm diameter ring core is also etched from Vitrovac 6025 amorphous ribbon. The sensor resolution is 40 nT, and the linearity error in the 400-μT range is 0.5%. A parallel-mode two-axis integrated fluxgate magnetometer by the same authors was developed for a low-power watch compass [134]; the sensor is shown in Figure 3.19.

A common weak point of integrated flat coils is that they cannot sufficiently saturate the core to erase the perming effect for two reasons: (1) the weaker coupling to the core than is present in the solenoid coil and (2) low metallization layer thickness, resulting in high coil resistance and low limit for current amplitude. High coil resistances are also the reason that the sensor cannot be tuned, neither in the excitation nor at the output. One possible improvement is a double-side core structure, which has an almost closed magnetic circuit [135]. The two-layer metallization process can form a solenoid around the core [136]; the noise was 40 nT p-p for a 5-mm-long sensor. A similar sensor was developed at Fraunhofer Institute [137] and has became a part of a CMOS integrated magnetometer [138].

The advantages of miniature fluxgate sensors (besides the small size) are small weight, high geometrical selectivity, low cost in mass-production, and the possibility of integration of on-chip electronics. They have a fast response, thanks to the high operation frequency. The sensitivity and the

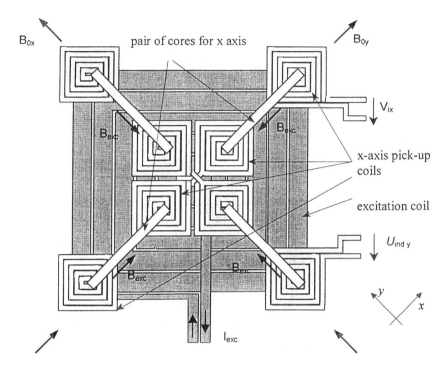

Figure 3.19 Planar fluxgate compass sensor (from [132]).

stability of miniature fluxgates are higher than that of any semiconductor sensors, but their competitors are AMR sensors. AMRs of the same size working in the flipping mode (see Chapter 4) may have similar parameters, including power consumption.

3.13 ac Fluxgates

There are three basic approaches to increasing the frequency range of the fluxgate: (1) increasing the excitation frequency, (2) exploiting the direct induction effect in the pickup coil, and (3) using the ac error signal of the dc loop [139, 140].

Marshall [141] measured the integral value of field impulses caused by lightning flashes by a ballistic method. Aroca [142] constructed a fast open-loop magnetometer with a 1% linearity error in the $\pm 2\mu T$ range. Kono designed the feedback fluxgate for a spinning sample rock magnetometer: The instrument has 450 Hz bandwidth and a linear phase response so that the measured signals with frequencies up to 200 Hz have a constant time delay [143, 144]. In a Molspin system, a 55-mm-inner-diameter fluxgate measures the magnetic moment of a rock sample rotating inside the core [145]. Ioan reached 3 kHz bandwidth even with feedback [146].

It is also possible to measure ac fields by induction coils (Chapter 2), which generally are simpler devices. But still there may be a good reason to use a fluxgate for ac field measurements, either when the fluxgate magnetometer is already a part of the system or for weak fields, in which a fluxgate in fact transforms the signal spectrum to a higher frequency, where the noise of the input amplifier is lower.

3.14 Multiaxis Magnetometers

The typical application of a two-axial fluxgate magnetometer is for a magnetic compass. The ring-core sensor with double cross-shaped pickup coil can measure the field in two directions simultaneously, as shown in Figure 3.20(a). Such sensors are used in simple magnetic compasses for automotive applications [147]. If a two-axis compensated sensor is used as a magnetic compass, the short-time angular accuracy may be 5 minutes of arc [43]. Figure 3.20(b) shows the tubular core sensor that measures all three components of the field. While in the x and y directions the parallel fluxgate effect is used, the z-output works as a perpendicular fluxgate. Afanasiev suggests making the core in the form of loops of tape, which results in a sensor having similar sensitivity in all three directions [148]. Single-core configurations have an advantage in low power and small dimension of the sensor head, but the long-term and temperature stability is limited by coil dimensional stability.

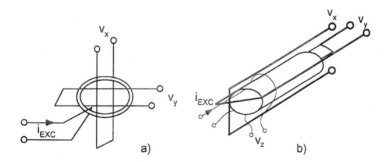

Figure 3.20 (a) Single-core dual-axis ring-core fluxgate sensor; (b) three-axis tubular sensor.

Another three-axial single-core sensor uses the crystalline anisotropy of its core [28]; although the principle is sophisticated, the device is unlikely to be stable enough for practical applications.

3.14.1 Three-Axial Compensation Systems

Three-axial fluxgate magnetometers usually have three single-axis sensors mounted perpendicularly in the sensor head. A large three-axial feedback system creating complete magnetic vacuum is preferred to three separate coils compensating the measured field in only one direction for each sensor. Three orthogonal circular or rectangular Helmholtz coils or more complex coil systems of both circular and rectangular shape are frequently used for the compensation system. Primdahl and Jensen [149] have constructed a spherically shaped three-axial coil system for rocket and satellite applications, in which the size of the sensor is strictly limited. The three coils have identical center points, each consisting of nine sections approximating the ideal spherical coil, which generates a uniform field (Figure 3.21). Although the outer diameter of the compact spherical coil (CSC) is just 90 mm, the homogeneity was proved to be sufficiently high for 40-mm-long sensors.

Figure 3.21 CSC magnetometer. (Courtesy of the Danish Technical University.)

The main advantage of keeping all three orthogonal sensors of the magnetometer in the center of the three-dimensional feedback system is that the sensors are kept in a very low field. That is important for long-term stability and low-noise operation, but the main advantage is that the system is free of errors caused by the crossfield effect (Section 3.10). The measuring axes are defined only by the feedback coil system (the exact position of the individual sensors is not critical) and thus may be easily determined and kept very stable. The CSC system was successfully used for the Oersted satellite magnetometer. The coil support was made from epoxy filled with glass microballoons and carbon fibers: This material has a thermal linear expansion coefficient of only 32 ppm/°C, while the density is just 750 kg/m^3. The temperature coefficients of the sensitivity in three axes were 31 to 36 ppm/°C, corresponding well to the expansion coefficient of the coil support material. Later the structural stability was even improved by using C-SiC material for the supporting shell. C-SiC is a compound quasi-ceramic material with low electric conductivity and high thermal conductivity [150]. It has excellent dimensional stability and low thermal expansion (2 ppm/°C at room temperature, near zero at 150°C). The temperature sensitivity coefficient of the CSC was then further reduced to 10 ppm/°C. The Oersted fluxgate magnetometer linearity was found to be below 1 ppm in the Earth's field, and the temperature coefficient of the deviation angles was 0.07 arc-sec/°C. The in-flight calibration of the Oersted CSC magnetometer has shown the instrument's stability for a 47-day period: The scatter was 0.3 nT for the offsets, less than 10 ppm for the scale values, and less than 1.5 arc-sec for the nonorthogonalities.

3.14.2 Individually Compensated Sensors

Although the CSC system is known to be the technically best solution, it is considered to be rather exotic because of its complexity and high price (e.g., some of the 27 coils should be wound after mounting the three sensors inside the sphere). Simpler and much cheaper are the systems consisting of sensors with individual feedback coils; however, these systems have many problems arising from the crossfield effect.

Three-axial systems also face additional problems from the crosstalk between the sensors. The vicinity of other sensor cores also changes the feedback coil sensitivity because the sensor core represents a region of higher magnetic conductivity for the neighboring sensor.

The sensor head geometry and the temperature stability of its dimensions are important. Sensors with individual feedback should be mounted symmetrically and at a maximum distance. Figure 3.22 shows a possible sensor arrangement. The sensor z should have a larger distance to the other, to keep the sensor sensitivities alike: The magnetic path of the z sensor includes the core of the x sensor in low demagnetization direction, while all the other couplings are much weaker [116].

The sensors are usually excited from the same generator. If the excitation amplitude allows, the excitation coils are simply connected serially. In the Oersted magnetometer, the excitation generator output was stepped up by a transformer, which also allowed use of a balanced drive, which reduces the capacitive couplings [20].

The precise MAGSAT satellite instrument had three single-axis compensated sensors [151]. Later a three-axial magnetometer was developed for the Swedish satellite Astrid-2, which has three closely mounted 17-mm ring-core fluxgates. The magnetometer construction and the test results are well documented in [152, 153]. All the critical mechanical parts are made of machinable ceramics MACOR [154]. The dimension of the sensor head is 32 by 47 by 55 mm. The close vicinity of the sensors increases the nonlinearities caused by the crosstalk and crossfield effect up to 8 nT p-p in the 60,000-nT range. Because of the low temperature expansion of MACOR (9.4 ppm/°C), the temperature drift of the sensitivity for the individual sensors is between 10 and 13 ppm/°C. The offset drift was 0.01 and 0.02 nT/°C for the x- and y-axes, respectively, but it degraded to 0.45 nT/°C for the z-axis, which is more magnetically coupled to the other sensors because of the holder geometry. Again, that effect is caused by the

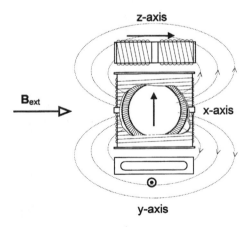

Figure 3.22 Three-axial sensor head (sensors compensated individually). (Courtesy of the Danish Technical University.)

crossfield sensitivity and can be suppressed by increasing the sensor distance or decreasing the core diameter. The temperature drift of the angles between the sensors was 0.17, 0.33, and 0.5″/°C [155].

Three individually gimbaled sensors were used in the attitude sensor used in an autonomous underwater vehicle [156]. The pitch and roll information is provided by air coils fixed to the vehicle frame.

3.15 Fluxgate Gradiometers

The traditional way to measure the field gradient is to use two sensors and subtract their reading. The sensor distance is called the gradiometric baseline. The gradient is thus approximated by the difference; error arises when higher order gradients are present. The measurement of the gradient is often used when the measured field source is at a short distance. In such a case, the disturbing fields from distant sources that are more homogeneous (such as the Earth's field) and their variations can be effectively suppressed, as shown in Figure 3.23.

Gradiometric nondestructive sensors are used for biomagnetic measurements, detection, and location of ferromagnetic objects, magnetic testing, and

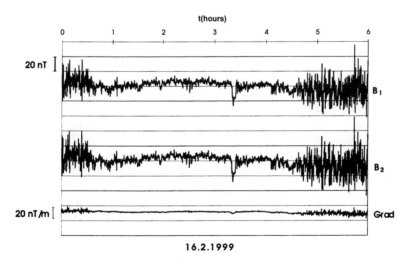

Figure 3.23 Magnetic field in the laboratory located in the city. Subway (underground) and trams do not operate between 12:30 A.M. and 4:30 A.M. B_1 and B_2 are the outputs of two fluxgate sensors measuring in the same direction in 50-cm distance. Grad is their difference recalculated to the magnetic gradient, in nanoteslas per meter.

other applications. The dual-sensor gradiometer can have a single feedback controlled by the master sensor so that the compensation current is the same for both sensors. In such a case, the slave sensor works in the open loop, and its output is directly the field difference. The alternative is to use one long solenoid for compensation of both sensors. Another possibility is to compensate the sensors individually and subtract their reading (preferably numerically). The disadvantage of the analog processing is that the compensation coils should have exactly the same constants and their axes should be perfectly aligned, which is difficult to guarantee in a wide temperature range. Digital processing needs two A/D converters, but the adjustment for the sensitivity mismatch can be made automatically. The latter configuration requires a larger sensor distance, to avoid interference between the sensors. Arrays of low-noise fluxgate sensors are used for detection systems for military and security applications. The sensor signals are processed numerically to compensate for crosstalk, individual sensitivities, and temperature coefficients.

Berkman has shown that field gradient can be measured by a single-core rod-type sensor [157]. Race-track single-core fluxgate gradiometers are described in [158, 159].

References

[1] Primdahl, F., "Bibliography of Fluxgate Magnetometers," *Publications of the Earth Physics Branch*, Vol. 41, 1970, No. 1.

[2] Serson, P. H., and F. Primdahl, "Bibliography of Magnetometers," *Publications of the Earth Physics Branch*, Vol. 43, No. 8, 1972.

[3] Primdahl, F., and Coles, R. L., "A Bibliography of Magnetometers," *Adv. Space Res.*, Vol. 8, 1988, pp. 79–83.

[4] Thomas, H. P., Direction Responsive System, U.S. Pat. No. 2,016,977, 1935.

[5] Aschenbrenner, H., and G. Goubau, "Eine Anordung zur Registrierung rascher magnetischer Storungen," *Hochfrequenztechnik und Elektroakustik*, 47, 1936, pp. 177–181.

[6] Dyal, P., and D. Gordon, "Lunar Surface Magnetometers," *IEEE Trans. Magn.*, Vol. 9, 1973, pp. 226–231.

[7] Neubauer, F. M., et al., "The Giotto Magnetic-Field Investigation," *Space Sci. Rev.*, Vol. 23, 1988, p. 250.

[8] Stuart, W. F., "Earth's Field Magnetometry," *Reports on Progress in Physics*, Vol. 35, 1972, pp. 803–881.

[9] Niblett, E. R., and R. L. Coles, "Scientific Requirements for Geomagnetic Observatory Data," *Proc. Intl. Workshop on Magnetic Observatory Instruments*, Ottawa, Canada, July 30–Aug. 9, 1988, pp. 11–15.

[10] Förster, F., "A Method for the Measurement of dc Field Differences and Its Application to Nondestructive Testing," *Nondestruct. Test.*, Vol. 13, 1955, pp. 31–41.

[11] Moore, W. J. M., and P. N. Miljanic, *The Current Comparator*, London, UK: Peter Peregrinus, IEE, 1988.

[12] Gordon, D. I., and R. E. Brown, "Recent Advances in Fluxgate Magnetometry," *IEEE Trans. Magn.*, Vol. 8, 1972, pp. 76–82.

[13] Primdahl, F., "The Fluxgate Magnetometer," *J. Phys. E: Sci. Instrum.*, Vol. 12, 1979, pp. 241–253.

[14] Kolachevski, N. N., *Fluktuacionnye Processy v Feromagnetikach (Fluctuation Processes in Ferromagnetic Materials)*, Moscow, Russia: Nauka, 1985.

[15] Afanasiev, Ju. V., *Ferroprobes* (in Russian), Moscow, Russia: Energia, 1969.

[16] Saito, T., *A Review of 70 Recent Papers on Development of Various Kinds of Magnetometers for Geophysical Use* (in Japanese with English abstract), Onagawa Magnetic Observatory, Tohoku University, Sendai, Japan, 1980.

[17] Boll, R., and K. J. Overshott (eds.), *Magnetic Sensors*, Vol. 2 of *Sensors*, Veiden, Germany: VCH, 1989, Chap. 12, pp. 477–485.

[18] Gibbs, M. R. J., and P. T. Squire, "A Review of Magnetic Sensors," *Proc. IEEE*, Vol. 78, 1990, pp. 973–989.

[19] Ripka, P., "Review of Fluxgate Sensors," *Sensors and Actuators A*, Vol. 33, 1992, pp. 129–141.

[20] Nielsen, O. V., et al., "Development, Construction and Analysis of the 'Orsted' Fluxgate Magnetometer," *Meas. Sci. Technol.*, Vol. 6, 1995, pp. 1099–1115.

[21] "Fluxgate Magnetometry," *Electronics World*, Sept. 1991, pp. 726–732.

[22] Alldredge, L. R., Magnetometer, U.S. Pat. No. 2,856,581, 1951.

[23] Primdahl, F., "The Fluxgate Mechanism, Part 1: The Gating Curves of Parallel and Orthogonal Fluxgates," *IEEE Trans. Magn.*, Vol. 6, 1970, p. 376.

[24] Seitz, T., "Fluxgate Sensor in Planar Microtechnology," *Sensors and Actuators A*, Vol. 21–23, 1990, pp. 799–802.

[25] Gise, P. E., and R. B. Yarbrough, "An Improved Cylindrical Magnetometer Sensor," *IEEE Trans. Magn.*, Vol. 12, 1977, pp. 1104–1106.

[26] Schonstedt, E. O., Saturable measuring device and magnetic core, U.S. Pat. No. 2,916,696, 1959.

[27] Nielsen, O. V., et al., "Miniaturisation of Low-Cost Metallic Glass Flux-Gate Sensors," *J. Magn. Magn. Mater.*, Vol. 83, 1990, pp. 405–406.

[28] Perlov, A. Ya., et al., "Three Component Magnetic Field Measurements Using the Cubic Anisotropy in (111) YIG-Films," Salford University Business Services, Salford, UK, 1997.

[29] Rabinovici, R., and B. Z. Kaplan, "Low Power Portable DC Magnetometer," *IEEE Trans. Magn.*, Vol. 25, 1989, pp. 3411–3413.

[30] Sonoda, T., and R. Ueda, "Distinctive Features of Magnetic Field Controlled Type Magnetic Field Sensor," *IEEE Trans. Magn.*, Vol. 25, 1989, pp. 3393–3395.

[31]   Heinecke, W., "Magnetfeldmessung mit Saturationskernsonden," *Technisches Messen*, Vol. 48, 1981, p. 6.

[32]   Takeuchi, S., and K. Harada, "A Resonant-Type Amorphous Ribbon Magnetometer Driven by an Operational Amplifier," *IEEE Trans. Magn.*, Vol. 20, 1984, pp. 1723–1725.

[33]   Ghatak, S. K., and A. Mitra, "A Simple Fluxgate Magnetometer Using Amorphous-Alloys," *J. Magn. Magn. Mater.*, Vol. 103, Nos. 1–2, 1992, pp. 81–85.

[34]   Zhang, H., et al., "A Novel Co-Based Amorphous Magnetic Field Sensor," *Sensors and Actuators A*, Vol. 69, 1998, pp. 121–125.

[35]   Moldovanu, A., et al., "Functional Study of Fluxgate Sensors With Amorphous Magnetic Materials Core," *Sensors and Actuators A*, Vol. 59, 1997, pp. 105–108.

[36]   Moldovanu, A., et al., "Performances of the Fluxgate Sensor With Tensile Stress Annealed Ribbons," *Sensors and Actuators A*, Vol. 81, 2000, pp. 189–192.

[37]   Moldovanu, A., et al., "The Applicability of Vitrovac 6025 X Ribbons for the Parallel-Gated Configuration," *Sensors and Actuators A*, Vol. 81, 2000, pp. 193–196.

[38]   Saito, T., et al., "Magnetometers for Geophysical Use, Part 1. Fluxgate Magnetometer With a 0.5m Length Two-Core Sensor," Science reports of Tohoku Univ., Sendai, Japan, Ser. 5, Vol. 27, Dec. 1980, pp. 85–93.

[39]   Weyand, K., and V. Bosse, "Fluxgate Magnetometer for Low-Frequency Magnetic Electromagnetic Compatibility Measurements," *IEEE Trans. Magn.*, Vol. 46, 1997, pp. 617–620.

[40]   Ripka, P., F. Jireš, and M. Macháček, "Fluxgate Sensor With Increased Homogeneity," *IEEE Trans. Magn.*, Vol. 26, 1990, pp. 2038–2041.

[41]   Ripka, P., "Race-Track Fluxgate Sensors," *Sensors and Actuators A*, Vol. 37–38, 1993, pp. 417–421.

[42]   Gordon, D. I., R. H. Lundsten, and R. A. Chiarodo, "Factors Affecting the Sensitivity of Gamma-Level Ring-Core Magnetometers," *IEEE Trans. Magn.*, Vol. 1, 1965, pp. 330–337.

[43]   Ripka, P., "Improved Fluxgate for Compasses and Position Sensors," *J. Magn. Magn. Mater.*, Vol. 83, 1990, pp. 543–544.

[44]   Ripka, P., "Race-Track Fluxgate Sensor," Czech Pat. No. CZ 286657, UV 8650/99, Int. appl. PCT/CZ00/00005.

[45]   Ripka, P., "Race-Track Fluxgate With Adjustable Feedthrough," *Sensors and Actuators A*, Vol. 85, 2000, pp. 227–234.

[46]   Ripka, P., "Contribution to the Ring-Core Fluxgate Theory," *Physica Scripta*, Vol. 40, 1989, pp. 544–547.

[47]   Kaplan, B. Z., "Duality of the Electric Covering Fieldmill and the Fluxgate Magnetometer," *IEEE Trans. Magn.*, Vol. 34, 1998, pp. 2306–2315.

[48]   Kaplan, B. Z., "Experimental Proof That Fluxgates Operation Is Directly Related to Electric Antennas Theory," *Sensors and Actuators A*, Vol. 69, No. 3, 1998, pp. 226–233.

[49] Kaplan, B. Z., "Treatment of Extremely Low Frequency Magnetic and Electric Field Sensors Via the Rules of Electromagnetic Duality," *IEEE Trans. Magn.*, Vol. 34, 1998, pp. 2298–2305.

[50] Primdahl, F., et al., "Demagnetising Factor and Noise in the Fluxgate Ring-Core Sensor," *J. Phys. E: Sci. Instrum.*, Vol. 22, 1989, pp. 1004–1008.

[51] How, H., L. Sun, and C. Vittoria, "Demagnetizing Field of a Ring Core Fluxgate Magnetometer," *IEEE Trans. Magn.*, Vol. 33, 1997, pp. 3397–3399.

[52] Burger, J. R., "Theoretical Output of a Ring Core Fluxgate Sensor," *IEEE Trans. Magn.*, Vol. 8, 1972, pp. 791–796; comments MAG-9, p. 708.

[53] Nielsen, O. V., et al., "Analysis of a Fluxgate Magnetometer Based on Metallic Glass Sensor," *Meas. Sci. Technol.*, Vol. 2, 1991, pp. 435–440.

[54] Moldovanu, B., C. Moldovanu, and A. Moldovanu, "Computer Simulation of the Transient Behaviour of a Fluxgate Magnetometric Circuit," *J. Magn. Magn. Mater.*, Vol. 157/158, 1996, pp. 565–566.

[55] Scouten, D. C., "Sensor Noise in Low-Level Flux-Gate Magnetometers," *IEEE Trans. Magn.*, Vol. 8, 1972, pp. 223–231; comments pp. 797–798.

[56] Gordon, D. I., et al., "A Fluxgate Sensor of High Stability for Low Field Magnetometry," *IEEE Trans. Magn.*, Vol. 4, 1968, pp. 397–405.

[57] Hoffman, G. R., "Some Factors Affecting the Noise in a Thin-Film Inductance Variation Magnetometer," *IEEE Trans. Magn.*, Vol. 17, 1981, pp. 3367–3369.

[58] Baltag, O., and D. Constadache, "Sensor With Ferrofluid for Magnetic Measurement," *IEEE Trans. Inst. Meas.*, Vol. 46, 1997, pp. 629–631.

[59] Gershenson, M., "High Temperature Superconductive Flux Gate Magnetometer," *IEEE Trans. Magn.*, Vol. 27, 1991, pp. 3055–3057.

[60] Chaplygin, Y. A., et al., "Experimental Research on the Sensitivity and Noise-Level of Bipolar and CMOS Integrated Magnetotransistors and Judgment of Their Applicability in Weak-Field," *Sensors and Actuators A*, Vol. 49, 1995, pp. 163–166.

[61] Mohri, K., "Review of Recent Advances in the Field of Amorphous Sensors and Transducers," *IEEE Trans. Magn.*, Vol. 20, 1984, pp. 942–947.

[62] Shirae, K., "Noise in Amorphous Magnetic Materials," *IEEE Trans. Magn.*, Vol. 20, 1984, pp. 1299–1301.

[63] Narod, B. B., et al., "An Evaluation of the Noise Performance of Fe, Co, Si and B Amorphous Alloys in Ring-Core Fluxgate Magnetometers," *Can. J. Phys.*, Vol. 63, 1985, pp. 1468–1472.

[64] Nielsen, O. V., et al., "Metallic Glasses for Fluxgate Applications," *Anales de Fisica*, Vol. B86, 1990, pp. 271–276.

[65] Tejedor M., B. Hernando, and M. L. Sanchez, "Magnetization Processes in Metallic Glasses for Fluxgate Sensors," *J. Magn. Magn. Mater.*, Vol. 140–144, 1995, pp. 349–350.

[66] Primdahl, F., et al., "The Sensitivity Parameters of the Short-Circuited Fluxgate," *Meas. Sci. Technol.*, Vol. 2, 1991, pp. 1039–1045.

[67] Bordin, G., et al., "Influence of the Chemical Etching on Magnetic Properties of Co-Rich Amorphous Sheets," *J. Magn. Magn. Mater.*, Vol. 133, 1994, pp. 259–261.

[68] Vertesy, G., A. Gasparics, and Z. Vertesy, "Improving the Sensitivity of Fluxset Magnetometer by Processing of the Sensor Core," *J. Magn. Magn. Mater.*, Vol. 167–169, 1999, pp. 333–334.

[69] Nielsen, O. V., et al., "Nanocrystalline Materials as the Magnetic Core in High Performance Fluxgate Sensors," *Proc. Nanomagnetic Devices*, NATO Adv. Res. Workshop, Sept. 1992, NATO ASI Series E Appl. Sci., Vol. 247, 1993, pp. 27–31.

[70] Nielsen, O. V., J. R. Petersen, and G. Herzer, "Temperature Dependence of the Magnetostriction and the Induced Anisotropy in Nanocrystalline FeCuNbSiB Alloys, and Their Fluxgate Properties," *IEEE Trans. Magn.*, Vol. 30, 1994, pp. 1042–1044.

[71] Benyosef, L. C. C., "Improvements With Amorphous and Nanocrystalline Ribbons To Be Applied as Fluxgate Sensor Cores," *4. Congresso Int. da Soc. Brasileira de Geofisica*, Rio de Janeiro, Brazil, 1995.

[72] Cruz, J. C., H. Trujillo, and M. Rivero, "New Kind of Magnetometer Probe With Enhanced Electronic Processing," *Sensors and Actuators A*, Vol. 71, 1998, pp. 167–171.

[73] Ripka, P., "Coherent Signal Processing in Sensor Technology," *Proc. Eurosensors VII Conf.*, Budapest, Hungary, 1993, p. 184.

[74] Primdahl, F., "The Short-Circuited Fluxgate," *Electronic Horizon*, Vol. 53, 1992, pp. 95–97.

[75] Primdahl, F., et al., "Digital Detection of the Fluxgate Sensor Output Signal," *Meas. Sci. Technol.*, Vol. 5, 1994, pp. 359–362.

[76] Auster, H., et al., "Concept and First Results of a Digital Fluxgate Magnetometer," *Meas. Sci. Technol.*, Vol. 6, 1995, pp. 477–81.

[77] Henriksen, J. P., et al., "Digital Detection and Feedback Fluxgate Magnetometer," *Meas. Sci. Technol.*, Vol. 7, 1996, pp. 897–903.

[78] Pedersen, E. B., et al., "Digital Fluxgate Magnetometer for the Astrid-2 Satellite," *Meas. Sci. Technol.*, Vol. 10, 1999, pp. N124–N129.

[79] Kawahito, S., et al., "A Delta-Sigma Sensor Interface Technique With Third-Order Noise Shaping and Its Application to Fluxgate Sensor System," *Transducers Conf.*, Sendai, Japan, 1999, pp. 824–827.

[80] Koga, S., et al., "Micro Fluxgate Magnetic Sensor Interface Circuits Using Sigma-Delta Modulation," *Trans. IEE of Japan*, Vol. 117E 2, 1997, pp. 84–88.

[81] Marshall, S. V., "A Gamma-Level Portable Ring-Core Magnetometer," *IEEE Trans. Magn.*, Vol. 7, 1971, pp. 183–185.

[82] Robertson, P. A., "Miniature Magnetic Sensor With a High-Sensitivity and Wide Bandwidth," *Electronic Letters*, Vol. 33, No. 5, 1997, pp. 396–397.

[83] Robertson, P. A., "Miniature Fluxgate Magnetic Field Sensors," *Sensor & Transducer Conf. MTEC '99*, Birmingham, England, p. 28.

[84] Rhodes, M. H., "Magnetic Field Detection by Differential Phase Lag," U.S. pat. 4,321,536, Mar. 1982.

[85] Praslicka, D., "A Relax-Type Magnetometer Using Amorphous Ribbon Core," *IEEE Trans. Magn.*, Vol. 30, 1994, pp. 934–935.

[86] Son, D., "A New Type of Fluxgate Magnetometer Using Apparent Coercive Field Strength Measurement," *IEEE Trans. Magn.*, Vol. 25, 1989, pp. 3420–3422.

[87] Kim, H. C., and C. S. Jun, "A New Method for Fluxgate Magnetometers Using the Coupling Property of Odd and Even Harmonics," *Meas. Sci. Technol.*, Vol. 6, No. 7, 1995, pp. 898–903.

[88] Mohri, K., "Magnetometers Using Two Amorphous Core Multivibrator Bridge," *IEEE Trans. Magn.*, Vol. 19, 1983, pp. 2142–2144.

[89] Kozak, M. Z., E. Misiuk, and W. Kwiatkowski, "A Converter-Type Magnetometer Using Amorphous Ribbon or Wire," *J. Appl. Phys.*, Vol. 69, 1991, pp. 5023–5024.

[90] Acuña, M. H., "Fluxgate Magnetometers for Outer Planet Exploration," *IEEE Trans. Magn.*, Vol. 10, 1974, pp. 519–523.

[91] Berkman, R. Ja., "The Effect of Higher Even Harmonics in the Excitation Circuit of Magnetic Modulators" (in Russian), *Avtomatika i telemekhanika*, Vol. 26, 1965, pp. 384–387.

[92] Berkman, R. Ja., and B. L. Bondaruk, "Feroresonansnyj rezhim vozbuzhdenia magnitnych modulatorov i ferozondov" ("Ferroresonance Mode of Excitation of Magnetic Modulators and Fluxgate Sensors"), *Geofiziczeskaya Apparatura*, 1972, No. 50, p. 20.

[93] Ripka, P., and S. Kawahito, "Processing of the Fluxgate Output Signal," *Imeko World Congress*, Osaka, Japan, 1999, Vol. 4, pp. 75–80.

[94] Serson, P. H., and W. L. W. Hannaford, "A Portable Electrical Magnetometer," *Can. J. Tech.*, Vol. 34, 1956, pp. 232–243.

[95] Russel, R. D., B. B. Narod, and F. Kollar, "Characteristics of the Capacitively Loaded Fluxgate Sensor," *IEEE Trans. Magn.*, Vol. 19, 1983, pp. 126–130.

[96] Narod, B. B., and R. D. Russell, "Steady-State Characteristics of the Capacitively Loaded Flux Gate Sensor," *IEEE Trans. Magn.*, Vol. 20, 1984, pp. 592–597.

[97] Gao, Z., and R. D. Russel, "Fluxgate Sensor Theory: Sensitivity and Phase Plane Analysis," *IEEE Trans. Geosci.*, Vol. 25, 1987, pp. 862–870.

[98] Player, M. A, "Parametric Amplification in Fluxgate Sensors," *J. Phys. D*, Vol. 21, 1988, pp. 1473–1480.

[99] Primdahl, F., and P. A. Jensen, "Noise in the Tuned Fluxgate," *J. Phys. E: Sci. Instrum.*, Vol. 20, 1987, pp. 637–642.

[100] Ripka, P., and W. Billingsley, "Fluxgate: Tuned vs. Untuned Output," *IEEE Trans. Magn.*, Vol. 34, 1998, pp. 1303–1305.

[101] Pedersen, E. B., "Digitalisation on Fluxgate Magnetometer," Ph.D. thesis, Academy of Technical Sciences, Copenhagen, Denmark, 2000.

[102] Primdahl, F., et al., "The Short-Circuited Fluxgate Output Current," *J. Phys. E: Sci. Instrum.*, Vol. 22, 1989, pp. 349–353.

[103] Petersen, J. R., et al., "The Ring Core Fluxgate Sensor Null Feed-Through Signal," *Meas. Sci. Technol.*, Vol. 3, 1992, pp. 1149–1154.

[104] Ripka, P., and F. Primdahl, "Tuned Current-Output Fluxgate," *Sensors and Actuators A*, Vol. 82, 2000, pp. 160–165.

[105] Burger, J. R., "Comments on D.C. Scouten, Noise in Low-Level Flux-Gate Magnetometers," *IEEE Trans. Magn.*, Vol. 8, 1972, pp. 797–798 (original paper published in the same volume on pp. 223–231).

[106] Berkman, R. Ja., "Sobstvennye schumy ferrozondov i metodika ich issledovania" ("Own Noises of the Fluxgate Sensors and Methods of Their Observation"), *Geofyzicyeskoe priborostroenie*, No. 7, 1960, p. 25.

[107] Afanasenko, M. P., and R. Ja. Berkman, "Vlianie konstruktivnych i elektricheskich parametrov na uroven sobstvennych schumov magnitnych modulatorov" ("Effect of Constructional and Electrical Parameters on the Magnetic Modulator Own Noise Level") (in Russian), *Otbor i Peredacha Informacii*, No. 42, 1974, p. 61.

[108] Moldovanu, C., et al., "The Noise of the Vacquier Type Sensors Referred to Changes of the Sensor Geometrical Dimensions," *Sensors and Actuators A*, Vol. 81, 2000, pp. 197–199.

[109] Nielsen, O. V., et al., "Selection and Processing of Metallic Glass Materials for Fluxgate Applications," in A. Conde and M. Mill (eds.), *Trends in Non-Crystalline Solids*, Singapore: World Scientific, 1992, pp. 417–420.

[110] Coles, R. L. (ed.), Geological Survey of Canada, paper 88-17, *Proc. Intl. Workshop on Magnetic Observatory Instruments*, Ottawa, Canada, July 30–Aug. 9, 1986.

[111] Kauristie, K., C. Sucksdorff, and H. Nevanlinna, *Proc. Intl. Workshop on Geomagnetic Observatory Data Acquisition and Processing*, Nurmijarvi, Finland, May 15–25, 1989, Finnish Meteorological Institute Geophysical Publication No. 15, 1990.

[112] Diaconu, E. D., et al., "Offset Thermostability of the Tfs-3 Magnetometric Sensors," *Sensors and Actuators A*, Vol. 59, 1997, pp. 109–112.

[113] Primdahl, F., "A Pedestrian Approach to Magnetic Cleanliness," *Danish Space Research Institute Research Report 1990*, Lyngby, Denmark, 1990.

[114] Ripka, P., "AC-Excited Magnetoresistive Sensor," *J. Appl. Phys.*, Vol. 79, No. 8, 1996, pp. 5211–5213.

[115] Primdahl, F., H. Luhr, and E. K. Lauridsen, "The Effect of Large Uncompensated Transverse Fields on the Fluxgate Magnetic Sensor Output," *Danish Space Research Institute Research Report 1992*, Lyngby, Denmark, 1992.

[116] Acuña, M. H., "MAGSAT—Vector Magnetometer Absolute Sensor Alignment Determination," *NASA Technical Memorandum 79648*, Sept. 1981.

[117] Brauer, P., et al., "Transverse Field Effect in Fluxgate Sensors," *Sensors and Actuators A*, Vol. 59, 1997, pp. 70–74.

[118] Ripka, P., and W. Billingsley, "Crossfield Effect at Fluxgate," *Sensors and Actuators A*, Vol. 81, 2000, pp. 176–179.

[119] Scarzello, J. F., and G. W. Usher, "A Low Power Magnetometer for Vehicle Detection," *IEEE Trans. Magn.*, Vol. 13, 1977, pp. 1101–1102.

[120] Ripka, P., and P. Kašpar, "Portable Fluxgate Magnetometer," *Sensors and Actuators A*, Vol. 68, 1998, pp. 286–289.

[121] Jankowski, J., and C. Sucksdorff, *Guide for Magnetic Measurements and Observatory Practice*, IAGA: Warsaw, Poland, 1996.

[122] Rasmussen, O., and E. K. Lauridsen, "Improving Baseline Drift in Fluxgate Magnetometers Caused by Foundation Movements, Using Band Suspended Fluxgate Sensor," *Phys. Earth Planet. Int.*, 1990, Vol. 59, pp. 78–81.

[123] Takahashi, Y., et al., "High Stability Three Axis Magnetometer for Observation of Geomagnetic Field" (in Japanese with English abstract), *Shimadzu Review*, Vol. 41, 1984, pp. 115–121.

[124] White, A., "A Sea Floor Magnetometer for the Continental Shelf," *Marine Geophys. Res.*, Vol. 4, 1979, pp. 105–114.

[125] Segawa, J., et al., "A New Model of Ocean Bottom Magnetometer," *J. Geomag. Geoelectr.*, Vol. 35, 1983, pp. 407–421.

[126] Siemens, "Magnetic Sensors," brochure, 1998.

[127] Dezuari, O., et al., "Printed Circuit Board Integrated Fluxgate Sensor," *Sensors and Actuators A*, Vol. 81, 2000, pp. 200–203.

[128] Vincueria, I., et al., "Flux-Gate Sensor Based on Planar Technology," *IEEE Trans. Magn.*, Vol. 30, 1994, pp. 5042–5045.

[129] Choi, S. O., et al. "An Integrated Micro Fluxgate Magnetic Sensor," *Sensors and Actuators A*, Vol. 55, 1996, pp. 121–126.

[130] Choi, S. O., et al., "A Planar Fluxgate Magnetic Sensor of On-Chip Integration," *Sensors and Materials*, Vol. 9, 1997, pp. 241–252.

[131] Schneider, M., et al., "High Sensitivity CMOS MicroFluxgate Sensor," *Proc. IEDM 97 Conf.*, IEEE, 1997, pp. 36.5.1–36.5.4.

[132] Kejik, P., "Contactless Measurement of Current by Using Fluxgate Method" (in Czech), Ph.D. thesis, Czech Technical University, Prague, 1999.

[133] Kejik, P., et al., "A New Compact 2D Planar Fluxgate Sensor With Amorphous Metal Core," *Sensors and Actuators A*, Vol. 81, 2000, pp. 200–203.

[134] Chiesi, L., et al., "CMOS Planar 2D Micro-Fluxgate Sensor," *Sensors and Actuators A*, Vol. 82, 2000, pp. 174–180.

[135] Ripka, P., et al., "Symmetrical Core Improves Micro-Fluxgate Sensors," *Eurosensors 2000*, submitted to *Sensors and Actuators* for publication.

[136] Kawahito, S., et al., "High-Resolution Micro-Fluxgate Sensing Elements Using Closely Coupled Coil Structures," *Sensors and Actuators A*, Vol. 54, 1996, pp. 612–617.

[137] Sauer, B, et al., "CMOS-Compatible Integration of Thin Ferromagnetic Films," *Sensors and Actuators A*, Vol. 42, 1994, pp. 582–585.

[138] Gottfried, R., et al., "A Miniaturized Magnetic-Field Sensor System Consisting of a Planar Fluxgate Sensor and a CMOS Readout Circuitry," *Sensors and Actuators A*, Vol. 54, 1996, pp. 443–447.

[139] Primdahl, F., et al., "High Frequency Fluxgate Sensor Noise," *Electronic Letters*, Vol. 30, No. 6, 1994, pp. 481–482.

[140] Ripka, P., et al., "AC Magnetic Field Measurement Using the Fluxgate," *Sensors and Actuators A*, Vol. 46–47, 1995, pp. 307–311.

[141] Marshall, S. V., "Impulse Response of a Fluxgate Sensor—Application to Lightning Discharge Location and Measurement," *IEEE Trans. Magn.*, Vol. 9, 1973, pp. 235–238.

[142] Aroca, C., et al., "Spectrum Analyzer for Low Magnetic-Field," *Review of Scientific Instruments*, Vol. 66, No. 11, 1995, pp. 5355–5359.

[143] Kono, M., et al., "A New Spinner Magnetometer," *Geophys. J. R. Astr. Soc.*, Vol. 67, 1981, pp. 217–227.

[144] Kono, M., M. Koyanagi, and S. Kokubun, "A Ring-Core Fluxgate for Spinner Magnetometer," *J. Goemag. Geoelectr.*, Vol. 36, 1984, pp. 149–160.

[145] Stephenson, A., and L. Mollyneu, "A Versatile Instrument for the Production, Removal and Measurement of Magnetic Remanence at Different Temperatures," *Meas. Sci. Technol.*, Vol. 2, 1991, pp. 280–286.

[146] Ioan, C., et al., "Extension of the Frequency Range of Fluxgate Magnetometers," *J. Magn. Magn. Mater.*, Vol. 158, 1996, pp. 567–568.

[147] Peters, T. J., "Automobile Navigation Using a Magnetic Flux-Gate Compass," *IEEE Trans. Vehic. Technol.*, Vol. 35, 1986, pp. 41–47.

[148] Afanasiev, Ju. V., and L. J. Bushev, "Three Component Ferroprobe" (in Russian), *Pribori i sistemi upravlenia*, 1978/1, 29–31.

[149] Primdahl, F., and P. A. Jensen, "Compact Spherical Coil for Fluxgate Magnetometer Vector Feedback," *J. Phys. E: Sci. Instrum.*, Vol. 15, 1982, pp. 221–226.

[150] Nielsen, O. V., et al., "A High-Precision Triaxial Fluxgate Sensor for Space Applications: Layout and Choice of Materials," *Sensors and Actuators A*, Vol. 59, 1997, pp. 168–176.

[151] Langel, R., et al., "The Magsat Mission," *Geophys. Res. Lett.*, Vol. 9, 1982, pp. 243–245.

[152] Brauer, P., "The Ring Core Fluxgate Sensor," Ph.D. thesis, Technical University of Denmark, Lyngby, Denmark, June 1997.

[153] Brauer, P., and O. V. Nielsen, "Fluxgate Sensor for the Vector Magnetometer Onboard the Astrid 2 Satellite," *Sensors and Actuators*, Vol. 81, 2000, pp. 184–188.

[154] Macor, product of Corning Inc., fabricated by AstroMet and Accuratus, www.accuratus.com.

[155] Merayo, J., et al., "Astrid-2 EMMA Magnetic Calibration at Lovoe Magnetic Observatory, May 15–16, 1997," Final Report, Department of Automation, Technical University of Denmark, Lyngby, Denmark, 1998.

[156] Fowler, J. T., and A. D. Little, "New Technology Magnetic Attitude Sensors in Autonomous Underwater Vehicle Applications," *Proc. IEEE Symp. Autonomous Underwater Vehicle Technology*, Washington, D.C., 1990, pp. 263–269.

[157] Berkman, R., "About a New Type of a Magnetomodulation Transducer for the Measurement of a Magnetic Field Gradient" (in Russian), *Avtomaticheski kontrol i inzmeritelnaya technika*, Vol. 4, 1960, pp. 157–162.

[158] Ripka, P., K. Draxler, and P. Kašpar, "Race-Track Fluxgate Gradiometer," *Electronic Letters*, Vol. 29, 1993, pp. 1193–1194.

[159] Ripka, P., and P. Navratil, "Fluxgate Sensor for Magnetopneumometry," *Sensors and Actuators A*, Vol. 60, 1997, pp. 76–79.

4

Magnetoresistors
Hans Hauser and Mark Tondra

This chapter covers the effects of electron scattering, depending on the spin direction with respect to the spontaneous magnetization in ferromagnetic materials. The anisotropic magnetoresistance (AMR) is utilized in thin films, and the giant magnetoresistance (GMR) is a phenomenon based on exchange coupling in sandwiches of ultrathin magnetic and nonmagnetic layers (superlattices). The recently found colossal magnetoresistance (CMR) will also have potential for future sensor applications.

The magnetoresistive effect was discovered in 1857 by Thomson [1], but only the last three decades of research and development have enabled its application in industrial sensors and read-heads for data storage devices. That recent progress is based on modern microelectronics technology and the demands of miniaturization.

Magnetoresistive sensors are well suited for medium field strengths, for example, earth field navigation and position measuring systems. They can be manufactured (also with on-chip electronics) by the packaging technology of integrated circuits at small sizes and low costs, which are the main prerequisites for mass-market acceptance.

This chapter is based on review books and articles [2–4] and the references cited therein, all of which are well suited for further reading. A general description of the AMR phenomenology and materials characterization and fabrication aspects are followed by linearization and stabilization techniques. The sensor layout is examined through selected examples of industrial products.

4.1 AMR Sensors

Three different physical effects in particular contribute to the influence of magnetic fields on the electrical resistivity of solid state conductors.

- The Hall effect is based on the Lorentz force of (1.15). An increase of the resistance is obtained in wide and short conductors[1] due to the deflection of the current paths by the electrical Hall field strength. It is proportional to the electron mobility μ_e and B^2. Because of a low μ_e in metals, it is negligible in this case.
- Another contribution is found in paramagnetic and diamagnetic semiconductors and metals, for example, bismuth. It is caused by band bending at the Fermi surface and is also proportional to B^2.
- The third effect appears distinctly in ferromagnetic and ferrimagnetic thin films (Chapter 1) with uniform orientation of the spontaneous magnetization that is parallel to the easy axis of minimum uniaxial anisotropy energy in the absence of an applied field. It is the AMR.

4.1.1 Magnetoresistance and Planar Hall Effect

The AMR effect is based on the anisotropic scattering of conduction electrons of the band with uncompensated spins (e.g., 3d orbit for the first transition metals Fe, Co, and Ni) in this exchange split band: The energies of the two states of the magnetic spin moment ($\pm \mu_B$) differ by the quantum-mechanical exchange energy. These electrons are responsible for ferromagnetism and ferrimagnetism (Chapter 1).

Theoretical analyses of AMR are given in terms of the electron density-of-states diagram and the Fermi level. The difficulty is that the anisotropic part of the resistance depends on the exact three-dimensional shape of the Fermi surface (3d envelope of the Fermi level), which is not precisely known except for a very few magnetic materials. Theorists, therefore, have not succeeded in calculating the effect to better than one order of magnitude. As a consequence, all materials data have to be found empirically [5]. The magnitude and the sign of the AMR cannot be predicted easily. Most of the materials have positive AMR coefficients, which means that the high resistivity state occurs when spontaneous magnetization $\mathbf{M}_s$ and current density $\mathbf{J}$ are parallel.

---

1. Field plates are sometimes called magnetoresistive sensors. They are made of InSb semiconductors and can exhibit a resistance variation of 1:10 at strong fields ($\approx$ 1T).

The description of the complex behavior of a general magnetoresistor can be simplified by dividing the problem into two parts: (1) the relation between resistivity ρ and the direction of $\mathbf{M}_s$ and (2) the relation between the applied field H and the magnetization direction.

4.1.1.1 Resistance and Magnetization

In soft magnetic thin films of a single domain state, the AMR can be described phenomenologically very simply as a two-dimensional problem. Figure 4.1 shows the dimensions (length l, width b, and thickness d) of the rectangular thin ferromagnetic film (AMR element) and the coordinate system. An applied field H_y rotates M_s out of the easy axis into the hard axis direction of the uniaxial anisotropy. The resistivity depends on the angle $\theta = \varphi - \psi$ between $\mathbf{M}_s$ and $\mathbf{J}$. With

$$\rho(\theta) = \rho_o + (\rho_p - \rho_o)\cos^2\theta = \rho_o + \Delta\rho\cos^2\theta \qquad (4.1)$$

and $\rho = \rho_p$ for $\mathbf{M}_s$ parallel $\mathbf{J}$, and $\rho = \rho_o$ for $\mathbf{M}_s$ and $\mathbf{J}$ orthogonal, the quotient $\Delta\rho/\rho_o$ is the magnetoresistive coefficient, which may amount to several percentage points. With the resistance

$$R(\theta) = \rho(\theta)\frac{l}{bd} = R + \Delta R\cos^2\theta \qquad (4.2)$$

(Figure 4.2), the voltage in x direction is

$$U_x = I\frac{l}{bd}(\rho_o + \Delta\rho\cos^2\theta) \qquad (4.3)$$

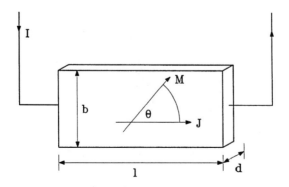

Figure 4.1 Current density and spontaneous magnetization in a single-domain thin-film strip.

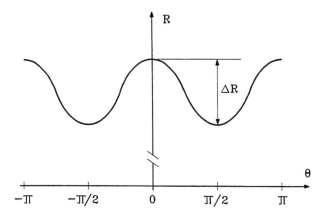

Figure 4.2 Dependence of the resistance R on the angle θ between $\mathbf{M}_s$ and $\mathbf{J}$.

Another effect related to AMR is determined by the tensor property of ρ. Perpendicular to the electrical field E_x causing the current density J_x is an electrical field

$$E_y = J_x \Delta \rho \sin \theta \cos \theta \qquad (4.4)$$

Because of its direction, the effect is known as the planar or extraordinary Hall effect. It must not be confused with the (ordinary) Hall effect because of different physical origins with respect to both $\mathbf{B}$ relations and materials. Depending on $\operatorname{sgn}(\theta)$, the planar Hall voltage

$$U_y = I \frac{\Delta \rho}{d} \sin \theta \cos \theta \qquad (4.5)$$

resulting from (4.4) will be much lower than U_x because of $b \ll l$ in usual geometrical AMR sensor designs. The planar Hall effect can be considered for sensors measuring fields at very small dimensions with high spatial resolutions, for example, for magnetic recording applications.

4.1.1.2 Magnetization and Applied Field

It should be noted that all magnetic sensors measure the flux density $\mathbf{B}$. The physical effect is the Lorentz force acting on moving electrical charges and the torque acting on magnetic moments (as spins in magnetic materials). As $\mathbf{B}$ relates to the applied field $\mathbf{H}$ in general outside the sensor ($\mathbf{B} = \mu_0 \mathbf{H}$), $\mathbf{H}$ is used conventionally for the following considerations. Furthermore, magnetic sensors can cause a distortion of the field to be measured. Therefore, the applied, previously homogeneous field is homogeneous only a certain distance from the sensor, depending on its geometric dimensions due to demagnetizing effects.

The theory of coherent rotation in thin films of the single-domain state was briefly described in Chapter 1. If **H** is acting in y direction, the angle

$$\theta = \arcsin\frac{H_y}{H_0} \tag{4.6}$$

between $\mathbf{M}_s$ and $\mathbf{J}$ (for $-1 < H_y/H_0 < 1$) leads to the field dependence of the resistance

$$R(H_y) = R_0 + \Delta R\left[1 - \left(\frac{H_y}{H_0}\right)^2\right] \tag{4.7}$$

for $|H_y| \leq H_0$ and $R(H_y) = R_0$ for $|H_y| > H_0$, shown in Figure 4.3. The characteristic field H_0 is the sum of the demagnetizing field H_d and the fictitious anisotropy field H_k, which relates to the anisotropy constant K_1 as $H_k = 2K_1/M_s$ (Chapter 1).

The magnetic field **H**, flux density **B**, and magnetization **M** are homogeneous only in ellipsoids. Therefore, H_d causes a variation of θ with respect

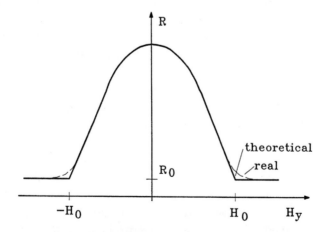

Figure 4.3 Dependence of the resistance R on the applied field H_y; theoretical (solid line) and of a 20-nm thin Permalloy film with $H_c \approx 0$ (dashed line).

to the y coordinate in general. If H_k is negligible compared with H_d and H_y is not strong enough to rotate M_s completely into the y direction, an analytical solution for θ in rectangular samples is

$$\theta(y) = \arctan\left(\frac{2H_y}{dM_s}\sqrt{\left(\frac{b}{2}\right)^2 - y^2}\right) \quad (4.8)$$

That effect is also shown in Figure 4.3. Even if $\theta \approx 60$ degrees in the middle of the sample ($y = 0$), there still is almost no deflection at the edges ($y = -b/2$ and $y = +b/2$). For a first approximation, the demagnetizing matrix is

$$\underline{N_d} \approx \left(\frac{d}{b}, \frac{d}{l}, 1\right) \quad (4.9)$$

as long as $d \ll b \ll l$ is valid.

4.1.2 Magnetoresistive Films

The most important properties of AMR materials are large coefficients $\Delta\rho/\rho$ at a large ρ (high signal at a certain resistance within a small area), low temperature dependence of ρ and l (low offset drift), low anisotropy field H_k (high sensitivity), low coercivity H_c in the hard axis direction (high reproducibility), zero magnetostriction ($\lambda_s \approx 0$, magnetic properties independent of mechanical stress), and long-term stability of those properties (no thermally activated effects, e.g., diffusion, recrystallization).

4.1.2.1 Materials

Besides amorphous ferromagnets, which are characterized by very low H_k and high ρ but only $\Delta\rho/\rho \approx 0.07\%$, the most established materials are crystalline binary and ternary alloys of Ni, Fe, and Co, mainly Permalloy 81Ni/19Fe. This material is characterized by $\mu_0 M_s = 1.1$T at 300K, $\rho = 2.2 \cdot 10^{-7}$ Ωm, $\Delta\rho/\rho = 2 \ldots 4\%$, thermal resistivity coefficient 0.3%/K, $H_k \approx 100 \ldots 1{,}000$ A/m, $H_c < 10$ A/m, and $\lambda_s \approx 0$. To achieve a material with simultaneously vanishing anisotropy and magnetostriction, it is necessary to add about 4% Mo (Supermalloy), but $\Delta\rho/\rho$ is reduced in that case. Therefore, most AMR films are made of Permalloy because of both very low H_k and λ_s at considerable $\Delta\rho/\rho$.

Materials with higher AMR coefficients are 90Ni/10Co (4.9%), 80Ni/20Co (6.5%), 70Ni/30Co (6.6%), and 92Ni/8Fe (5.0%), but they are not suitable for sensor applications due to either high H_k or high λ_s. Other low-anisotropy materials are 50Ni/50Co (2.2%) and Sendust, which is used for magnetoresistive heads because of the mechanical hardness requirements.

4.1.2.2 Film Processing

The two established methods for deposition of Permalloy layers are vacuum evaporation and cathode sputtering at very low oxygen partial pressure. The main advantage of the latter procedure is the good corresponding composition of the film with the target alloy, which can be vacuum melted or sintered. Both elevated target and substrate temperatures have been proven to yield an AMR coefficient of $\Delta\rho/\rho = 3.93\%$ in a 50-nm thin film [6], which is almost the bulk value of about 4%.

The thermally activated crystallite growth or the incorporation of residual gas atoms at grain boundaries, which reduces ρ, can also be done by an annealing process in vacuum or hydrogen, but that can increase coercivity. As the resistivity increases with decreasing film thickness, the AMR coefficient decreases at constant $\Delta\rho$ to 3.5% at 20 nm (which is thought to be the optimum thickness for AMR films [7]). That leads to a tradeoff between sensor resistance and AMR effect.

To define the easy axis orientation, the deposition and/or annealing has to be done in a homogeneous magnetic bias field of some kiloamperes per meter (kA/m), providing a spatial aligning of Ni and Fe atom pairs. That leads to an additional induced anisotropy, which can be used to either decrease or increase the intrinsic H_k, depending on the bias field direction. To avoid unwanted anisotropy and texture, the amorphous substrates (glass or oxidized silicon) have to be very smooth.

The further processing of the film with contact and passivation layers are well-known microelectronic packaging technologies. Because the dimensional tolerances for AMR sensors are smaller than for other devices—due to bridge unbalance—the masks are made by high-precision electron beam lithography.

4.1.2.3 Measurements

The magnetic properties of the film can be determined by measuring the hysteresis loops (component M_H of M_s in H direction versus H) in different directions of the homogeneous applied field, which can be provided by Helmholtz arrangements or cylindrical coils. Suitable methods are vibrating coil magnetometry (VCM) or reflected light modulation by the magneto-optical Kerr effect. Figure 4.4 shows such a hysteresis loop [8] parallel and

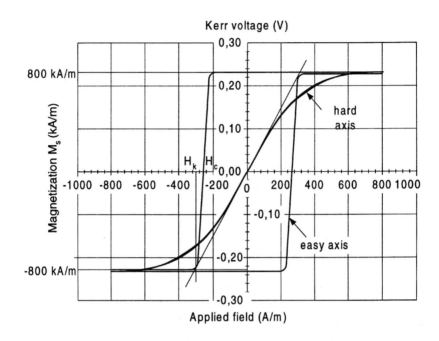

Figure 4.4 Hysteresis loops $M_H(H)$ of a 20-nm thin Permalloy film ($\Delta\rho/\rho = 3.4\%$).

perpendicular to the easy axis, which corresponds well to the theory of coherent rotation (Chapter 1).

Furthermore, ρ and $\Delta\rho(H)$ are determined by a four-point probe, arranged in a square or rectangle, under an applied field of variable direction and strength. Figure 4.3 shows the resistance dependence on H_y for the sample given here.

4.1.3 Linearization and Stabilization

The simple magnetoresistive element in Figure 4.1 has several drawbacks without additional measures: The resistance change at low fields is about zero; the characteristics are nonlinear, even in the vicinity of the inflection point; and the film would exhibit a multidomain state that could cancel out the AMR effect or at least give rise to magnetic Barkhausen noise. Therefore, linearization and stabilization have to be done by several biasing techniques.

Generated either by permanent magnets or currents in simple coils, an external magnetic field can be used for both purposes, but it is advantageous only if the field is also needed for the given application, for example, position sensing via field distortion. In this section, only internal, integrated solutions are discussed.

4.1.3.1 Perpendicular Bias

Perpendicular bias is used to establish an angle $\theta > 0$ between $\mathbf{M}_s$ and $\mathbf{J}$ in absence of $\mathbf{H}$. That can be done either by applying a magnetic bias field H_B in y direction or by rotating the current path out of the easy axis (which is discussed separately as geometric bias because of its practical importance). Replacing H_y in (4.7) by $H_y + H_B$ yields

$$R(H) = R_0 + \Delta R \left(\frac{2H_y H_B}{H_0^2} + \frac{H_y^2 + H_B^2}{H_0^2} \right) \quad (4.10)$$

which is linear for $H_y \ll H_0$.

Permanent Magnetic Films

Besides using an external magnet, a sandwich structure with a hard magnetic film (thickness d_m) can provide a bias field by simple magnetostatic coupling, as shown in Figure 4.5. Such films can be processed by well-established thin-film recording technologies and are made of CoPt, CoNiPt, or CoCrPt. Their remanence should relate to $\mathbf{M}_s$ of the AMR film as

$$\mathbf{M}_r = \mathbf{M}_s \frac{d}{d_m} \sin \theta \quad (4.11)$$

To prevent exchange coupling, electrical shunting, and diffusion processes, a nonmagnetic, electrically insulating layer should be provided. It is typically 10–20 nm of SiO_2 or Al_2O_3. Disadvantages are the magnetic stray field of the sensor itself and the fact that postfabrication adjustment of the bias field is impossible.

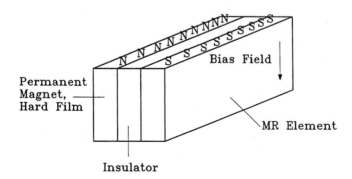

Figure 4.5 Perpendicular bias field generation by hard magnetic films.

Exchange Coupled Films

The basic idea is to induce a unidirectional anisotropy by exchange coupling of the AMR film with an antiferromagnetic or a ferromagnetic layer (thickness d_a) in contrast to the uniaxial anisotropy of the AMR film. The direction of anisotropy can be determined by deposition in a field or by subsequent field annealing above the Néel or Curie temperature, respectively. Figure 4.6 illustrates the principle.

The interfacial exchange coupling depends on the submicroscopic details of the interface. The coupling strength is usually described by an effective exchange field

$$\mathbf{H}_E = \frac{J_E}{\mu_0 \mathbf{M}_s d} \qquad (4.12)$$

with respect to the interfacial exchange coupling energy J_E (typically 1 mJ/m^2) and the AMR film parameters $\mathbf{M}_s$, K_1, and d. Only if $J_E < K_1 d_a$, the $\mathbf{M}(H)$ curve is shifted unidirectionally because of the effect of the exchange anisotropy. Otherwise, only coercivity is increased.

Examples of antiferromagnetic films are MnFe, CoO, and TbCoFe. The materials for the previously described permanent magnet biasing are used for ferromagnetic exchange coupled films. Because of corrosion problems, the latter method is preferred (increasing coercivity) to yield a long-term stability. Besides the complex materials science problem of exact exchange control on the atomic level, the main advantage is that the antiferromagnetic bias field cannot be reset or changed accidentally.

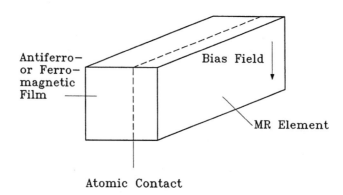

Figure 4.6 Exchange coupling of ferromagnetic and antiferromagnetic films cause unidirectional anisotropy represented by an effective exchange (bias) field.

Shunt Bias

Several methods utilize shunting of currents or magnetic fields. Figure 4.7(a) shows the basic arrangement of an AMR layer, an insulating layer, and a nonmagnetic conductor. The magnetomotive force (1.6) of the conductor produces the bias field. A disadvantage of the insulator thickness is the comparatively high shunt current

$$I_B = M_s d \tag{4.13}$$

of about 30–100 mA in this case. A further disadvantage of thick insulators is the reduction of the bias field at the ends of the AMR film (underbiased region). An advantage of this scheme is that the conductor can be used for servo bias or field compensation purposes in a feedback mode. Omitting the insulator layer causes a partial reduction of the resistance variation, and problems can arise from diffusion of the conductor layer material into the AMR film.

Other arrangements are sandwiches of two magnetoresistors of opposite bias that can be used to form bridge circuits. The current of the scheme in Figure 4.7(b) is reduced by the factor of total sandwich thickness or width with respect to (4.13). The double element biasing scheme of Figure 4.7(c) utilizes the sensing currents of each AMR layer for mutual biasing.

The soft adjacent layer bias arrangement is shown in Figure 4.7(d). Consisting of a high permeability material (Permalloy or Sendust), the soft magnetic shunt layer is magnetized by the field of the AMR sensor current. According to Figure 1.9, the magnetic poles at the surface ($\nabla \cdot M \neq 0$) are the origin of a magnetic stray field H_d, which acts as a bias field for the AMR layer.

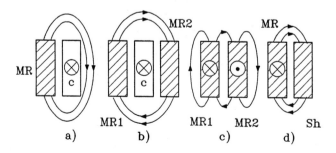

Figure 4.7 Shunt bias field generation by (a) AMR film and conductor, (b) double AMR films and conductor, (c) double element biasing, and (d) soft adjacent layer bias.

A disadvantage of the soft adjacent layer bias scheme is the shunting of the magnetic field $\mathbf{H}_y$ (to be measured) because of the two flux paths. That drawback is eliminated by the double element biasing scheme. Because of the nonlinear mutual magnetostatic interaction between the two soft magnetic layers, the design problems must be solved by micromagnetic numerical techniques.

4.1.3.2 Longitudinal Bias

The theoretical dependence of resistivity change dependence on the applied field is valid under the condition of a single-domain state of the film. Experimental investigations [9] have shown that the magnetization may divide into several domains. That is caused by the large stray field at the ends of a single domain and by favored location of domain walls at imperfections of the film (Chapter 1). The main effects are a smaller $\Delta\rho/\rho$ due to antiparallel domain magnetization and hysteresis with Barkhausen noise due to irreversible domain wall displacements.

The deviations of $\mathbf{M}_s$ from the easy axis (ripple) have been described in terms of internal stray fields, taking into account exchange coupling and different orientation of crystallites. Those stray fields are responsible for blocking the free rotation of $\mathbf{M}_s$, which leads to restrictions for both direction and amplitude of $\mathbf{H}$ to yield a characteristic without hysteresis. The theoretical limit for $\mathbf{H}$ is determined by the Stoner-Wohlfarth astroid (Chapter 1).

Longitudinal bias is necessary to stabilize the single-domain state against perturbation by external fields and both thermal and mechanical stresses. That is achieved by pinning the end-zone domain walls (Figure 4.8) in their initial position or by avoiding those domains by special geometric designs,

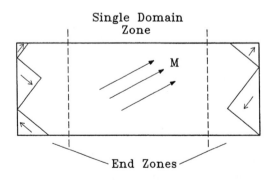

Figure 4.8 AMR film with a single-domain center and multidomain end zones.

for example, triangular endings. Besides adapting the techniques described for perpendicular bias, some other methods are discussed briefly.

Hard Magnetic Films

Thin films with high coercivity (like CoCr, CoPt, SmCo, and ferrites) are sputtered beneath both ends of the AMR film, as shown in Figure 4.9. Their in-plane magnetization provides a longitudinal bias field and a reduction of the end-zone domains. The performance of this method strongly depends on the microscopic details of the junction between the hard film and the AMR film.

A much simpler possibility is also illustrated in Figure 4.9. Overcoating the ends of the AMR film with the conductor material provides the current to flow in the single-domain center zone only. A further improvement can be achieved by progressing the AMR film toward the end zones, which increases the coercivity in that region and contributes to the domain wall pinning. This method reduces noise in the current signal, but it cannot prevent blocking of magnetization at strong fields. An alternative method is screen printing SmCo or hard magnetic ferrite powder onto the rear or the AMR film substrate, which also provides a longitudinal bias field to stabilize the single-domain state.

Exchange Tabs

The end-zone domains are pinned by exchange coupling (as described for perpendicular bias) to either a ferromagnetic layer or an antiferromagnetic (e.g., MnFe) layer. In the zone of strong exchange coupling (Figure 4.10), the spins are prevented from direction changes; therefore, neither domain wall displacements nor rotation of domain magnetization occurs. The direction of exchange anisotropy can be induced by deposition in a magnetic field or by subsequent field annealing above the Néel temperature.

4.1.3.3 Geometric Bias by Herringbone and Barber-Pole Structures

An alternative way to achieve linearity is to rotate the current direction by an angle ψ with respect to the easy axis. The two possible solutions for this

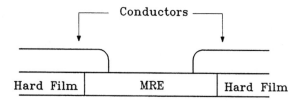

Figure 4.9 Hard magnetic films providing a longitudinal bias field.

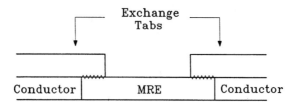

Figure 4.10 Exchange coupling in the tabs providing a longitudinal bias field.

kind of geometric bias have been known since the beginning of industrial AMR sensor production.

Herringbones

The rectangular resistive element is inclined by an angle ψ (two possibilities) to the easy axis, and the current still flows in the longitudinal direction, as shown in Figure 4.11. The spontaneous magnetization is in the direction of the minimum of the total anisotropy energy, as described in Chapter 1. The initial direction of $\mathbf{M}_s$ (two possibilities) has to be defined by an additional stabilizing field. The voltage signal relates to the planar Hall effect of (4.5) with $\theta = \phi \pm 45°$; using (4.2), (1.56), and (1.60), we find the ideal case of resistance dependence on the applied field H_y:

$$R(H_y) = R_0 \pm \Delta R \frac{H_y}{H_0} \sqrt{1 - \left(\frac{H_y}{H_0}\right)^2} \qquad (4.14)$$

As shown in Figure 4.12, the linearity of this characteristic is better than 5% for $H_y < H_0/2$.

A Wheatstone bridge can be realized by using AMR elements of different inclination sign (see Figure 4.11). Problems of this kind of geometric bias

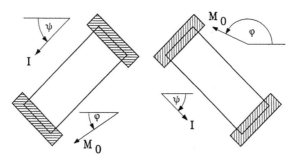

Figure 4.11 AMR elements with inclined easy axis by 45 degrees with respect to the longitudinal direction.

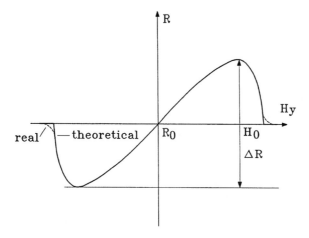

Figure 4.12 Characteristic of AMR elements with $\psi = 45°$ due to geometrical bias; theoretical (solid line) and of a 20-nm thin Permalloy film with $H_c \approx 0$ (dashed line).

are some uncertainty of the angles ψ and ϕ due to misalignment, lithographic limits, and strain gauge effects.

Barber Poles

Entirely different from the methods described so far, the barber pole[2] biasing scheme is used successfully to establish the correct angle θ between $\mathbf{M}_s$ and $\mathbf{J}$. The AMR film is covered with stripes made of materials of high electrical conductivity (Al, Au, Ag, Cu), slanted by an angle of about 45 degrees with respect to the easy axis of the film. If the resistance of the strips is much lower than that of the AMR film and if the contact resistance between barber pole and AMR film is negligible, the current is inclined by 45 degrees with respect to $\mathbf{M}_s$, as shown in Figure 4.13. Therefore, the resistance characteristic also can be described by (4.14).

Imperfections of this linearization are caused by edge effects, in which the current is still flowing parallel to the longitudinal direction of the film, and by a voltage drop along the barber-pole strips due to their finite conductivity. Therefore, the average current angle is less than 45 degrees, which can be considered by correction terms. The resistance

$$R_0 = \rho \frac{l}{bd} \frac{a}{s+a} \frac{f}{2} \quad (4.15)$$

---

2. The name comes from the rotating signs of barbershops.

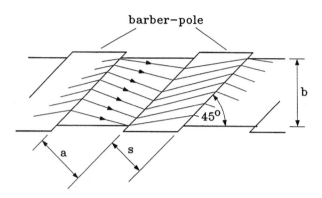

Figure 4.13 Barber-pole structure and directions of **J** in the AMR layer.

is decreased by two to four times compared to other AMR sensor types of the same area due to partially shunting the film by the barber poles (length $b\sqrt{2}$, width s, gap a). The correction factor $f \approx 1.2$ considers edge effects and the resistivity of the barber poles. The geometric design optimization is a tradeoff between high $\Delta\rho/\rho$ and linearity, which leads to the relation $d/b \leq 1/20 \cdot s/a$.

The current flowing in the barber poles generates a magnetic field with an x component

$$H_x = \frac{I}{2b} \frac{a}{a+s} \qquad (4.16)$$

which can be used for stabilization by longitudinal bias [10]. The equal field directions of the stripes restrict some possible combinations of current directions, for example, normal conductors have to be provided for current reversal in AMR meander structures.

The resistance of a bridge consisting of AMR elements with different barber-pole slopes (±45 degrees) depends slightly on the applied field. But the main disadvantage of the barber-pole scheme is that only about 60% of the AMR film width contributes to the active sensor area. Nevertheless, the technique is successfully used for industrial sensor fabrication because of the main advantage: The angle of $\mathbf{M}_s$ and **J** is set directly by geometry only and there are no difficult magnetostatic complications.

4.1.4 Sensor Layout

Besides sensitivity and resistance, it is mainly the type of linearization—longitudinal bias by magnetic fields or geometrical bias by slanted stripes or barber poles—that determines the layout. General restrictions exist for the lower limit of the film thickness (≈ 5 nm) due to decreasing $\Delta \rho / \rho$ and for the upper limit of the current density ($J_{max} \approx 10^{10}$ A/m^2 for Permalloy).

A simple stripe arrangement (see Figure 4.1) is used for very small sensor volumes (e.g., read-heads for data storage) or if only fields above a certain range have to be detected. To reduce thermal effects and to increase the signal voltage, full or half bridges are used for industrial general purpose sensors. Furthermore, both temperature limits and small gradients require low-power dissipation $I^2 R$.

4.1.4.1 Sensitivity and Measuring Range

Due to the large demagnetizing field in z direction, the sensitivity[3]

$$S = \frac{dU}{dH_y} \frac{U_{max}}{\mu_0 U} \tag{4.17}$$

can be defined as the variation dU of the bridge voltage U with respect to the applied field H_y (U_{max} is the maximum permitted operating voltage), which yields the convenient unit V/T (or $\mu V/\mu T$). Considering a full-bridge barber-pole circuit and using (4.17) and (4.14), the maximum sensitivity is

$$S_{max} = \frac{U_{max}}{\mu_0} \frac{\Delta \rho}{\rho} \frac{1 - 2\frac{H_y^2}{(H_0 + H_y)^2}}{(H_0 + H_x)\sqrt{1 - \frac{H_y^2}{(H_0 + H_y)^2}}} \tag{4.18}$$

and for $H_y \ll H_0$, $H_x \approx 0$, it is

$$S_{max} = \frac{\Delta \rho}{\rho} \frac{U_{max}}{\mu_0 H_0} \tag{4.19}$$

The measuring range is also determined by the characteristic field H_0 and the linearity limit. Following (4.14) and Figure 4.3, the output signal of the bridge is zero at $H_y = H_0$ and $H_x = 0$. Therefore, the maximum field is

---

3. A different definition is $S_0 = S/\mu_0 U_{max}$ which is measured in (mV/V)/(kA/m). Other definitions relate the sensitivity also to the AMR film current.

$$H_{max} \approx \frac{H_0 + H_x}{2} \qquad (4.20)$$

and the sensitivity and the maximum field are inversely proportional. The demagnetizing field can be reduced by arranging parallel AMR strips (spacing in the range of the film thickness) connected electrically in series, thus providing magnetically $l \approx b$. The bridge output signal is proportional to the operating voltage U_{max}, considering the maximum power dissipation P_{max} for the tolerable temperature rise.

The lower detection limit H_{min} is determined by the noise level of the bridge output voltage. Main contributions are the thermal noise of resistors ($4kTR\Delta f$ in bandwidth Δf) and the Barkhausen noise (about 4 pT/$\sqrt{Hz}$ in the best case).

Offset drift, for example, due to thermal gradients or thermal mismatch of the bridge circuit, can be minimized by operating the sensor at constant temperature and low power or by special layout. The frequency bandwidth of magnetization reversal in thin films is theoretically from dc to the gigahertz range, but some 100 MHz already have been achieved.

4.1.4.2 Examples

Various designs of AMR read-heads have been developed for magnetic recording, but only some characteristic layout possibilities for industrial AMR sensors are discussed here. The principal schemes of spatial arrangements of full-bridge circuits are illustrated in Figure 4.14. The + and − signs in the rectangles (schematical resistors, distanced by Δx and Δy) denote the relative change of the resistance (increase or decrease) with respect to the applied field

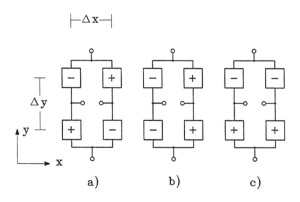

Figure 4.14 Bridge circuits: (a) magnetometer; (b) and (c) gradiometers.

$$H(x, y) = H_m + \frac{dH}{dx}\Delta x + \frac{dH}{dy}\Delta y \qquad (4.21)$$

which is linearized around the medium field H_m, assuming small spatial variance.

Figure 4.14(a) is a magnetometer, and the bridge output voltage is proportional to the field H_m in the center. Figures 4.14(b) and (c) are gradiometers, measuring dH/dx and dH/dy, respectively. The field at each resistor should not exceed the limit, H_{max}. Furthermore, the design can be made sensitive to fields of periodic spatial variation by structuring the AMR elements with the same wavelength.

Herringbone Full Bridge

Figure 4.15 shows a classic design with slanted elements [11]. The black areas represent a NiFeCo AMR film. Four meandered resistors are connected to a Wheatstone bridge. The contact pins for supply and output are indicated by + signs. The strips (width about 50–100 μm) of the meandered resistors are slanted by ±45 degrees with respect to the easy axis, established by induced anisotropy perpendicular to the arrow (measurement direction).

Half Bridge

A typical half bridge with perpendicular AMR elements is shown schematically in Figure 4.16. The output signal at pin b, therefore, depends only on the direction of **H** at $H \gg H_0$. By adding a known constant bias field, we can also measure the field strength.

General-Purpose Full Bridge

A full bridge with meandered resistors using the barber-pole scheme is shown in Figure 4.17 [12]. Depending on the stabilizing field H_x, the sensitivity S has been varied between 100 μV/μT and 10 μV/μT at maximum fields

Figure 4.15 Full-bridge herringbone sensor.

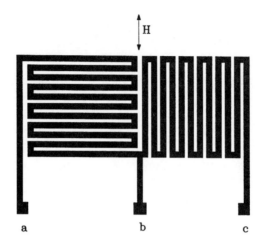

Figure 4.16 Half-bridge sensor with perpendicular elements.

$\mu_0 H_{max}$ between ±0.6 mT and ±9 mT, respectively. A typical operation voltage is 10V. The bridge offset can be adjusted by laser trimming of integrated resistor networks.

Sensors for Weak Fields

If the sensor is used as a magnetometer, for example, to measure deviations of the Earth's magnetic field, a method of switching the spontaneous magnetization between the two stable states in the AMR film is advantageous. The flipping procedure can be done by applying field pulses with an amplitude larger than H_0. The typical characteristics of different slope but equal bridge offset are shown in Figure 4.18. Using a phase-sensitive detector (PSD) or similar electronic circuits, all offset, drift, and noise of sensor and amplifier can be eliminated or reduced.

Figure 4.19 shows a differential switched bias sensor layout [13]. It consists of a pair of identical AMR stripes and overlaid bias conductors. The bias field is generated by the same bias current I_B, but it has the opposite sign for each AMR film. The square wave I_B switches the characteristic of each stripe in different directions (see Figure 4.18), the differential output signal is a square wave voltage, and its amplitude is propotional to the applied field (zero without field). A sensitivity of 24 mV/nT at Δf = 250 Hz and a resolution of 0.1 nT have been achieved for a prototype.

To avoid blocking of magnetization rotation, this kind of sensor can work without a stabilizing field H_x at small fields only (up to 200–300 A/m). That disadvantage can be overcome by the well-known field

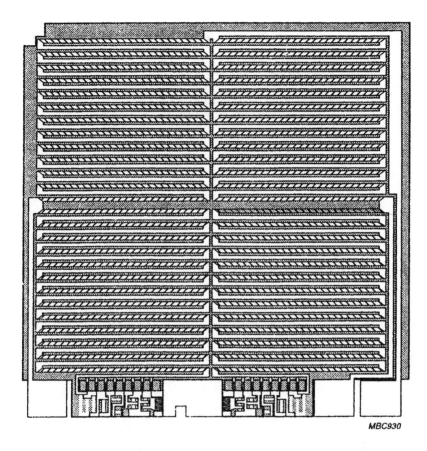

Figure 4.17 General purpose full bridge with barber-pole bias. (Courtesy of Philips.)

compensation method. Because of its high sensitivity, the AMR sensor can be used as a zero field detector, providing further accuracy enhancement.

Figure 4.20 shows schematically a sensor layout with flipping and compensation in-plane circuits [14, 15]. The current of the compensation conductor is the output signal, which is controlled to keep the bridge output at zero. A maximum sensitivity of 400 mV/mT and a resolution of 1 nT have been achieved for industrial sensors [16] (Figure 4.21).

Finally, further improvement of sensitivity and noise can be achieved by designing the whole sensor area for homogeneous fields and low shape anisotropy in the sensitive direction. An overall ellipsoidal shape of the active area with the long ellipse axis perpendicular to the easy axis of the AMR film is a possible approach, increasing sensitivity by two times and reducing noise (caused by field deviations at the end zones of the sensor area) by about three times [8] compared to rectangular sensor designs.

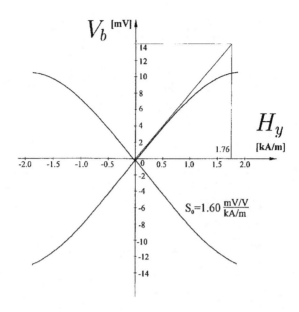

Figure 4.18 Mirror characteristics of a bridge, normal and flipped.

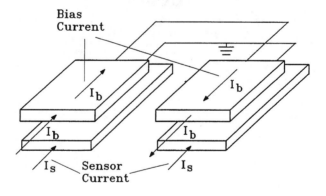

Figure 4.19 Switched bias differential sensor.

4.2 GMR Sensors

4.2.1 Introduction

Many applications in magnetic sensing and detection require high sensitivity, high speed, and low power transducers that can be integrated with silicon circuitry. GMR materials constitute a versatile foundation for these uses.

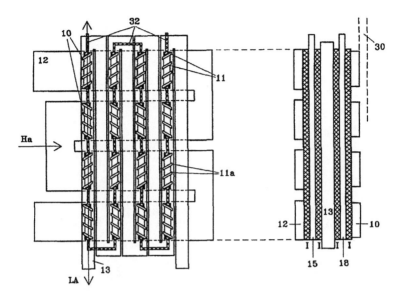

Figure 4.20 Full bridge with barber-pole bias for weak field measurements; conductor films for flipping field pulses and compensation for H_y are sputtered below and above the AMR film, respectively.

They can be made to have a wide range of saturation fields, resistance values, shapes and sizes, and can be formed directly on the surface of an integrated circuit. Because of their high sensitivity, they generally result in the smallest sensor size for a given magnetic sensing application. Because these sensors can be fabricated using the techniques and equipment developed for semiconductor manufacturing, they can be made small and cheaply. An excellent example of this is the ubiquitous integrated read-head for hard disk drives.

The GMR effect was discovered in the late 1980s by several different research groups [17–19], and its fundamental properties were studied extensively [20–22]. As a young technology, it is relatively undeveloped compared to other magnetic sensing technologies like the Hall effect, AMR, and fluxgate magnetometers. But because of some key advantages over other types of sensing materials, a great deal of effort has been put into developing GMR sensors for the full spectrum of magnetic sensing applications. The pace of development is fastest for applications with the largest markets (i.e., read-heads) [23]. It is slower for applications with low volumes and very specific sensing requirements (e.g., ultra low field magnetometers) [24, 25]. The state of GMR sensors is still changing rapidly for some applications. Consequently, this section will try to address not only the state of the art in

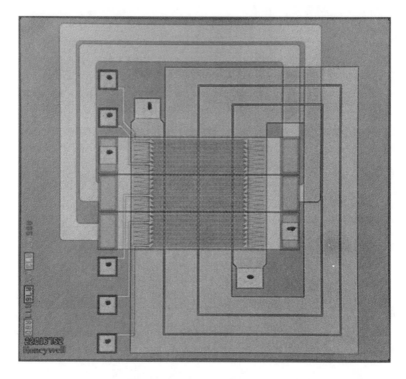

Figure 4.21 Industrial full-bridge layout with conductor strips for control of offset, field compensation, and flipping (set/reset) for weak field measurements. The on-chip straps for set/reset and offset fields are patented by Honeywell. (Courtesy of Honeywell.)

GMR sensor technology, but also some of the fundamental aspects of GMR materials that will help the readers assess the feasibility of using GMR in a particular project.

Because the output of GMR sensors is directly related to the magnetic state of thin films of ferromagnetic metals, they have much in common with both AMR and fluxgate sensors. In particular, the design constraints revolve around the need for smooth and reproducible magnetization behavior. This section looks at three aspects of GMR sensors: the physical origins of the sensing capability, the technical aspects of design and construction of the sensors, and some popular applications for the finished GMR products.

4.2.2 Spin Valve Effect Basics

4.2.2.1 Simple Spin Valves

A straightforward way of understanding GMR is to look at the resistive properties of the simplest GMR structure: the spin valve. It consists of two ferromagnetic metallic layers separated by a nonferromagnetic layer. To further simplify the magnetics, one of the ferromagnetic layers is "pinned" with an adjacent antiferromagnetic layer so that only one of the two magnetic layers is free to respond to an externally applied field. These layers are typically just a few nanometers thick, and the total stack thickness comes to 10 nm to 20 nm. The resistance is lowest when the layers' magnetizations are parallel and highest when their magnetizations are antiparallel. This is true for all GMR structures.

Cross sections of common GMR structures (spin valve, sandwich, and multilayer) are shown in Figure 4.22. The vertical dimension is overstated so that the layered structure is visible. Typical dimensions of a GMR resistor in the film plane are 2 μm wide by hundreds of μm long, depending on the desired resistance value.

This structure has resistance versus field characteristics that look very much like the B versus H loops of the unpinned FM layer (Figures 4.23 and 4.24).

If the soft layer's magnetization is rotated in the film plane rather than irreversibly switched, the resistance varies smoothly as a function of the angle between magnetizations of adjacent layers. Assuming the magnetization within each magnetic layer can be made uniform, this dependence of the resistance on the angle between magnetizations is:

$$R(\theta) = R_{par} + (\Delta R/2)[1 - \cos(\theta)] \qquad (4.22)$$

Here, ΔR is the magnitude of the maximum change of resistance as the angle θ between the two FM layers goes from parallel to antiparallel, and

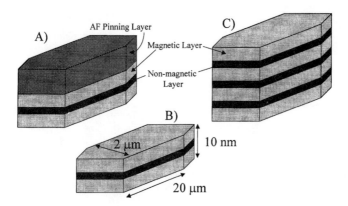

Figure 4.22 Cross sections of common GMR structures: (a) spin valve, (b) sandwich, and (c) multilayer.

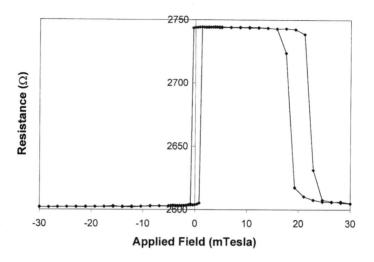

Figure 4.23 The high field (major loop) response of a spin valve when the applied field is parallel to the easy axis of the soft layer and pinning directions. The soft layer switches at about 1 mT, while the pinning of the hard layer is overcome at about 20 mT.

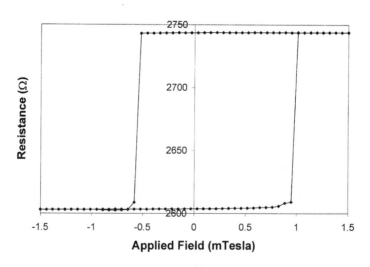

Figure 4.24 Low field response of the spin valve. Only the soft layer is switching.

R_{par} and R_{anti} are the resistances when the layers are parallel and antiparallel, respectively. Also,

$$\Delta R = R_{anti} - R_{par} = R_{max} - R_{min} \quad (4.23)$$

One can see, by differentiating, that the maximum slope for $R(\theta)$ is at $\theta = 90°$ (Figure 4.25). Consequently, when the soft layer of a spin valve element is magnetically biased orthogonally to the hard layer, the element's $R(\theta)$ is at its most sensitive point for very small external fields. The total GMR is usually calculated as a percentage:

$$\text{GMR} = \Delta R/R_{min} = (R_{anti} - R_{par})/R_{par} \quad (4.24)$$

4.2.2.2 Origin of Spin Dependent Scattering

A common term to describe the physical mechanism of GMR is *spin dependent scattering*. As with AMR, the GMR effect is due to differences in the conduction properties of spin-up and spin-down conduction electrons. These differences arise because certain metals (primarily the transition metals Ni, Fe, and Co) have a sizable mismatch in the density of states for up and down

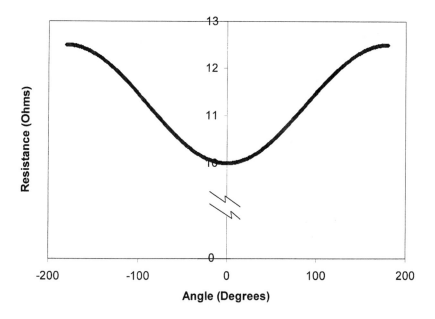

Figure 4.25 The resistance of a spin valve as a function of angle between the pinned and free layers.

electrons at the Fermi energy. When two magnetic thin films are separated by a nonmagnetic film thin enough to allow electrons to pass from one magnetic layer to another, those electrons probe the density of states of the other magnetic layer. When the two magnetic layers' magnetizations are parallel, their state densities are spatially matched and no unusual scattering occurs. But when the two magnetizations are antiparallel, the electrons traversing the nonmagnetic layer experience an inversion of the three-dimensional density of states upon reaching the other magnetic layer. In other words, electrons that were spin-up become spin-down, and vice versa. The mismatched densities of states of the two antiparallel magnetic layers result in increased scattering.

The same electrons that do the conduction in these metals are also the ones that give them their ferromagnetic characteristics. They exhibit the AMR for similar reasons. In a well-constructed GMR film, AMR is a smaller effect, but does contribute to the overall resistance versus field signal of a GMR sensor. The GMR and AMR in a single magnetoresistive film can be deconvolved by making measurements with the current passing in different directions. The GMR is isotropic with current direction, while the AMR will vary (as described in Section 4.1).

There are varying opinions on the detailed causes of the spin dependent scattering that leads to GMR. Some argue that the effect is localized in a very thin region at the interfaces between the magnetic and nonmagnetic layers. Others claim that the spin dependent scattering extends well into the bulk of the ferromagnetic films. The correct explanation likely depends on the particular system being studied (Fe/Cr, Co/Cu, and so forth). The net observable effect, however, is always an increase in the measured resistivity in the antiparallel state versus the parallel state.

In order to observe the spin dependent transport effect, several conditions must be met.

1. It is necessary for a significant fraction of conduction electrons to be able to go from one ferromagnetic layer to another before scattering. This requires very thin nonmagnetic metal spacers (< 50A).
2. There must be a way to rotate the relative magnetizations of the ferromagnetic layers from parallel to nonparallel.
3. Some fraction of electrons going from one ferromagnetic layer to the other must be spin polarized, and maintain that polarization. The magnitude of the GMR effect is qualitatively proportional to the polarization of the ferromagnetic material at the interfaces of the system.

Good lattice matching between the ferromagnetic and nonmagnetic layers in the GMR system seems to be important for maximizing GMR. The most common base systems are Co/Cu and Fe/Cr. In practice, the structures are more complicated. For instance, a pinned sandwich might have the following structure:

[substrate] NiFeCo 3/CoFe 1.5/Cu 3/CoFe 4.5/IrMn 10 (in nm)

The NiFeCo and CoFe alloys are chosen for their high polarization with low magnetostriction. The bottom NiFeCo layer behaves better as a sensing layer (lower coercivity, higher anisotropy) while the CoFe has higher polarization. Thus, the bottom ferromagnetic layer is designed to have the soft magnetic properties of the NiFeCo portion while the dusting of CoFe at the interface with the Cu increases the overall magnetoresistance.

A very large challenge in the construction of GMR films is to make the nonmagnetic layer thin enough without allowing magnetic pinholes to ferromagnetically connect the two magnetic layers. These pinholes do not necessarily change the nature of spin dependent scattering, but they do make it very difficult to align antiparallelly the two magnetic layers (more on this in the micromagnetic design section).

It is interesting to note that the net flow of electrons is parallel to the plane of the films in the standard thin film resistor construction. Only a fraction of the conduction electrons have enough out-of-plane energy to cross the nonmagnetic layer regardless of their spin states. The measured GMR effect is much larger for a given structure when current is passed perpendicular to the film plane. Unfortunately, the vertical resistance of a thin film material is so small that superconducting contacts are required to measure it unless submicron diameter pillars are made. Such structures are still in the exploratory phase.

In practical commercial films, the GMR can routinely be made to be 8% in simple sandwich structures and 15% in multilayer structures. The GMR can be increased to more than 100% by reducing the thickness of the nonmagnetic layer to about 1 nm. However, these structures exhibit extremely large antiferromagnetic coupling, and the resulting high saturation fields (greater than 1T). A useful figure of merit for GMR materials is sensitivity expressed in %/Oe which is calculated by taking the GMR and dividing by the saturation field. This parameter is on the order of 1%/mT in commercial GMR films.

Having described the magnetoresistive response of a simple spin valve and examined the physical origin of this effect, a few varieties of GMR structures are discussed.

Unpinned Sandwich A variation of the spin valve is the GMR sandwich, a spin valve without the pinning layer [Figure 4.22(b)]. While this structure has a simpler layer composition, its magnetoresistive response is harder to control. This is because both magnetic layers are free to rotate, adding an additional degree of freedom. Sensors using unpinned sandwich material must be carefully designed to achieve a useful response [26, 27]. This response is typically unipolar, is linear in a narrow range, and has some hysteresis (Figure 4.26).

Multilayer If the basic ferromagnet/nonferromagnet/ferromagnet structure is repeated several times in a GMR, a multilayer is constructed. Multilayer structures typically have higher GMR and higher saturation fields. One significant advantage is they can be made to have linear output versus field (Figure 4.27).

4.2.2.3 Spin Dependent Tunneling

A new and very promising magnetoresistive structure is the spin dependent tunneling (SDT) device. It is made by using an insulating material (usually Al_2O_3) in place of the nonmagnetic metal in a sandwich or spin valve [28]

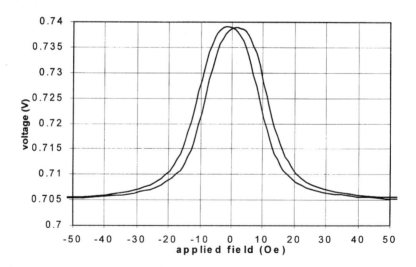

Figure 4.26 GMR sandwich structure output: resistance versus field.

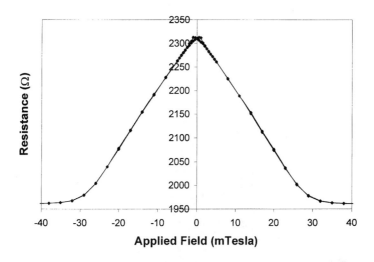

Figure 4.27 GMR multilayer structure output: resistance versus field.

(Figure 4.28). The measured resistance is then inversely proportional to the tunneling probability across the thin insulating barrier. While the details of the conduction mechanisms are quite different, the resistance versus field response of SDT devices look exactly like those of a sandwich or spin valve. (See the MR response for a pinned SDT device in Figure 4.29.)

SDT materials have not appeared in any commercial products to date. However, because they have significantly higher sensitivity than even the best GMR spin valves, product developments are imminent.

4.2.2.4 Micromagnetic Design

Making a useful sensor out of a GMR material requires careful micromagnetic design. Several types of micromagnetic interactions create the total magnetic environment of the GMR sensor:

- Magnetostatic interlayer edge coupling;
- Magnetostatic interlayer coupling (Néel or "orange peel" coupling);
- Sense current field;
- Interlayer exchange coupling;
- Intralayer demagnetizing fields;
- Intralayer anisotropy energies.

All of these fields must be managed to give the resulting magnetoresistive structure high sensitivity to external fields while still having low hysteresis.

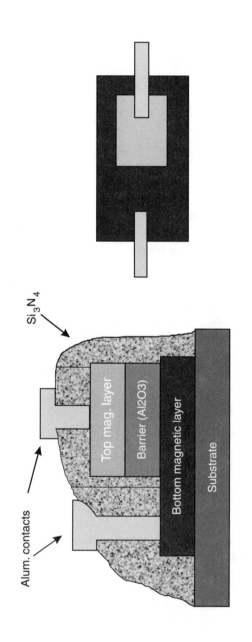

Figure 4.28 A cross section of the SDT structure. The vertical scale is exaggerated so the thicknesses are visible. The lateral dimensions of the tunnel junctions range from 0.1 μm to 1,000 μm.

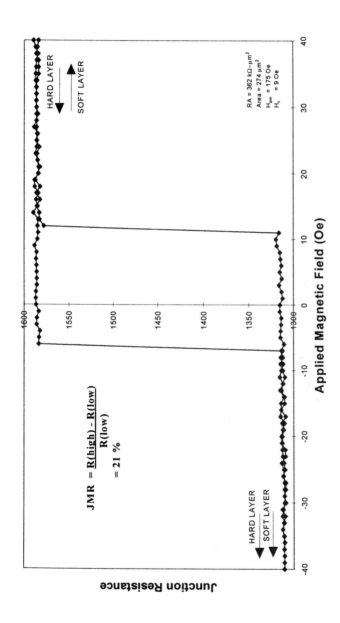

Figure 4.29 The magnetoresistive response of a pinned magnetic tunnel junction. The field is parallel to the easy axis of the free and pinned layers. Note that the characteristic shape is identical to that of the standard GMR spin valve, but the magnitude of the SDT magnetoresistance can be greater than 40%, compared to approximately 10% for spin valves.

The design parameters include GMR stripe width, current density, deposition conditions, and material composition.

Of all the interactions listed, interlayer exchange coupling is most unique to GMR systems. The GMR effect was first discovered by researchers looking for evidence of antiparallel interlayer coupling [17]. This coupling causes nearest ferromagnetic layers in an alternating nonmagnetic/ferromagnetic metallic stack to be antiparallelly aligned. Recalling that the resistance maximum in GMR materials occurs when nearest magnetizations are antiparallel, it is easy to see why the GMR effect is often associated with antiferromagnetic interlayer coupling. This coupling is only observed for interlayer thicknesses less than about 5 nm, and actually has an oscillitory behavior with thickness for thinner layers. It can be very strong (greater than 1T). In many GMR sensor material designs, there would be no way to get the magnetic layers antiparallel (and, hence, maximize the resistance) without antiparallel coupling.

It has only recently been possible to control the deposition of thin metallic films to the precision required for incorporating antiparallel coupling effects into the design of magnetoresistive material.

4.2.2.5 GMR Material Processing Techniques

There are two critical steps in the GMR material fabrication process: material deposition and resistor patterning.

Film Deposition For commercial applications, GMR materials are deposited in vacuum systems. The most common arrangements are RF diode sputtering and ion beam sputtering. Many other types of depositions are done in research laboratory environments including evaporation, electrodeposition, MBE, and dc magnetron sputtering. Making metallic layers whose thicknesses must be controlled to within fractions of an angstrom is challenging for any deposition system.

Patterning Techniques Standard photolithography techniques are used to pattern the wafers full of blank GMR material into resistors of various values. The material itself can be etched using dilute solutions of acid or ion milling. Reactive ion etching has not yet been adapted successfully to ferromagnetic metals. Because the magnetic behavior of thin ferromagnetic films is sensitive to the shape and texture of the edges, the detailed profile of the etching process is critical. Tapers, "mouse-bites," oxidation, and corrosion of the magnetic thin film edges change the magnetization dynamics of the sensor material and potentially ruin the sensor.

4.2.3 Sensor Construction

Having good magnetoresistive material is a start towards a good magnetic field sensor, but it is only the beginning. Typically, a finished sensor is made of several GMR resistors in a Wheatstone bridge configuration. In order for the bridge to have a nonzero response, something must be done to make two of the resistors behave differently than the other two in an external field. In GMR sensors, this is most often accomplished with flux concentrators (Figure 4.30). These are two relatively thick (15 μm) formations of Permalloy plated directly on the GMR sensor chip. The flux concentrators are positioned so that two of the resistors are in the gap between the two concentrators while the other two resistors are underneath the concentrators. The gap resistors experience a multiplication of the external field by a factor of the length-to-gap ratio. The resistors underneath the flux concentrators are effectively shielded from the external field, so their resistance does not change for moderate fields (Figure 4.31).

4.2.3.1 Bipolar Response Using Biasing Coils

Another technique for generating a nonzero output from a bridge is with on-chip planar biasing coils. These coils can be used to create a localized field in different directions for different GMR resistors on the chip. This technique can be used to make a sensor with a bipolar response out of material with a unipolar MR characteristic (Figure 4.32).

Bipolar Material GMR material (or SDT material) can be made to have a bipolar MR response, as is typical with the spin valve structure. A bipolar spin valve must be perpendicularly biased to achieve a linear nonhysteretic output rather than a square, open-loop type response.

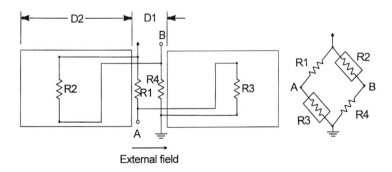

Figure 4.30 A schematic top view of a GMR sensor with flux concentrators, GMR resistors both under and between the concentrators, and external interconnects.

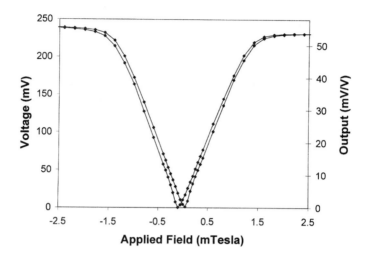

Figure 4.31 Data from an NVE AA00-02 GMR multilayer bridge sensor.

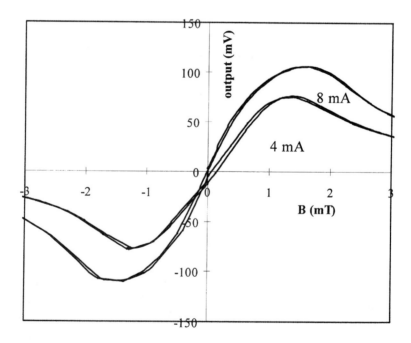

Figure 4.32 The output from a field-biased sandwich material bridge sensor. There are no flux concentrators on this sensor.

A biased SDT resistor is shown in Figure 4.33. Because the basic magnetoresistive response is the same, a similar technique can be used with a GMR spin valve. However, the SDT elements can be any shape and size while the spin valve material must be patterned into long, narrow stripes to get a measurable resistance. Consequently, it is easier to avoid undesirable demagnetizing fields in the SDT structure. The dashed line here has a slope of about 800 mV/mT.

A common magnetic tool for getting a linear response from almost any material is to incorporate a feedback coil and measure the current in this coil that is required to keep the sensor's output at a certain value.

Gradiometer It is also useful to create a bridge whose opposite resistor legs are identical, but separated in space by a relatively large distance. Then the bridge output is nonzero only in the case where the field varies from one end of the sensor to the other. This type of sensor measures the field gradient rather than the absolute field, and is very useful for detecting small nearby objects with relatively low magnetizations. Several examples are given below.

Temperature Characteristics GMR materials have been shown to be able to operate in environments above 225°C [29]. This is particularly important for applications in the automotive industry. The output of all magnetoresistive material has some sensitivity to temperature. These effects come from two sources: the usual increase in R with increasing T for metals; and the decrease in the magnetic moment, which ultimately leads to a decrease in the GMR

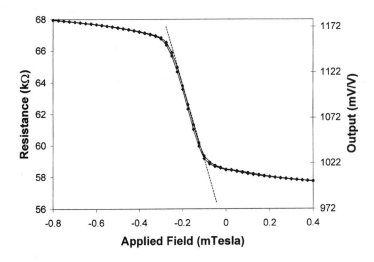

Figure 4.33 A biased SDT resistor.

signal, with temperature. The resistance increase with temperature is about 10%/100°C. The GMR decrease with temperature is not consistent from one type of structure to the other, but is on the order of 0.1%/°C at room temperature.

Cross Axis Sensitivity Another operating parameter to consider is the cross axis sensitivity—how much an x-axis sensor reacts to fields in the y and z directions. This parameter can be specified by dividing the off axis sensitivity by the x-axis sensitivity. For example, if a 1 mT field along the x-axis generates a 10 Volt signal while a 1 mT field along the z-axis generates a 0.01 Volt signal, the sensor would have a 0.001 (0.01/10) off-axis sensitivity to z-axis fields. GMR sensors have very little off-axis sensitivity to z-axis fields due to their thin film (x-y plane) nature (demagnetizing fields are very strong in the z direction). However, some designs of magnetoresistive sensors have significant off-axis sensitivity to y-axis fields (10%). Flux concentrators keep off-axis sensitivity to under 1%.

Trim Sites Automated wafer test equipment dramatically improves the productability of GMR sensors. Sensors are designed with trim sites, which are specially designed links of GMR material connected in parallel with the other bridge resistors. These links are burned out with a short laser beam burst, resulting in an increase in the total resistance of a bridge leg. The trimming process takes place with the sensor powered up and exposed to a known field.

Packaging GMR material has adequate thermal stability to survive most standard integrated circuit packaging processes. But it is very important to choose a package with minimal magnetic material. This involves careful selection of the lead frame material to which the sensor is bonded, and which becomes the pins for the finished package. GMR sensors can be made to fit inside the smallest available package type; 8 pin DIP, 8 pin SOIC, and 4 pin TSOP are common GMR sensor packages.

4.2.4 Applications

GMR sensors are applied to many practical problems. This section briefly discusses some of the best uses for GMR sensors. For each application, the feature of GMR sensors that make them desirable is mentioned.

Vehicle Detection A potentially high-volume application for low-power, high-sensitivity integrated field sensors is for roadway vehicle detection and classification [30].

Cylinder Position Sensing Automated pneumatic cylinders are prevalent in industrial facilities. These devices have pistons that move from one end of the cylinder to the other when actuated by air pressure. An easy way to electronically determine a cylinder head's location is with a magnetic sensor, which when positioned outside the cylinder, detects the field from a magnet mounted internally on the cylinder head. GMR sensors enable this function to be performed using little power, and with a detector that mounts in a small groove on the side of the cylinder.

Gear Tooth Sensing This application is common in automotive and industrial environments where measurements of the angular position and velocity of rotating parts are needed (antilock braking systems, transmission sensors, and so forth). GMR sensors are particularly well suited for these applications because of their high bandwidth, high sensitivity, and good temperature stability. The motion and position of gear teeth can be detected with a gradiometer and a back-biasing magnet. The magnet is placed perpendicular to the sensitive plane of the GMR sensor and near the object to be detected. Thus, a large field can be applied to magnetize the target without saturating the magnetization of the sensor material.

Nondestructive Evaluation The very high bandwidth of GMR sensors makes them ideal for detecting eddy currents. An excellent example of this type of application is nondestructive evaluation (NDE) of metallic structures (boats, aircraft, pipe, and so forth). NDE is used to detect microscopic and/or hidden defects (cracks, pits, oxidation, or other chemical corrosion) in metal layers [31]. The principle is that a time-varying source field induces eddy currents in the target metal layer. These eddy currents generate their own time-varying fields, which can be detected by a magnetic field sensor. Defects on or near the metal surface will change the response of the GMR detector. Coils are often used as the field detector. However, GMR sensors have two distinguishing advantages over coils: They have a flat sensitivity versus a frequency characteristic, and they can be made much smaller and are more easily put in sensor arrays. Having a response at low frequencies is very valuable in NDE because the depth of sensing is inversely related to the excitation frequency.

Unexploded Ordnance A closely related application is unexploded ordnance detection (land mines, shells, and so forth). In this case, eddy currents are induced in metallic objects buried in the ground. The fields from these eddy currents are detectable with an array of GMR sensors, and the collected data can be used to create an image of the buried object [32].

Geomagnetic Another ground searching application is called MagnetoTellurics. This technique measures the response of the ground to electric and magnetic fields. An effective resistance of a given layer of earth or rock is calculated by dividing the E by the B field at a given frequency. The higher the frequency, the shallower the layer. GMR sensors may be useful in this application because of their relatively low cost, power requirements, and small size.

Currency Detection Magnetic ink is used in most currencies, checks, and other financial instruments. Magnetic sensors can detect these inks, help sort currency, and detect bogus currency. The main advantage here of GMR is the signal size in a small area. Because the sensor separation from the currency may be 1 mm or more and the amount of magnetic ink is very small, magnetic signal is at a premium. The ability to put an array of GMR sensors in a small area is an advantage for currency sensing.

Linear and Rotary Position/Motion Sensors GMR sensors are well suited for these applications because of their high output in small sizes. The position detection systems often use coded permanent magnets having regions alternately magnetized up and down, or left and right at a known spacing (Figure 4.34). A series of GMR detectors are placed at appropriate positions to read the magnetization from the nearby magnetic target. A multitrack target, each track having different spatial resolutions, can be used to get absolute position detection using a multisensor detection head. Even more precision may be obtained by using a Vernier-type arrangement where 11 GMR sensors are put in the same space as 10 target transitions. These types of arrangements can achieve spatial resolution under 1 μm (a read-head is an excellent example).

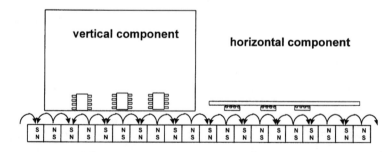

Figure 4.34 A permanent magnet target with about a pitch of a few millimeters. The GMR sensors can be configured to pick up either the vertical or horizontal component of the target.

Signal Isolator/Current Sensors   To this point, the only purposes mentioned for on-chip planar coils have been for generating biasing and feedback fields. But a very important use of these coils is in an application that completely ignores the external field of the GMR device. Note that the field generated is exactly proportional to the current passing through one of these coils, and a GMR element with coils can easily be used as a current sensor. Two on-chip coils can be arranged such that the fields from input currents on one coil are exactly canceled by fields from the output currents in the other coil. This arrangement results in a highly linear current mirror. If the signals are digital rather than linear, the GMR resistance switches in response to discrete field pulses generated by the input coil. These resistance switches can be amplified and passed on by a separate electrically isolated output circuit in whatever form is required by external circuitry. This very useful configuration is a digital signal isolator.

Memory   A digital memory circuit can be made by fabricating an array of bits, each of which is a tiny GMR sensor. This type of memory, magnetoresistive random access memory (MRAM), has the important feature of being nonvolatile, which means it does not lose its information when the power is removed from the memory. It can be thought of as a hard disk on a chip. The military is the primary user of MRAM, although several major commercial developments are under way.

Hard Disk Drive Read-Heads   Although read-heads are the highest-volume GMR sensor application, we will not discuss them here. However, it is useful to note how much GMR read-heads have been developed as an indication of what is technically feasible for other applications. The state-of-the-art in read-heads is using a GMR spin valve sensor to detect bits whose surface area is 50 nm × 500 nm. These bits speed by at a rate on the order of 10^8 bits/second. The bit density and detection rate are increasing at an incredible pace, and GMR read-heads are the best bit sensor for the foreseeable future.

References

[1] Thomson, W., "On the Electro-Dynamic Qualities of Metals—Effects of Magnetization on the Electric Conductivity of Nickel and Iron," *Proc. Royal Society,* London, Vol. A8, 1857, pp. 546–550.

[2] Dibbern, U., "Magnetoresistive Sensors," in W. Göpel, J. Hesse, and J. N. Zemel (eds.), *Sensors,* Vol. 5, *Magnetic Sensors* (vol. eds. R. Boll and K. J. Overshott), Weinheim: VCH, 1989, pp. 342–379.

[3] Mallinson, J. C., *Magneto-Resistive Heads*, San Diego, CA: Academic Press, 1996.

[4] Mapps, D. J., "Magnetoresistive Sensors," *Sensors and Actuators A,* Vol. 59, No. 1, 1997, pp. 9–19.

[5] McGuire, T. R., and R. I. Potter, "Anisotropic Magnetoresistance in Ferromagnetic 3d Alloys," *IEEE Trans. Magn.,* Vol. 11, No. 4, 1975, pp. 1018–1038.

[6] Aigner, P., G. Stangl, and H. Hauser, "Cathode Sputtered Permalloy Films of High Anisotropic Magnetoresistive Effect," *J. Physique,* Vol. 8, 1998, pp. 461–464.

[7] Song, Y. J., and S. K. Joo, "Magnetoresistance and Magnetic Anisotropy of Permalloy Based Multilayers," *IEEE Trans. Magn.,* Vol. 32, 1996, pp. 5–8.

[8] Hauser, H., et al., "Anisotropic Magnetoresistance Effect Field Sensors," *J. Magn. Magn. Mater.,* Vol. 216, 2000, pp. 788–791.

[9] McCord, J., et al., "Domain Observation on Magnetoresistive Sensor Elements," *IEEE Trans. Magn.,* Vol. 32, No. 6, 1996, pp. 4803–4805.

[10] Feng, J. S. Y., L. T. Romankiw, and D. A. Thompson, "Magnetic Self-Bias in the Barber-Pole MR Structure," *IEEE Trans. Magn.,* Vol. 13, 1977, pp. 1466–1468.

[11] Hebbert, R. S., and L. J. Schwee, "Thin/Film Magnetoresistance Magnetometer," *Rev. Sci. Instru.,* Vol. 37, 1966, pp. 1321–1323.

[12] Philips KMZ10.

[13] Mapps, D. J., et al., "A Double Bifilar Magnetoresistor for Earth's Field Detection," *IEEE Trans. Magn.,* Vol. 23, No. 5, 1987, pp. 2413–2415.

[14] Dettmann, F., and U. Loreit, Magnetfeldsensor, aufgebaut aus einer Ummagnetisierungsleitung und einem oder mehreren magnetoresistiven Widerständen, German Patent Application DE 43 19 146 A1, 1994.

[15] Hauser, H., Magnetic field sensor, Austrian Patent Application 1595/96, 1996.

[16] Honeywell HMC1001.

[17] Grunberg, P., et al., "Layered Magnetic Structures: Evidence for Antiferromagnetic Coupling of Fe Layers Across Cr Interlayers," *Phys. Rev. L,* Vol. 57, 1986, pp. 2442–2445.

[18] Baibich, M. N., et al., "Giant Magnetoresistance of (001)Fe/(001)Cr Magnetic Superlattices," *Phys. Rev. L,* Vol. 61, 1988, pp. 2472–2475.

[19] Binash, G., et. al., "Enhanced Magnetoresistance in Layered Magnetic Structures With Antiferromagnetic Interlayer Exchange," *Phys. Rev. B,* Vol. 39, 1989, pp. 4828–4830.

[20] Barthelemy, A., et al., "Magnetic and Transport Properties of Fe/Cr Superlattices," *J. Appl. Phys.,* Vol. 67, 1990, pp. 5908–5913.

[21] Dieny, B., et al., "Giant Magnetoresistive in Soft Ferromagnetic Multilayers," *Phys. Rev. B,* Vol. 43, 1991, pp. 1297–1300.

[22] Parkin, S. S. P., N. More, and K. P. Roche, "Oscillations in Exchange Coupling and Magnetoresistance in Metallic Superlattice Structures: Co/Ru, Co/Cr, and Fe/Cr," *Phys. Rev. L,* Vol. 64, 1990, pp. 2304–2307.

[23] Tsang, C., et al., "Design, Fabrication and Testing of Spin-Valve Read-Heads for High Density Recording," *IEEE Trans. Magn.,* Vol. 30, 1994, pp. 3801–3806.

[24] Daughton, J., and Y. Chen, "GMR Materials for Low Field Applications," *IEEE Trans. Magn.*, Vol. 29, 1993, pp. 2705–2710.

[25] Daughton, J. M., et al., "Magnetic Field Sensors Using GMR Multilayer," *IEEE Trans. Magn.*, Vol. 30, 1994, pp. 4608–4610.

[26] Daughton, J. M., et al., "Giant Magnetoresistance in Narrow Stripes," *IEEE Trans. Magn.*, Vol. 28, 1992, pp. 2488–2493.

[27] Daughton, J. M., "Weakly Coupled GMR Sandwiches," *IEEE Trans. Magn.*, Vol. 30, 1994, pp. 364–368.

[28] Moodera, J., and L. Kinder, "Ferromagnetic-Insulator-Ferromagnetic Tunneling: Spin-Dependent Tunneling and Large Magnetoresistance in Trilayer Junctions," *J. Appl. Phys.* Vol. 79, No. 8, 1996, pp. 4724–4729.

[29] Wang, D., J. Anderson, and J. M. Daughton, "Thermally Stable, Low Saturation Field, Low Hysteresis, High GMR CoFe/Cu Multilayers," *IEEE Trans. Magn.*, Vol. 33, 1997, pp. 3520–3522.

[30] "Tiny Sensor Measures Vehicles Speed," nu-metrics News Release, 1998, www.nu-metrics.com.

[31] Dogaru, T., and S. T. Smith, "A GMR Based Eddy Current Sensor," (to be published in *IEEE Trans. Magn.*).

[32] Wold, R. J., et al., "Development of a Handheld Mine Detection System Using a Magnetoresistive Sensor Array," *Proc. of SPIE Detection and Remediation Technologies for Mines and Minelike Targets IV*, Orlando, FL, Apr. 5–9, 1999, pp. 113–123.

5

Hall-Effect Magnetic Sensors
Radivoje S. Popovic, Christian Schott, Ichiro Shibasaki, James R. Biard, and Ronald B. Foster

Magnetic field sensors based on the Hall effect are probably the most widely used magnetic sensors. Interestingly, Hall magnetic sensors are relatively rarely used to measure just a magnetic field. They are much more used as a key component in contactless sensors for linear position, angular position, velocity, rotation, and electrical current. From the trend shown in Figure 5.1, we estimate that more than 2 billion Hall magnetic sensors will be sold worldwide in 2000. There is hardly a new car in the world without a dozen Hall magnetic sensors, used mostly as position sensors; millions of ventilators and personal computer disc drives use brushless motors with Hall magnetic sensors inside; and millions of current sensors in various products also depend on Hall magnetic sensors. Moreover, the world production of Hall magnetic sensors is increasing, and the application area is becoming ever broader.

Apart from their simplicity and good characteristics, the importance of Hall magnetic sensors is due to their almost perfect compatibility with microelectronics technology. The optimal material characteristics, device structures and dimensions, and fabrication processes are similar to those readily available in semiconductor industry. Therefore, the development in Hall magnetic sensors does not require much specific investment in fabrication processes, contrary to all other magnetic sensors.

In terms of physical parameters, Hall magnetic sensors usually perform well in the following areas: at magnetic flux densities higher than 1 mT, temperatures between −100°C and +100°C, and frequencies from dc to

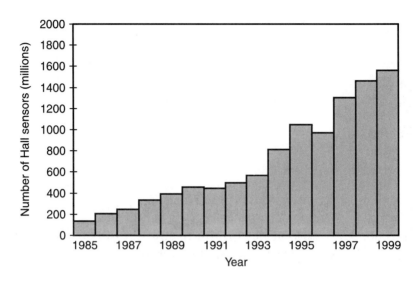

Figure 5.1 World market of Hall sensors.

30 kHz. Of course, the exact values of those parameters depend on the material and the design of the Hall device and may be considerably different in particular cases.

Currently, most applied Hall magnetic sensors are low-cost discrete devices. However, an ever increasing proportion come in the form of integrated circuits. The integration offers an opportunity to apply the system approach to improve the performance in spite of the mediocre characteristics of the basic Hall cells. Moreover, the integrated Hall magnetic sensors are "smart:" they usually incorporate means for biasing, offset reduction, temperature compensation, signal amplification, signal level discrimination, and so on.

A note on the terminology: A basic device exploiting the Hall effect in the form similar to that in which it was discovered is usually called a Hall device, Hall element, or Hall cell. In Japanese literature, the term *Hall element* is used for a discrete Hall device applied as a magnetic sensor. To stress the application of a Hall device, we usually refer to a *Hall magnetic sensor* or just a *Hall sensor*. For magnetic field measurement, a Hall magnetic field sensor, packaged in a suitable case, is normally referred to as a *Hall probe*. In German literature, we often see the expression *Hall generator,* which gives a hint on one aspect of its operation. The term *Hall plate* reflects a conventional form of a Hall device. An integrated circuit, incorporating a combination of a Hall device with some electronic circuitry is usually called an *integrated Hall-effect magnetic sensor* or just a *Hall IC*.

This chapter first briefly discusses some basics of the Hall effect and Hall magnetic sensors (Section 5.1). The three subsequent sections present the three areas of, in our opinion, the highest practical importance in the field: high-mobility discrete Hall plates (Section 5.2), integrated Hall sensors (Section 5.3), and the emerging technology of nonplatelike Hall sensors (Section 5.4).

5.1 Basics of the Hall Effect and Hall Devices

This section is a summary of the physics of Hall magnetic sensors and their basic characteristics. The depth of the explanations are limited to providing just enough of a basis for the following sections. Interested readers can find more detailed treatment of the subject in a monograph on Hall-effect devices [1].

5.1.1 The Hall Effect

The Hall effect is the best known among the physical effects arising in a condensed matter carrying electrical current in the presence of a magnetic field. The effect is named after E. H. Hall, who discovered it in 1879 [2]. A first report on the application of a semiconductor Hall-effect device as a magnetic field sensor was published in 1948 [3].

The Hall effect shows up in its classic and simplest form when a long current-carrying strip is exposed to a dc magnetic field (Figure 5.2). All charge carriers in the strip are then affected by the Lorentz force:

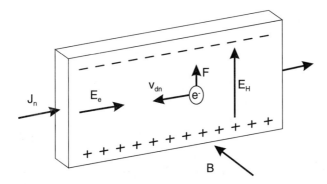

Figure 5.2 The Hall effect in long samples of N-material. The magnetic forces press the electrons toward the upper boundary of the strip so that a Hall voltage appears between the charged edges of the strips.

$$\mathbf{F} = e \cdot \mathbf{E} + e(\mathbf{v} \times \mathbf{B}) \tag{5.1}$$

Here e denotes the electrical charge of a carrier, $\mathbf{E}$ is the local electrical field, $\mathbf{v}$ is the velocity of a charge carrier, and $\mathbf{B}$ is the magnetic flux density, which we take to be perpendicular to the strip plane.

Let us assume that the strip material is a strongly extrinsic N-type semiconductor. We neglect the presence of holes. Along the length of the strip, in the x-direction, an external electrical field E_e is applied. Most of the electrical field $\mathbf{E}$ in (5.1) is due to that external field. The electrons respond to the external electrical field by moving along the strip with the average drift velocity

$$\mathbf{v}_{dn} = \mu_n \cdot \mathbf{E}_e \tag{5.2}$$

μ_n being the drift mobility of electrons. The associated current density is given by

$$\mathbf{J}_n = q \cdot n \cdot \mu_n \cdot \mathbf{E}_e \tag{5.3}$$

where q is the elementary charge.

The carrier velocity $\mathbf{v}$ in (5.1) is due to thermal agitation and drift. Let us neglect for a moment the thermal motion. Then the magnetic part of the Lorentz force (5.1) is given by

$$\mathbf{F}_{mn} = q \cdot \mu_n [\mathbf{E}_e \times \mathbf{B}] \tag{5.4}$$

That force pushes the electrons toward the upper edge of the strip. Consequently, the electron concentration at the upper edge of the strip increases and that at the lower edge decreases. Because of those space charges, an electrical field appears between the strip edges. This electrical field, denoted in Figure 5.2 as $\mathbf{E}_H$, acts on the electrons by a force

$$\mathbf{F}_{en} = -q \cdot \mathbf{E}_H \tag{5.5}$$

That force tends to decrease the excess charges at the edges of the strip. At steady state, the two transverse forces $\mathbf{F}_{mn}$ and $\mathbf{F}_{en}$ balance. By equating (5.4) and (5.5), we find

$$\mathbf{E}_H \simeq \mu_n [\mathbf{E}_e \times \mathbf{B}] \tag{5.6}$$

The transverse electric field $\mathbf{E}_H$ is called the Hall electric field. We use the approximate sign because we neglected the thermal agitation of the charge carriers. Nevertheless, (5.6) is a surprisingly close approximation of the accurate result. Without neglecting the thermal agitation of electrons, instead of (5.6) we obtain

$$\mathbf{E}_H = -\mu_{Hn}[\mathbf{E}_e \times \mathbf{B}] \qquad (5.7)$$

Here μ_{Hn} denotes the Hall mobility of electrons. The Hall mobility differs a little from the drift mobility: It is given by

$$\mu_{Hn} = r_H \cdot \mu_n \qquad (5.8)$$

where r_H is the Hall scattering factor. This is a numerical factor that reflects the influence of the thermal motion of carriers and their scattering on the Hall effect. In most cases, r_H differs less than 20% from unity.

According to (5.6) and (5.7), the Hall electrical field is proportional to the externally applied electrical and magnetic fields. The proportionality coefficient is the carrier mobility. That gives us a first idea about a suitable material to build a magnetic sensor based on the Hall effect: It should be a high-mobility material. Because the mobility of electrons is about three times higher than the mobility of holes, it is better to use an N-type than a P-type semiconductor.

Another useful expression for the Hall electrical field is obtained when the external electrical field in (5.6) is expressed by the current density (5.3):

$$\mathbf{E}_H = -R_H[\mathbf{J} \times \mathbf{B}] \qquad (5.9)$$

Here, R_H denotes the Hall coefficient, in this case given by

$$R_H \approx \frac{1}{q \cdot n} \qquad (5.10)$$

where n is the density of free electrons.

Here again we use the approximate sign because we neglected the thermal agitation of charge carriers. Without neglecting their thermal agitation, instead of (5.10) we obtain

$$R_H = \frac{r_H}{q \cdot n} \qquad (5.11)$$

That gives us another criterion for the choice of the material for a Hall magnetic sensor: it should be a relatively low-doped semiconductor material.

When more than one type of charge carrier is present and/or the material is anisotropic, the Hall coefficient takes a more complicated form. However, if one type of carrier is predominant in terms of the product concentration times mobility, then an equation like (5.11) gives a good approximation for the Hall coefficient.

The most tangible thing associated with the Hall effect is the appearance of a measurable voltage between the edges of the strip. This voltage is known as the Hall voltage. With reference to Figure 5.2, let us choose two points, M and N, at the opposite edges of the strip so that the potential difference between them is zero when $B = 0$. Then the Hall voltage is given by

$$V_H = \int_{S_1}^{S_2} E_H ds \qquad (5.12)$$

In this particular case, we find that

$$V_H = \mu_{Hn} \cdot E_e \cdot B \cdot w \qquad (5.13)$$

where w denotes the width of the strip. We assume here that the magnetic flux density vector **B** is perpendicular to the strip plane; otherwise, we would have to replace B in (5.13) with the perpendicular component of **B**, which we denote by $B_\perp$.

We can obtain another useful expression for the Hall voltage when we combine (5.3) and (5.13) and take into account that the current density in the strip is given by

$$J = \frac{I}{t \cdot w} \qquad (5.14)$$

Here, I denotes the current in the strip and t is the thickness of the strip. So the Hall voltage is also given by

$$V_H = \frac{R_H}{t} \cdot I \cdot B \qquad (5.15)$$

which gives us an idea about a suitable geometry of a Hall magnetic sensor: A Hall magnetic sensor usually has the form of a thin plate.

Let us estimate the value of the Hall voltage in a typical Hall sensor application: take $n = 10^{16}$ cm^{-3}, $t = 10$ μm, $I = 1$ mA, and $B_\perp = 100$ mT; then $V_H \sim 60$ mV.

5.1.2 Structure and Geometry of a Hall Device

In accordance with the conclusions drawn in Section 5.1.1, we can imagine the practical Hall magnetic sensor shown in Figure 5.3: a piece of a strip of N-type semiconductor material, fitted with four ohmic contacts at its periphery. The electrical energy is supplied to the device via two of the contacts, called the current contacts (CCs) or the input terminals. The other two contacts are placed at two equi-potential points at the plate boundary. Those two contacts are used to retrieve the Hall voltage. They are called the voltage contacts, the sense contacts (S), or the output terminals. Some early Hall devices really had such a form, with dimensions (length and width) of several millimeters. Most modern Hall magnetic sensors are much smaller, but they still have a general structure reminiscent of that in Figure 5.3. That is why a Hall device is often called a Hall plate.

Apart from the simple rectangular shape shown in Figure 5.3, many other shapes are possible, such as a square, an octagon, or a cross. It could be shown that all these shapes could be transformed into each other by conformal mapping. Therefore, in the ideal case, the basic characteristics of a Hall device are not dependent on its general shape. However, the sensitivity

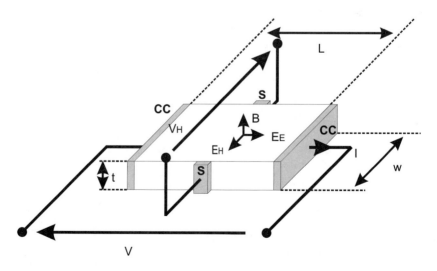

Figure 5.3 Conventional Hall sensor in the shape of a plate.

of a Hall device to parasitic effects, such as fabrication tolerances, may depend a lot on the basic shape of the device. The most commonly used shapes of Hall devices are presented later in this chapter, in the examples of technological realizations of Hall magnetic sensors.

A particularly important issue in the geometry of a Hall device is the relative size of the contacts. If a contact is very small, then the contribution of the semiconductor material resistance adjacent to the contact to the total device resistance may be too high. If an S covers an essential portion of the device periphery, it will short-circuit part of the bias current, and the Hall voltage will be lower than expected. Similarly, if a CC is large, it will short-circuit a part of the Hall voltage.

The influence of the geometry of a Hall plate, including the geometry of the contacts, can be represented by the so-called geometrical correction factor G. That factor describes the diminution of the Hall voltage in a Hall device due to the above-described short-circuiting effects. Let us denote the actually measured Hall voltage by V_H and that of a hypothetical corresponding point-contact Hall device by $V_{H\infty}$. Then the geometrical correction factor is defined as

$$G = \frac{V_H}{V_{H\infty}} \quad (5.16)$$

Theoretically, $0 < G < 1$. In practical Hall devices, we usually find the values of G between 0.7 and 0.9.

Using the notion of G, we can now correct (5.15), which was developed for the case of an infinitely long strip. In a real Hall device, the Hall voltage is given by

$$V_H = G \cdot \frac{R_H}{t} \cdot I \cdot B \quad (5.17)$$

That is probably the most often used equation in the field of Hall magnetic sensors.

5.1.3 Main Characteristics of Hall Magnetic Field Sensors

To be useful as a magnetic sensor, a Hall device must feature a set of characteristics adequate for the intended application. We now discuss a few characteristics that are decisive for the applicability of Hall magnetic sensors [4].

5.1.3.1 Sensitivity

The responsivity of the output voltage of a Hall device to a magnetic field can be characterized by three figures of merit, that is, absolute sensitivity S_A, supply current-related sensitivity S_I, and supply voltage-related sensitivity S_V.

The absolute sensitivity S_A is defined by

$$S_A = \frac{V_H}{B_\perp}, \quad V_H = S_A \cdot B_\perp \tag{5.18}$$

Here V_H is the Hall voltage and $B_\perp$ is the normal (to the Hall plate) component of the magnetic induction.

Supply current-related sensitivity (in short, sensitivity S_I) is defined by

$$S_I = \frac{S_A}{I}, \quad V_H = S_I \cdot I \cdot B_\perp \tag{5.19}$$

where I is the supply (or bias) current of the Hall device. For a strongly extrinsic Hall plate [see (5.17)],

$$S_I = G \cdot \frac{R_H}{t} \tag{5.20}$$

where G denotes the geometrical correction factor ($G \leq 1$), R_H is the Hall coefficient, and t is the thickness of the plate.

In most currently used Hall magnetic sensors, one finds the values of the sensitivity S_I of the order of 100 V/AT. It was estimated that a maximum value of the current-related sensitivity of about 3,000 V/AT could be reached in an integrated Hall device [1].

In low-voltage applications, the absolute sensitivity of a Hall magnetic sensor is often limited by the available supply voltage V. The relevant parameter is then the supply voltage-related sensitivity (in short, sensitivity S_V), defined by

$$S_V = \frac{S_A}{V}, \quad V_H = S_V \cdot V \cdot B_\perp \tag{5.21}$$

The value of S_V is particularly important in low-voltage applications of Hall devices.

For a strongly extrinsic Hall plate,

$$S_V = \mu_H \cdot \frac{w}{l} \cdot G \tag{5.22}$$

where μ_H is the Hall mobility of the majority carriers, w/l is the width-to-length ratio of the equivalent rectangle of the Hall plate, and G is the geometry correction factor.

The measured Hall voltage is then given as

$$V_H = \mu_H \cdot \frac{w}{l} \cdot G \cdot V \cdot B_\perp \tag{5.23}$$

The value of the term $(w/l) \cdot G$ is the largest in large-contact Hall devices, the limit sensitivity being

$$S_{V\text{max}} = 0.742 \cdot \mu_H \tag{5.24}$$

Sensitivity S_V depends strongly on the material used to fabricate a Hall device. While silicon, with its modest mobility, allows, at room temperature, $S_{V\text{max}} \approx 126$ V/VT, GaAs gives 0.67 V/VT, and InGaAs 0.78 V/VT. Therefore, one clear and important trend in the development of Hall devices is the search for and application of high-mobility materials.

5.1.3.2 Offset

The offset voltage of a Hall device is a quasi-static output voltage that exists in the absence of a magnetic field. With reference to Figure 5.3, in virtue of the symmetry, we would expect the output voltage of the Hall device V_H to be zero in the absence of the magnetic field. However, the symmetry of a Hall device is never perfect: there are always small errors in geometry and variations in doping density, surface conditions, contact resistance, and so forth. Also, a mechanical stress in the Hall device, in combination with piezoresistance effect, can produce an electrical nonsymmetry. The result is a parasitic component in the Hall voltage, which cannot be distinguished from the real quasi-static part of the Hall voltage. Therefore, the offset severely limits the applicability of Hall devices when nonperiodic or low-frequency magnetic signals have to be detected.

The offset of a Hall device is best characterized by the offset-equivalent magnetic induction B_{off}. Using (5.21) and (5.22), we find

$$B_{\text{off}} = \frac{1}{S_V} \cdot \frac{V_{\text{off}}}{V} \approx \frac{1}{\mu_H} \cdot \frac{V_{\text{off}}}{V} \qquad (5.25)$$

where we take $(w/l)G \approx 1$. This equation demonstrates once again the importance of a high Hall mobility of the material used for Hall devices.

When microelectronics technology is used to fabricate a Hall device, the offset voltage amounts to usually less than 0.1% of the voltage applied between the input (current) contacts. Inserting this value into (5.25), we find $B_{\text{off}} \approx 10$ mT, 1 mT, 0.1 mT for Si, InGaAs, and InSb Hall devices, respectively.

It is important to note that the offset voltage is not stable. It varies with temperature and time. Even if all other influences are somehow eliminated, there remain long-term (over a period of more than an hour) fluctuations of the output voltage due to $1/f$ noise. In high-quality silicon Hall devices, those fluctuations correspond to a $B_{\text{off}} \approx 10$ μT.

5.1.4 Other Problems

The applicability of Hall magnetic sensors also depends on the following nonideal characteristics.

- Long-term stability of all characteristics, particularly of sensitivity and offset. Long-term instability due to the surface effects and piezo-resistive and piezo-Hall effects are fairly well understood. We think that some bulk effects may also play a role, but practically nothing is published in the open literature on the subject. The best published long-term stability of the sensitivity S_I [5] is

$$\frac{\Delta S_I}{S_I} = 10^{-4}/\text{year} \qquad (5.26)$$

- Noise. Noise is a limiting factor in low-level magnetic measurements, such as in current sensing. Usually, $1/f$ noise is the most disturbing. If perfect materials and buried structures are used, $1/f$ noise can be decreased by several orders of magnitude. In a good silicon Hall sensor, the noise equivalent magnetic induction in the frequency range from 0.1 Hz to 10 Hz is about 1 μT [5].
- Temperature cross-sensitivity of a Hall device is undesirable sensitivity of its characteristics, such as magnetic sensitivity, to temperature.

In the carrier density saturation range of the semiconductor material used for the Hall device, the temperature cross-sensitivity of the magnetic sensitivity S_I is about 0.1%/K. By a simple compensation, even that can be reduced by a factor of 10. But outside the saturation range, namely, in the intrinsic range and the freeze-out range, the temperature dependence of S_I becomes exponential. That renders a Hall device useless in some applications. To extend the operating range to higher temperatures, wide bandgap semiconductors are used (i.e., GaAs up to 175°C).

5.2 High Electron Mobility Thin-Film Hall Elements

5.2.1 Introduction to Thin-Film Hall Elements

This section describes the fabrication and characteristics in practical and very important thin-film InSb and InAs Hall elements.

Referring to (5.23), the Hall output voltage is proportional to the electron mobility, and so a high-electron mobility material is suitable for fabricating Hall elements. Referring to (5.17), to obtain a Hall element that is highly sensitive to a magnetic field, the active layer must be very thin and have a low carrier density. Therefore, III-V thin-film semiconductors such as InSb, InAs, GaAs, and similar materials have been used for practical Hall elements because they have high electron mobility and their carrier (electron) densities can be easily controlled. Moreover, the thin-film technology is very well suited for fabricating high electron mobility thin films.

The offset voltage V_u between the output electrodes cannot be easily subtracted from the Hall output voltage in practical applications; therefore, reducing the V_u has become the most important problem to be solved in the production of practical Hall elements. The effect can be caused by unwanted asymmetry of the Hall element pattern or electrode and by nonuniformity of material properties of the thin film.

In 1947, Pearson used the Hall effect observed in germanium to measure a magnetic field [3]. That was the first application of the Hall effect for magnetic field sensing. The physical properties of InSb, such as its extremely high electron mobility, were first reported by W. Welker in 1952 [6]. Important progress in InSb thin-film technology for application to Hall effect devices was made by Guenther in 1958 [7, 8]. He proposed the effective vapor pressure control method, referred to as the "three temperatures method," for producing stoichiometric, high electron mobility InSb thin films by vacuum deposition. In 1960, Sakai and Ohsita also studied Hall effect devices and reported that the vacuum deposition method would be a key technology for the development and mass production of practical, low-cost, highly sensitive Hall elements [9]. Then, in 1975, Asahi Chemical Company developed a highly sensitive InSb thin-film Hall element with a novel device structure and a small plastic package using a novel vacuum deposition method [10, 11]. In 1974, the first design of magnetic field amplification was established. This Hall element has been mass produced and used as a magnetic sensor in the field of consumer electronics [11–13]. Molecular beam epitaxy (MBE) technology was found to be a key technology for growing thin film of III-V compound semiconductors having high electron mobility [14]. By using this technology, it is easy to epitaxially grow high electron mobility single-crystal InAs thin films, or more complex structures such as an InAs quantum well on GaAs single-crystal substrates, which are also used for making Hall elements [15]. The role of MBE in the development of practical Hall element technology will be described later.

5.2.2 Highly Sensitive InSb Hall Elements

This section describes the properties of practical InSb thin-film Hall sensors produced by vacuum deposition. In the early days, InSb Hall elements (Hall plates) were fabricated mainly from thin bulk single-crystal InSb, making them expensive and not suitable for mass production. Under the pressure of a strong demand for low-cost, highly sensitive Hall elements for use in electronic equipment, such as small-sized dc brushless motors, where Hall elements are used mainly as magnetic sensors for fine angular velocity control, highly sensitive InSb thin-film Hall elements with a novel structure and a small plastic package were developed [11–13].

5.2.2.1 Highly Sensitive InSb Hall Elements Produced by Vacuum Deposition

Novel production technology for InSb polycrystal thin films having a high electron mobility of 20,000 to 30,000 cm^2/Vs and a thickness of 0.8 μm grown on thin mica substrates was established by the multisource vacuum deposition method with time-dependent (variable) substrate heating using InSb as a source material. In this unique vacuum deposition method, the crystal stoichiometry is controlled by sequential evaporation of InSb from several source boats. The surface of mica is perfectly flat and stable under heating. By choosing suitable parameters, high electron mobility InSb thin films were obtained. Table 5.1 shows the basic properties of InSb thin films; Figure 5.4 shows the temperature dependence of electron mobility, Hall coefficient, and resistivity.

Table 5.1
Typical Properties of InSb Thin Film Formed by Vacuum Deposition (at 25°C)

	Dopant	Electron Mobility, μ_H	Electron Density, n	Thickness, t
InSb	None	20,000–30,000 cm^2/Vs	$2 \cdot 10^{16}$ cm^{-3}	0.8 μm

The InSb thin films deposited by vacuum deposition on mica substrates showed several new and important properties. The temperature dependence of their Hall coefficients was similar to that of single-crystal InSb. However, the electron mobility showed a very small temperature dependence near room temperature, which was different from single-crystal InSb. In the case of single-crystal InSb, the electron mobility increases at low temperature because of decreasing lattice scattering. InSb thin films grown on mica substrates are polycrystal and have many defects, and the scattering of electrons by crystal boundaries or defects is independent of temperature and does not decrease at low temperature. Moreover, this special electron transport phenomenon leads to a reduction of the electron mobility at low temperature and explains the temperature dependence shown in Figure 5.4(a). That led to the discovery of a new Hall element that had a Hall output voltage with a small temperature dependence, that is, the Hall element enabled us to drive it at constant voltage according to (5.23).

A new device structure of the InSb thin-film Hall element having a high sensitivity was also developed. The InSb thin film was removed from the thin mica substrate and sandwiched between a ferrite substrate and a small ferrite chip. The structure amplified the magnetic field in the gap between the ferrite substrate and chip by a factor of about 3 to 6 compared to the original magnetic field applied to the Hall element. Exact calculation of the amplification factor is complex. Because the InSb thin film in the gap experiences the amplified magnetic field, the Hall elements have ultrahigh sensitivity to the magnetic field. This special structure is shown in Figure 5.5.

Figure 5.6 is a photograph of the Hall element chip.

The standard production process for fabricating these Hall elements is shown in Figure 5.7.

5.2.2.2 Typical Characteristics of Highly Sensitive InSb Hall Elements

Table 5.2 shows typical characteristics (standard specification) of commercial highly sensitive InSb Hall elements [16]. The basic characteristics in a magnetic field (i.e., V_H-B characteristics) are shown in Figure 5.8 [11–13].

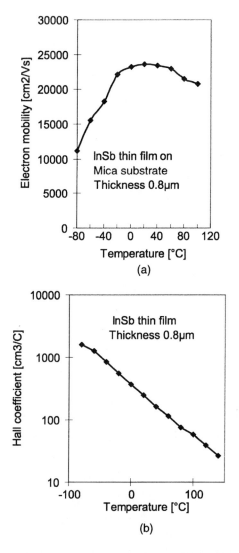

Figure 5.4 Temperature dependence of properties of InSb thin film formed by vacuum deposition (thickness $t = 0.8$ μm): (a) electron mobility; (b) Hall coefficient; and (c) resistivity.

The temperature dependencies of V_H at constant voltage drive and constant current drive are shown in Figure 5.9(a) and (b), respectively; the temperature dependence of these InSb Hall elements is 2.0%/°C [11–13].

To understand those important temperature characteristics, a brief discussion on the temperature dependence of Hall output voltage near room

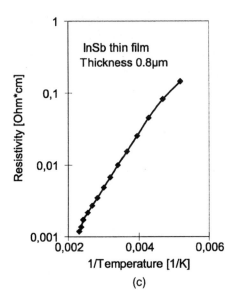

Figure 5.4 (continued).

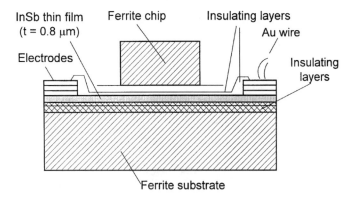

Figure 5.5 The highly sensitive InSb Hall element (cross section).

temperature is important. The temperature coefficient of the Hall output voltage V_H at constant voltage driving is easily derived from (5.13) or (5.23):

$$\frac{1}{V_H} \cdot \frac{dV_H}{dT} = \frac{1}{\mu_H} \cdot \frac{d\mu_H}{dT} \qquad (5.27)$$

For constant current driving, it is also derived from (5.11) and (5.15) or (5.17) and expressed as

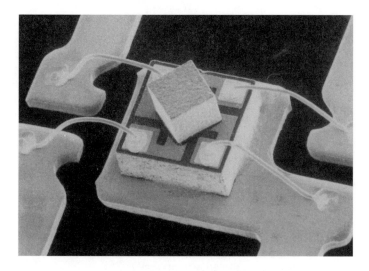

Figure 5.6 Photograph of InSb Hall element chip (Asahi Kasei Electronics: InSb Hall elements HW series).

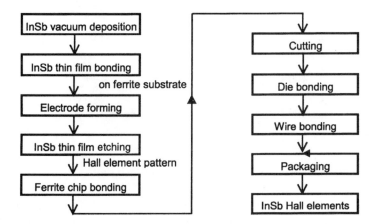

Figure 5.7 Production process of InSb Hall elements.

$$\frac{1}{V_H} \cdot \frac{dV_H}{dT} = -\frac{1}{n} \cdot \frac{dn}{dT} \qquad (5.28)$$

Early Hall elements made using single-crystal InSb plates were driven at constant current to avoid breakdown due to excess current because they had a very low input resistance of only a few ohms. Therefore, temperature dependence of V_H is approximately $-2\%/°C$ because of the large temperature

Table 5.2
Characteristics of InSb Hall Elements Formed by Vacuum Deposition

(a) Electrical and Magnetic Field Characteristics

	Driving Voltage, V_{in}	Hall Output Voltage, V_H (B = 0.05T)	Offset Voltage, V_{off} (B = 0T)	Resistance, R_{in}
InSb	1V	150–320 mV	<±7 mV	240–550 Ω

(b) Absolute Maximum Ratings

Driving Current, I_c	Max. 20 mA
Driving temperature Storage temperature	–40°C to +110°C –40°C to +125°C

Source: Asahi Kasei Electronics HW-300A [11–13, 16].

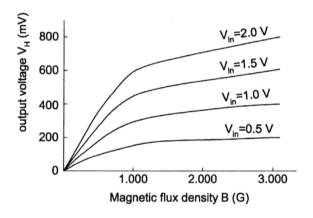

Figure 5.8 Magnetic field characteristics of high-sensitivity thin-film InSb Hall element at constant voltage driving (V_H-B characteristics, Asahi Kasei Electronics HW-300A).

dependence of carrier density n due to the narrow bandgap of InSb. That large temperature dependence was a major problem for applications of Hall elements made from InSb. However, the newly developed thin-film InSb Hall elements have a high input resistance of around 350Ω. Therefore, those Hall elements are stable under an input voltage of 1V to 2V and are driven at a constant voltage. Such constant voltage driving results in the Hall output voltage of highly sensitive InSb Hall elements having a very small or stable

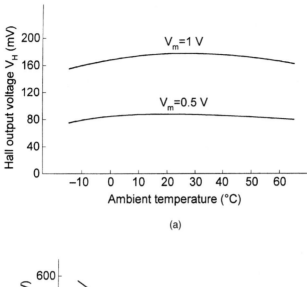

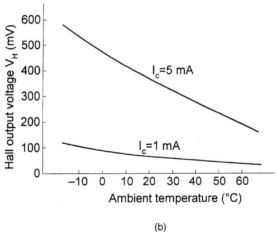

Figure 5.9 Temperature dependence of Hall output voltage at (a) constant voltage driving and (b) constant current driving.

temperature dependence near room temperature. Because the temperature dependence of V_H is the same as electron mobility, as shown in (5.27) and Figure 5.4(b), then this new driving technique reduces the temperature coefficient of the Hall output voltage from −2%/°C to ±0.1%–0.2%/°C near room temperature, as shown in Figure 5.9(a). That is one of the most important practical merits of these highly sensitive InSb thin-film Hall elements produced by the vacuum deposition process and is now a standard driving method for this type of Hall element. These Hall elements have ultrahigh sensitivity to magnetic fields, have practical reliability, and enable the design of small packages that have a wide range of applications.

This InSb Hall element was a commercial success and opened up a new area for brushless dc motor technology and later resulted in the large-scale application of Hall elements as magnetic sensors in small dc brushless motors. The first practical application in 1976 of the highly sensitive Hall element was as a magnetic sensor for music record (audio) player motors. Since then, Hall elements have been mass produced, and recent large-scale applications include dc brushless motors or Hall motors used in VCRs, FDD motors, CD-ROM drive motors of personal computers, and similar electrical equipment. In 1998, more than a billion InSb thin-film Hall elements were produced by vacuum deposition and used in many kinds of applications.

5.2.3 InAs Thin-Film Hall Elements by MBE

The only problem with the InSb Hall elements is their narrow operating temperature range, which is effectively restricted to being near room temperature and which originates from the large temperature coefficient for the input resistance of 2.0%/°C. The success of InSb Hall elements as magnetic sensors has led to many new applications for Hall elements. Some of those new applications impose more severe driving conditions on the Hall elements. A typical new application is automotive sensors, located in the engine compartment or outside the main body frame. This application requires a wide operating temperature range, from −40°C to +150°C; stability over a wide range of driving conditions is also very important. Therefore, a small temperature coefficient of the input resistance of the Hall element or an active layer having a wide bandgap is required. To obtain a Hall element suitable for such new applications, InAs thin-film Hall elements were developed using MBE [15, 17, 18].

Molecular beam epitaxy (MBE) is a technology for growing thin films on single-crystal substrates in an ultrahigh vacuum chamber. With this method, it is possible to fabricate thin films of InAs or III-V compound semiconductor on GaAs substrates (see [14, 15, 17]).

5.2.3.1 Properties of InAs Thin Film for Hall Elements

Bulk single-crystal InAs has a high electron mobility of more than 30,000 cm^2/Vs. The InAs thin films grown by MBE also have a high electron mobility and a bandgap energy of 0.36eV, which is larger than InSb of 0.17eV. Thus, InAs could be used to make a Hall element stable over a wide temperature range because the larger bandgap energy may reduce the temperature instabilities. However, there is a well-known large lattice mismatch between InAs and GaAs (about 7.4%). The fabrication of high-sensitivity Hall elements, using epitaxially grown InAs thin films by liquid phase epitaxy, was previously proposed. Later, by optimizing growth parameters in MBE, a condition was found to grow high-quality InAs thin films, and the large lattice mismatch did not result in any problems in Hall element applications [15, 17, 18].

The room temperature properties of Si doped and undoped InAs thin films grown directly on (100) GaAs substrates (2 degrees off) are shown in Table 5.3.

The temperature characteristics of the InAs thin films are shown in Figure 5.10(a) and (b).

To reduce the temperature dependence of the Hall output voltage for InAs Hall elements at higher temperatures, N-type impurity doping (i.e., Si doping to InAs) was found to be practically effective [15, 17, 20, 21]. However, because this is not trivial, the qualitative discussion that follows shows the effect of doping on the temperature dependence of V_H and input resistance near room temperature.

From (5.28), the temperature dependence of V_H at constant current drive is equal to $-1/n \cdot dn/dT$. Because the dn/dT may be a function of the bandgap energy of InAs, it does not vary much with temperature. Therefore, by doping we can easily increase n and produce a small value of $1/n \cdot dn/dT$. Equation (5.28) illustrates the effectiveness of N-type impurity doping to reduce the temperature dependence of the Hall output voltage at constant current drive. A similar qualitative argument is also valid for the temperature dependence of the resistivity ρ of InAs thin films and input resistance R_{in} of InAs Hall elements. The input resistance of the Hall element in Figure 5.3 is given by

$$R_{in} = \rho \cdot \frac{L}{W \cdot t} \qquad (5.29)$$

Table 5.3
Typical Properties of InAs Thin Film Formed by MBE (at 25°C)

	Dopant	Electron Mobility, μ_H	Electron Density, n	Thickness, t
InAs	None	9,000 cm^2/V · s	$2.2 \cdot 10^{16}$ cm^{-3}	1.2 μm
InAs	Si	11,000 cm^2/V · s	$8 \cdot 10^{16}$ cm^{-3}	0.5 μm

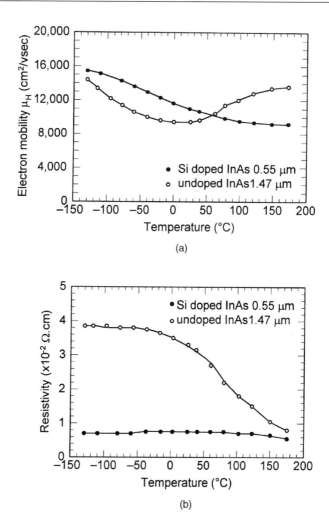

Figure 5.10 Temperature dependence of InAs thin films: (a) electron mobility and (b) resistivity.

Therefore, the temperature coefficient of R_{in} is equal to that of ρ. For a simple model of N-type conduction for the active layer of a Hall element, $\rho = 1/en\mu_H$, where e is the electron charge. Therefore, a simple calculation shows the temperature coefficient of ρ to be given by

$$\frac{1}{\rho} \cdot \frac{d\rho}{dT} = -\frac{1}{n} \cdot \frac{dn}{dT} - \frac{1}{\mu_H} \cdot \frac{d\mu_H}{dT} \qquad (5.30)$$

That simple result gives us an idea of how to reduce the temperature dependence of the input resistance and V_H of InAs Hall elements. As seen in Figure 5.10(a), the temperature dependence of electron mobility is drastically reduced by Si doping. The reason is the unique electron transport mechanism observed in InAs thin films grown directly on GaAs substrates. Such InAs thin films have a two-layerlike mobility structure in the depth direction. Near the interface region of the InAs layer and GaAs substrate, there is a low mobility layer, and far from the interface region, there is a high electron mobility layer. By Si doping of InAs, electrical conduction is always dominated by the high electron mobility layer without any transition from the low to high electron mobility layer; therefore, the Si-doped InAs thin films show a very small temperature dependence for electron mobility [15, 20]. Thus, the Hall output voltage V_H of Si-doped InAs thin-film Hall elements shows a very small temperature dependence at constant drive voltage, as seen from (5.27) and Figure 5.10(a). Moreover, because both terms on the left side of (5.30) are reduced by doping, there is also a very small temperature dependence for resistivity and thus for input resistance of InAs Hall elements. Therefore, Si-doped InAs thin films are suitable for practical Hall elements. The room temperature electron density and mobility of typical 0.5-μm-thick Si-doped InAs thin films for use in Hall elements are $8 \cdot 10^{16}/cm^3$ and 11,000 cm^2/Vs, respectively.

5.2.3.2 Design and Fabrication of InAs Hall Elements

InAs thin films doped with Si were used for designing practical InAs Hall elements with 0.36-mm^2 chip size. The InAs thin films were processed to form Hall elements by a specially developed procedure and assembled in a mass production line. The fabrication process is shown in Figure 5.11.

Figure 5.12 is a photograph of an InAs Hall element chip bonded on a lead island.

5.2.3.3 Typical Characteristics of InAs Hall Elements

Table 5.4 lists the standard specifications of InAs Hall elements [15, 16]. The typical Hall output voltage (or sensitivity) of this Hall element in a magnetic field is 100 mV/0.05T/6V. The characteristics are shown in Figure 5.13.

Good linearity of Hall output voltage for sensing a magnetic field is observed. The temperature dependence of the Hall output voltage is shown in Figure 5.14 and that of input resistance in Figure 5.15.

Because the input resistance of the InAs Hall element does not change much with temperature, practical applications over a wide temperature range

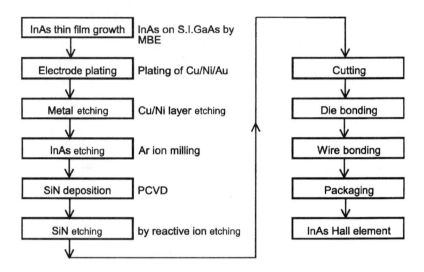

Figure 5.11 Fabrication process of InAs Hall elements.

Figure 5.12 Photograph of the InAs Hall element chip.

Table 5.4
Typical Characteristics of Si-Doped InAs Hall Elements Formed by MBE:
Electrical and Magnetic Field Characteristics

	Driving Voltage, V_{in}	Hall Output Voltage, V_H (B = 0.05T)	Offset Voltage, V_μ (B = 0T)	Resistance, R_{in}
InAs	6V	100 mV	<±16 mV	400Ω

Source: Asahi Kasei Electronics HZ-302C.

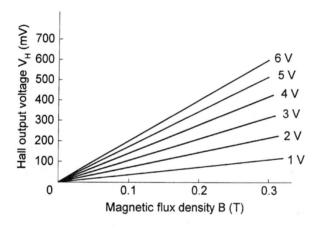

Figure 5.13 The magnetic field characteristics of Si-doped InAs Hall element (Asahi Kasei Electronics HZ-302C).

are possible. The temperature characteristics and stability of InAs Hall elements depend on the electron density in the active layer [20, 21]. The higher operation temperature (near 150°C) is attainable by optimizing Hall element design and Si doping. These Si-doped InAs Hall elements also work well at lower temperatures. It is also possible to fabricate a small package suitable for many kinds of applications.

The excellent characteristics of high stability, low offset drift, and low $1/f$ noise properties are special features of the InAs Hall elements. By using the thin-film InAs Hall element with heavy Si doping (electron density of $5 \cdot 10^{17}/cm^3$), a very small magnetic field of 0.003 mT was detected and the element showed a high sensitivity and low-noise properties. Moreover, the temperature dependence of the Hall output voltage and input resistance of heavily doped InAs Hall elements was very small, effectively near zero

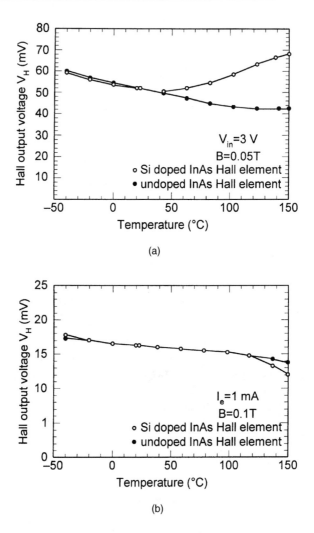

Figure 5.14 Temperature dependence of Hall output voltage for InAs Hall elements (Asahi Kasei Electronics HZ-302C): (a) constant voltage driving and (b) constant current driving.

over a wide temperature range (from −40°C to +160°C) [20]. These excellent properties of InAs Hall elements are promising for current sensor applications, magnetic field measurement, and other applications such as contactless sensors.

5.2.4 InAs Deep Quantum Wells and Application to Hall Elements

To achieve even higher sensitivity for InAs Hall elements, an even higher electron mobility and higher sheet resistance are required for the InAs active

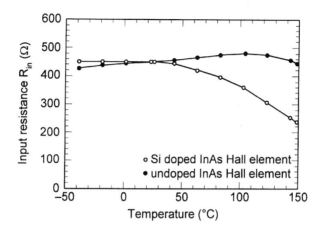

Figure 5.15 Temperature dependence of input resistance for InAs Hall elements (Asahi Kasei Electronics HZ-302C).

layer. That means an ultrathin InAs active layer with a higher electron mobility or quantum well structure is required. Next, we describe the extremely high potential of quantum well structures as magnetic sensors. To obtain a higher electron mobility, InAs deep quantum well (DQW) structures were studied. One new type of insulating layer is a quaternary material incorporating Sb having the same lattice constant as InAs and with a large bandgap energy of about 1.0 eV. This composition works well as a high potential barrier to form the InAs quantum well. For example, $Al_xGa_{1-x}As_ySb_{1-y}$ ($0 < x < 1$, $0 < y < 1$) is a suitable composition range. This layer absorbs many kinds of defects produced by the large lattice mismatch (7.4%) between the GaAs substrate and the InAs. Moreover, the defects are electrically inactive, and the layer acts as an insulating layer, pinning electrically active defects.

This insulating layer was used to form a DQW with a conductive InAs channel layer (i.e., InAs DQW) and applied to a Hall element [22–24]. Figure 5.16 illustrates a typical InAs DQW Hall element structure.

A 15-nm-thick InAs well was used as an active layer, and a structure comprising AlGaAsSb (35 nm)/InAs (15 nm)/AlGaAsSb(600 nm)/GaAs (at $x = 0.65$ and $y = 0.02$) was grown by MBE. The InAs DQW had a high electron mobility of 20,000–32,000 cm^2/Vs, as shown in Table 5.5.

The typical Hall output voltage from that DQW was 300 mV/0.05T/6V, as shown in Table 5.6.

The Hall output voltage was also proportional to the magnetic flux density of the applied magnetic field. The output voltage and temperature

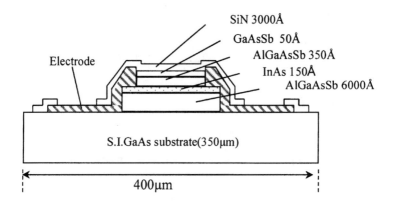

Figure 5.16 InAs DQW Hall element structure (cross section).

Table 5.5
Typical Properties of InAs DQW (at 25°C)

	Dopant	Electron Mobility, μ_H	Electron Density, n	Thickness, t
InAs DQW	None	20,000–32,000 cm²/Vs	$50 \cdot 10^{16}$ cm^{-3}	0.015 μm

Table 5.6
Typical Characteristics of InAs DQW Hall Elements

	Driving Voltage, V_{in} (V)	Hall Output Voltage, V_H (B = 0.05T)	Offset Voltage, V_μ (B = 0T)	Resistance, R_{in}
DQW	6V	250–300 mV	<±16mV	700Ω

dependence of InAs DQW Hall elements are compared with various kinds of Hall elements in Figure 5.17.

Figure 5.17 shows that InAs DQW Hall elements have a high sensitivity comparable to InSb thin-film Hall elements with magnetically amplified structure and good stability over a wide temperature range. The InAs DQW is applicable to many kinds of sensors and electronic devices, and InAs DQW Hall elements hold promise as magnetic sensors of the future.

5.2.5 Conclusion

Hall elements formed by thin-film technology have been utilized in important applications such as magnetic sensors in electronic equipment. InSb Hall

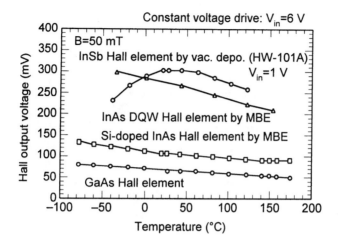

Figure 5.17 Temperature dependence of Hall output voltage for various kinds of Hall elements.

elements show ultrahigh sensitivity to magnetic fields and have resulted in revolutionizing dc brushless motor technology. The InAs Hall elements have practical characteristics of high sensitivity and temperature stability suitable for new applications. In the future, the InAs DQW Hall element will prove its usefulness with its ultrahigh sensitivity and stability over a wide range of operation temperatures. These thin-film Hall elements will open new areas of applications for magnetic sensors and will contribute to the advancement of electronic systems.

This section described only a selection of practical thin-film Hall elements that have been developed by Asahi Chemical Industry Co. Ltd. The GaAs Hall element was not described, but it also is important for practical applications. Moreover, important early applications of discrete Hall elements as contactless sensors (except from dc brushless motors) are summarized in [25, 26]. These applications may be still valuable in future.

5.3 Integrated Hall Sensors

5.3.1 Historical Perspective

The history of fully integrated Hall effect sensors closely parallels the development of linear analog bipolar integrated circuit technology. The reason for that is the natural fit of the Hall element requirements to the linear bipolar process. The Hall element can be fabricated in linear bipolar silicon with no additional process steps, resulting in simplicity and low cost. In addition, both high-quality amplification and temperature compensation are readily provided with linear bipolar circuit technology. Complexity and parasitics associated with interconnect are easily managed by full integration, resulting in low cost and high reliability. For many applications, a requirement for built-in protection from voltage transients can be added, again with little or no added process complexity. Early work on integrated Hall sensors is summarized in [27].

Figure 5.18 is an example of an early implementation of Hall elements in a keyboard switch. This first application made use of relatively high magnetic fields to provide an overall Hall sensitivity, which could overcome the large offset voltages of the single rectangular Hall element. The device employed a flip-chip mounting technique using lead-tin solder bumps to minimize mounting stresses.

Figure 5.19 is an example of a next-generation Hall element, which made use of dual Hall elements to cancel several of the components of Hall offset. The higher magnetic sensitivity made available by this approach made it possible to sense weaker magnetic fields. The technology resulted in a variety of magnetic position-sensing switches. The dual Hall element results in reduction in the cost of the magnet and often allows a physically smaller magnet to be used; the combination enabled further miniaturization.

Figure 5.20 is an example of a more modern magnetic sensor, one that features a quad of cross-connected Hall elements. In this design, the Hall elements are placed along thermal and mechanical stress centerlines on the

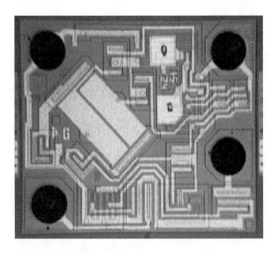

Figure 5.18 Early Hall sensor design used in keyboard switches.

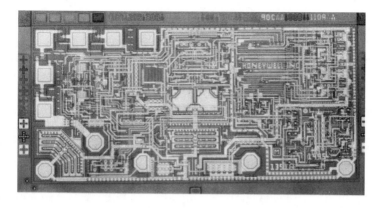

Figure 5.19 Improved Hall sensor using dual Hall elements.

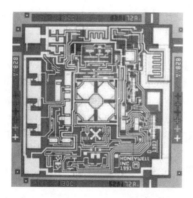

Figure 5.20 Magnetic sensor using four cross-connected Hall elements.

bipolar silicon integrated circuit. Although there is some tradeoff with current draw from the power supply each time an element is added, the resultant improvement in Hall sensitivity has made this approach worthwhile in many applications.

Alternatively, metal oxide semiconductor (MOS) integrated circuit technology has been slow to make inroads as the technology of choice for Hall effect integrated circuits. In addition to the potential for added process complexity, MOS-based linear amplifiers are fundamentally more difficult to apply, due to higher amplifier offset voltages, which in turn are due to variations in threshold voltages. In recent years, MOS-based Hall effect circuits have seen increasing application. As CMOS processing has become standardized worldwide, the associated wafer cost has been reduced. That gives CMOS sensor circuits a potential cost advantage. Through application of circuit techniques such as autozeroing and chopper stabilization, many of the circuit disadvantages of MOS can be overcome. In addition, the near-perfect analog switch available with CMOS has been applied with some success as part of a commutation strategy to reduce Hall offset voltages and thereby improve magnetic sensitivity and accuracy [28]. Even so, high-volume applications such as automotive engine sensors have been slow to accept CMOS Hall sensors because of their requirements for high reliability at elevated temperature and the ability to withstand repeated high-voltage transients. In such high-performance applications, bipolar Hall sensors have seen wider application.

An N-type epitaxial layer on a high-resistivity P-type substrate is typically used in linear bipolar silicon technology to achieve high-performance vertical NPN transistors. This N-type layer is ideally suited to the fabrication of Hall elements with no adjustments. Figure 5.21 shows how such a device can be realized in integrated technology.

All CCs and SCs are realized as n+ implants at the surface of the epitaxial layer (N). The lateral isolation of the active volume is implemented using deep p-diffusion (P). The surface is covered by a shallow p-layer (SP) to keep the active zone away from the surface and thus achieve higher stability. The final layer on top is a protecting oxide layer (OX). Between the pn-diodes is a depletion-layer (DL), isolating the two parts.

The final sheet resistance of this layer is roughly 1,000 to 5,000 Ω/square, which corresponds to an impurity concentration of about 8.0×10^{14}

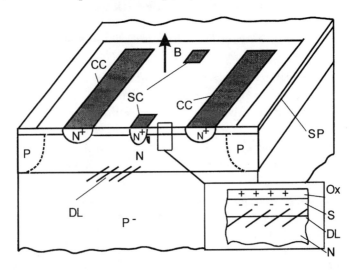

Figure 5.21 Cross section of a Hall device realized in integrated technology.

to 5.0×10^{15} atoms/cm^3. At those levels of impurity concentration, the mobility is high, being limited primarily by lattice scattering, resulting in good Hall responsivity. Hall elements made from P-type silicon with the same impurity concentration have about one-third the sensitivity because of the inherent mobility ratio of silicon. For low-supply current requirements, it is desirable to have a higher sheet resistance. Noise considerations, however, limit the sheet resistance of the Hall element at the upper end. Process reproducibility is also more challenging because impurity concentrations are reduced. The standard N-type epitaxial layer concentration for a typical high-voltage linear bipolar silicon process is nearly ideal in every respect for fabrication of Hall elements.

A key requirement for Hall elements is low drift. Nonrepeatable error in the form of drift cannot be compensated and therefore reduces overall accuracy. To fabricate a stable element, some form of protection from spurious surface charges must be incorporated into the design of the Hall element. Essentially, two techniques, both implementing what generally is described as a Faraday shield, have been applied successfully. The first is a metal field plate that covers the oxide over the Hall element and is tied to a stable voltage. Any surface charges that may appear on the surface of the metal are prevented from creating an image charge in the underlying silicon. The second approach is to add a P-type region over the N-type Hall element. Such a P-cap is tied to a known, stable circuit potential. Both approaches have seen broad usage.

The sensitivity of a magnetic sensor describes the minimum magnetic signal change that can be reliably detected. Sensitivity is a signal-to-noise ratio. The signal, in this case, is determined by the magnetic responsivity of the Hall element. In a direct-coupled sensor, the noise usually is dominated by offset in the Hall element and input amplifier. Direct-coupled integrated Hall sensors have been produced in high volume with standard linear bipolar silicon technology; they have a magnetic sensitivity of 5.0 mT over a temperature range of $-40°C$ to $+200°C$. Those same integrated sensors withstand repeated direct application of positive and negative voltage transients in excess of 100V on the supply and output pins.

5.3.2 CMOS Hall Elements

In the implementation of Hall elements in a CMOS circuit, two fundamental approaches can be applied. Most standardized CMOS processes begin with a heavily doped P-type substrate covered with a more lightly doped P-type epitaxial layer. To maintain high mobility and therefore high Hall responsivity, an N-type layer must be used for the Hall element. The Hall element can be accomplished as a simple N-type layer ion implanted into the P-type substrate (or P-type epitaxial layer). The implanted Hall element could be either common with an implanted, diffused N-well used for fabrication of PMOS transistors or a separate, custom N-type implant. Typically, there will not be common requirements for the CMOS N-well and the Hall element. The CMOS N-well is usually doped somewhat heavily to minimize circuit latch-up effects, to optimize the temperature coefficient of the threshold voltage, or to maximize field inversion thresholds. At the higher level of doping, the electron mobility is reduced, with resultant reduction in Hall responsivity. Assuming that a custom N-well is applied, production costs may increase relative to using the standard PMOS N-well. However, use of a custom N-well allows more freedom in the optimization of the Hall element.

The second solution may be to include an N-type ion implanted region over a previously implanted and diffused P-well region. In this case, the resulting Hall element impurity levels may be higher than in the first solution, making this approach less desirable because of a further reduction in mobility that results from greater impurity scattering. The lower electron mobility will reduce the Hall responsivity. In addition, because of process control considerations, it is difficult to actually fabricate a Hall element by first implanting and diffusing a P-well, then implanting an N-type layer. Nonuniformity in the two implants tends to be additive. As a result, this approach has seen little service.

In the case of a Hall element integrated with a CMOS process, a P+ source-drain ion implant can be included over the Hall element to provide a Faraday shield. The shallow P-cap eliminates surface accumulation; there are no compensating effects on overall responsivity to magnetic field. The Hall responsivity of an implanted/diffused layer generally is degraded by the relatively lower mobility of the N-type ion implanted profile near the surface. Using a shallow P-cap as a Faraday shield converts the more heavily doped N-type surface region, and the Hall element begins below the surface. That reduces the peak concentration of donor impurities and electrons in the Hall element and results in higher mobility. However, the introduction of the implanted/diffused acceptor concentration in the P-cap results in some compensation in the active N-type layer. For a well-designed process, there is little change in the responsivity of the Hall element as the P-cap is included.

5.3.3 Hall Offsets

Matching of the Hall effect to real sensing applications is limited by the accuracy of the sensor, which in turn is limited by the offset voltage of the

Hall element. Offset is the voltage at the sense terminals of the Hall element when the magnetic field is zero. Relative to many other sensing technologies, such as piezoresistive pressure sensors or magnetoresistive sensors, the Hall voltage is quite small for the magnitude of magnetic field that can be economically achieved. Therefore, rather high amplification is required in Hall effect sensors. For example, a maximum Hall output voltage of 1.0 mV is somewhat typical. Any dc offset in the Hall element also will be amplified, and that dc offset will fundamentally limit the usefulness of the amplified output. Of course, if the application allows for ac coupling, then the dc offset can be eliminated, and other circuit limitations will set the overall achievable accuracy.

The history of the Hall element development includes initial determination of the variables that affected the Hall offset. Following some rapid progress, there was little further improvement in reduction of the Hall offset until cancellation by making use of multiple Hall elements was employed. Usage of dual or quad Hall elements resulted in perhaps a tenfold reduction in the net Hall offset. At that point, another plateau was reached, and no further progress was made until the theory of spinning the current through the Hall element was fully developed. Today, some commercially available circuits make use of a spinning current, or commutation, technique that cancels a large component of the offset voltage within a single Hall element. A 90-degree rotation of the current requires the integration of eight CMOS switches (two switches for each of the four nodes of the element). The switches must have excellent characteristics, and four of the eight must carry the entire Hall element current. The outputs from the Hall element must then be averaged to achieve the desired analog output. All of this switching and averaging must be done at a small signal level (~1.0 mV) ahead of the first stage of amplification. All those requirements tend to increase the size, complexity, and cost of the resultant sensor. It is yet to be seen whether the cost of the added functionality will be broadly accepted in high-volume magnetic sensor applications.

Bellekom and Munter [29] have summarized the various components of Hall offset. It is instructive to consider each component separately and to discuss the ability of the different cancellation techniques to deal with each component of offset voltage. First, the origin of various components of offset voltage can be considered to be either systematic or statistical in nature. Systematic variables are correlated such that, if the gradients in one Hall element are known, the gradients in the adjacent element can be predicted with high accuracy. For example, a gradient in the resistivity across a wafer could be considered to be systematic. Two Hall elements placed side-by-side tend to have the same local gradients in resistivity. Alternatively, a statistical variable is considered to be purely random in nature. The fabrication-related defects that might be found in the two adjacent Hall elements typically are not correlated.

Fabrication variables such as photomask misalignment, gradients in epitaxial sheet resistance or ion implant resistivity, and misalignment to the crystal plane generally are systematic. By careful design of a dual Hall element, offsets due to those variables can be neatly canceled.

Variables such as absolute Hall contact position or size, impurity concentration or crystal defect-related variations in depletion layers, and point defects due to localized penetration of impurities are statistical in nature. While the dual Hall element provides some reduction in the effect of those variables, they are by no means canceled. In fact, assuming that true randomness is at work, the standard deviation of the Hall offset voltage due to statistical variables will be divided by the square root of 2 for a dual element and the square root of 4 for a quad element.

Multiple thermal effects influence the offset voltage. For a good thermal design in an integrated Hall sensor, the temperature gradients across the Hall elements due to power dissipation in other circuit elements are well behaved and do not contribute to offset voltage. However, there may be inhomogeneity in the thermal resistance path that is statistically distributed. For a dual or quad Hall element, inhomogeneity will produce a statistically distributed offset.

Systematic thermal variables are the interrelated Seebeck and Peltier effects. The two side contacts of the Hall element will be at slightly different temperatures, giving rise to a potential difference due to the Seebeck effect. In an integrated Hall sensor, the temperature difference may be due to power dissipation in circuit elements other than the Hall element. The sense contact current normally is quite low; however, any current drawn from the side contacts will result in heat being absorbed or released due to Peltier effect. It is generally assumed that these variables are small in a well-designed Hall sensor.

Finally, stress effects must be considered. Because silicon has a significant piezoresistance effect, a Hall element can be an effective strain transducer. Thus, any inhomogeneity in the stress distribution in the Hall element will result in a component of offset voltage that cannot be distinguished from a magnetic signal. Mechanical stress may be an intrinsic part of the silicon crystal growth process and may also result from impurity diffusions, which generate localized lattice strains. The largest stresses, however, are thought to arise from differences in thermal expansion coefficients between silicon and packaging materials. Because the piezoresistive coefficients of silicon are very orientation sensitive, the mechanical stress effects can be dramatically reduced by proper orientation of the silicon. Maupin and Geske [27] and Kanda [30] have shown that fabricating Hall elements in (100) silicon wafers with current flow in the [100] direction minimizes the effect of all of the piezoresistive π coefficients except π_{66}. The coefficient π_{66} has its minimum value for that orientation and produces an offset voltage in response to a properly oriented shear stress in the plane of the Hall element. Use of a dual or quad Hall element will result in the cancellation of the offset due to the symmetrical component of mechanical stress. However, unsymmetrical components of mechanical stress will produce offsets of a statistical nature.

Most typically, lead-frame technology results in the silicon die being attached via an adhesion layer to a copper alloy. Attachment is completed at an elevated temperature. Once the temperature has been reduced, a stress arises because of the mismatch in material properties. Although the overall stress effects can be minimized by symmetrical design, the remaining effects still are important enough to generate significant offset voltages. It is worth noting that the component of Hall offset voltage due to piezoresistive stress effects is the only component that is fundamentally a function of temperature. Any approach to compensate for the offset voltage by trim techniques will be limited by the fact that the offset voltage will reappear as the temperature is moved away from the trim temperature.

In the continuation of the lead-frame assembly, plastic material can be molded directly over the silicon die surface. The temperature-dependent stress associated with this material mismatch is also an important variable. Various approaches to buffer the stress have been attempted with varying success. Simplistically, by coating the silicon die surface with a material that will not transfer shear stress, the effect of expansion or contraction of the plastic will be minimized.

As has been mentioned, systematic variables normally can be canceled with a simple dual-element approach. Multiple elements will act to reduce—but not eliminate—the effects of statistical variables. In the spinning current approach, cancellation occurs for any component of Hall offset that is symmetrical with respect to the bias switching terminals. In the simple, practical case of using only two bias orientations that are 90 degrees apart, the performance of a commutated Hall element can be judged by comparison to a dual Hall element configuration. With commutation, the Hall element is being compared to itself; in the dual configuration, the cancellation is between two nearby but statistically different Hall elements. That means compensation for many statistical variables can be accomplished with 90-degree commutation. For instance, offset due to a diffused point defect of any kind will be opposite and equal for the two bias orientations and will average to zero. In fact, for two current orientations, most components of Hall offset average to zero, including the important temperature-dependent stress effects. However, an important statistical offset voltage is not canceled by the spinning current approach. As the applied bias voltage is rotated, new depletion regions are created. Offset voltages due to the nominal depletion effects are canceled by averaging. Inhomogeneity in the depletion region is statistical in nature, however, and thus will produce a statistical net offset voltage. For that reason at least, the two-current-orientation approach to commutation of offset voltages is less than perfect.

An additional concern that arises with the commutation approach is that the frequency of switching must be designed so as to not affect measurement accuracy. Such design is not trivial, because there are tradeoffs with power supply rejection, the need to retain higher order components of the signal, and bandwidth limitations of the circuit.

Because the cost effectiveness or other limitations of the commutation approach may limit applicability, there clearly is opportunity for further reduction in offset voltages by fundamental improvements. For instance, die-attach–related stress effects can be reduced by the use of new lead-frame materials that have a closer match in thermal expansion coefficients to silicon. In some cases, attempts have been made to reduce the die-attach effects by shaping the back surface of the silicon through micromachining techniques. Although progress certainly can be made in this manner, such solutions have not proved to be cost effective, and there has been little commercial application of the technique. Innovations in buffer materials to be applied to the top surface of the silicon will continue to be in demand. Finally, new trim techniques offer the prospect of allowing for further improvement in accuracy, as the offset voltage is compensated with reduced interaction with the amplifier circuits.

5.3.4 Excitation

Proper biasing of a silicon Hall element is important in achieving optimum performance from an integrated Hall sensor over the operating temperature range. From (5.11) and (5.17), the Hall element responsivity is given by

$$\frac{V_H}{B} = \frac{r_H}{qnt} IG \qquad (5.31)$$

For lightly doped N-type silicon Hall elements, the electron concentration n, thickness t, and geometrical correction factor G are constant over temperature. Thus, constant current bias in a Hall element gives a magnetic responsivity that has the temperature coefficient of the Hall factor, r_H. For a particular uniformly doped epitaxial Hall element with donor concentration of ~3.0 · 10^{15} cm^{-3}, the measured temperature coefficient of the Hall factor is 6.75 · 10^{-4}/°C. The temperature coefficient of r_H causes the magnetic responsivity to increase with increasing temperature. It follows that for this Hall element a constant magnetic responsivity over temperature can be achieved by biasing with a current source having a temperature coefficient of −6.75 · 10^{-4}/°C. For the temperature range from −40°C to +200°C, that represents a 16.2% decrease in bias current.

The resistance of the Hall element is given by

$$R = \frac{\rho}{t}\left(\frac{L}{W}\right)_e = \frac{1}{q\mu_c nt}\left(\frac{L}{W}\right)_e \quad (5.32)$$

where ρ/t is the sheet resistance of the Hall element, $(L/W)_e$ is the effective length-to-width ratio of the Hall element, μ_c is the conductivity mobility (drift mobility) of electrons, n is the electron concentration in the Hall element, and t is the thickness of the Hall element.

For the Hall element described above, the conductivity mobility changes rather dramatically with temperature. Measurements of this same particular N-type epitaxial Hall element have shown

$$\mu_c = \mu_{co}\left(\frac{T_0}{T}\right)^{2.268} \quad (5.33)$$

where μ_{co} is the value of electron mobility at T_0, T_0 is the reference temperature in K, and T is the absolute temperature in K.

Between −40°C and +200°C, the resistance of the Hall element changes by a ratio of ~5/1. With the negative temperature coefficient on the bias current required for constant magnetic responsivity, the voltage across the Hall element changes by a ratio of ~1/4.3 for the same temperature range.

Kanda [30] has shown that a slightly unbalanced Wheatstone bridge can represent the passive portion of the Hall element. Thus, the nominal offset voltage at the sense terminals is proportional to the bias voltage across the Hall element. Statistical inhomogeneity in the Hall element depletion layers introduces nonratiometric variations in the offset as the voltage bias changes with temperature. Package-induced stresses also cause variation of offset voltage with temperature. Biasing the Hall element with a constant current makes it difficult to trim the offset. Overall, constant current bias has been found to have a poorer signal-to-offset ratio than constant voltage bias.

When the Hall element is biased at constant voltage, the nominal offset voltage is independent of temperature. In that case, the temperature coefficient of the offset voltage is the result of packaging stress, and laser trimming can remove the nominal offset. For constant voltage bias, the offset changes due to statistical depletion effects are eliminated. With constant voltage bias, the magnetic responsivity is given by

$$\frac{V_b}{B} = V_b \mu_H G \left(\frac{W}{L}\right)_e \quad (5.34)$$

where μ_H is the Hall mobility given by $\mu_H = r_H \mu_c$, and V_b is the bias voltage.

The measured Hall mobility as a function of temperature for the same Hall element described above is given by

$$\mu_H = \mu_{H0} \left(\frac{T_0}{T}\right)^{2.044} \quad (5.35)$$

where μ_{H0} is the Hall mobility at T_0 in K.

For constant voltage bias over the temperature range from $-40°C$ to $+200°C$, the Hall mobility and magnetic sensitivity change by a ratio of $\sim 4.3/1$.

The elimination of statistical depletion effects with constant voltage bias results in a higher signal-to-offset ratio for that biasing scheme compared to constant current bias. As a result, constant voltage bias has been used most widely in commercial integrated Hall sensors.

Constant voltage bias also has been widely used in linear Hall sensors. In this application, the power supply is applied directly to the Hall element to provide a means of calibrating the magnetic responsivity. From (5.9), the magnetic responsivity is directly proportional to V_b.

For epitaxial Hall elements with low donor concentration, the Hall and drift mobility is determined almost entirely by lattice scattering for the temperature range of interest. Normal run-to-run variations in donor concentration encountered in a silicon bipolar process are small enough that there is little effect on the magnetic responsivity. In integrated Hall sensors, N-type epitaxial resistors and zero-temperature-coefficient thin-film resistors can be used together to compensate for the variation of drift mobility over temperature. Such compensation schemes do not correct for the temperature coefficient of the Hall factor, r_H. Compensation for the temperature variation of r_H must be provided by other circuit means to obtain constant magnetic responsivity over temperature.

5.3.5 Amplification

The Hall element generates a differential signal at the sense terminals at a common mode dc voltage of approximately 0.5 V_b. As a result, most integrated Hall sensors use a linear differential amplifier as the first stage in the on-chip signal conditioning circuitry. The first stage amplifier is a critical part of the design of an integrated Hall sensor because its input offset voltage and noise add directly to the offset voltage and noise of the Hall element to determine the magnetic sensitivity of the sensor. Linear differential amplifiers can be constructed from both bipolar and CMOS transistors. However, the characteristics of the two types of amplifiers are considerably different.

All bipolar transistors are characterized by a predictable exponential relationship between the emitter base bias voltage and the collector current:

$$I_C = I_S \exp\left(\frac{qV_{BE}}{kT}\right) \quad (5.36)$$

where I_C is the collector current, I_S is the saturation current, V_{BE} is the emitter base voltage, k is Boltzmann's constant, and T is absolute temperature in K.

The transconductance, g_m, of a bipolar transistor is given by

$$g_m = \frac{qI_C}{kT} \quad (5.37)$$

A well-designed differential amplifier made from closely spaced integrated bipolar transistors has a typical input offset voltage of less than 300 μV. Laser trimming can be used to further reduce the input offset voltage. When bipolar differential amplifiers are trimmed for zero offset at room temperature, they also have a zero temperature coefficient of offset. That is an inherent characteristic of the Boltzmann statistics of the minority carriers in the semiconductor.

The input amplifier in an integrated Hall sensor is a linear amplifier and must be biased continuously. As a result, in linear amplifiers, the use of CMOS offers no power or current reduction, as it does in digital logic. In a CMOS process, the first-stage differential amplifier normally employs NMOS transistors. For an MOS transistor, the transconductance is given by

$$g_m = \frac{2I_D}{(V_G - V_T)} \quad (5.38)$$

where I_D is the drain current, V_G is the gate-source voltage, and V_T is the threshold voltage.

To achieve the same transconductance at the same current as the bipolar amplifier, the MOS amplifier must operate at a $(V_G - V_T)$ value of about 52 mV. NMOS linear differential amplifiers operating in this mode require considerably more area than their bipolar counterpart. The exact area ratio depends on the minimum feature size of the bipolar process and the gate length of the MOS process used in the comparison.

A well-designed integrated NMOS amplifier using closely spaced transistors has a typical input offset voltage of 2 to 3 mV. For MOS differential amplifiers, the input offset is determined by the local variation of V_T. Offset trimming is not as productive with MOS as with bipolar amplifiers. MOS transistors lack the inherent matching offered by the Boltzmann statistics that govern the performance of bipolar transistors. As a result, with MOS amplifiers the trim for zero offset voltage does not in general give a zero temperature coefficient of offset.

The availability of ideal switches in the CMOS process makes it possible to chopper stabilize the linear NMOS amplifier. However, chopper stabilization is required in MOS to achieve acceptable input offset voltages. The use of chopper stabilization tends to increase the area and complexity of the integrated Hall sensor but provides input offset voltage of 10 to 20 μV.

Useful differential amplifiers can be made using both CMOS and bipolar processes. In both cases, good mechanical and thermal design techniques should be employed to achieve optimum performance over temperature. All the resistors and transistors used in the amplifier will produce input offset voltage in response to mechanical strain and temperature gradients.

The best input amplifier performance is obtained from a BiCMOS process. The combination allows the CMOS switches to be used to chopper stabilize a bipolar input amplifier. Chopper stabilization can be used to reduce the input offset voltage by about a factor of 100.

5.3.6 Geometry Considerations

Conformal mapping theory shows that all Hall elements with the same effective (L/W) will have the same magnetic responsivity. That means the specific geometry of the Hall element can be selected to minimize the piezoresistance effects and to comply with layout requirements of integrated Hall sensors.

Two popular geometries used in integrated Hall sensors are depicted schematically in Figure 5.22. It is evident that neither design has the simple rectangular geometry often used to describe the Hall effect. The (L/W) of these Hall elements cannot be determined by inspection. However, the effective (L/W) can be determined from

$$\left(\frac{L}{W}\right)_e = \frac{R}{\rho_s} \quad (5.39)$$

where R is the measured resistance between the bias or sense contacts, and ρ_s is the sheet resistance of the Hall element material, $\rho_s = \rho/t$.

The two geometries, as depicted in Figure 5.22, use contacts of significant size for both bias and sensing. That means two significant $(L/W)_e$ values must be used to describe their operation. In general, the bias and sense contacts do not have to be the same size. The bias (L/W) is given by

$$\left(\frac{L}{W}\right)_b = \frac{R_b}{\rho_s} \quad (5.40)$$

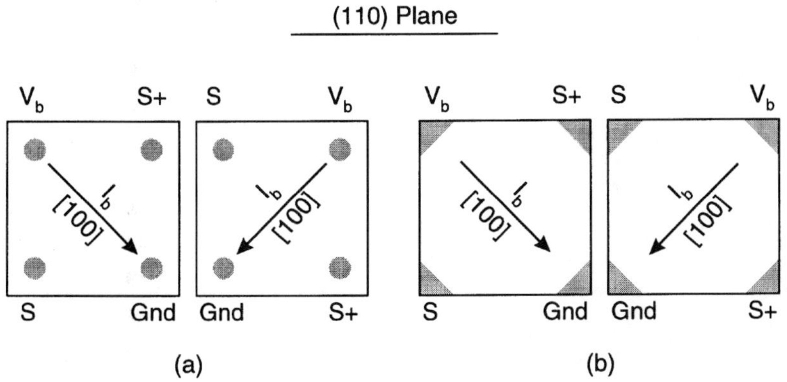

Figure 5.22 Hall element geometries.

and the sense (L/W) is given by

$$\left(\frac{L}{W}\right)_s = \frac{R_s}{\rho_s} \qquad (5.41)$$

where R_b and R_s are the resistance values measured between the bias contacts and sense contacts, respectively.

The expression for magnetic responsivity in (4.4) must be rewritten as

$$\frac{V_h}{B} = V_b \mu_H G_T \left(\frac{W}{L}\right)_b \qquad (5.42)$$

where G_T is the total geometry factor, which includes the shorting effect of both the bias and the sense contacts.

For (100) silicon wafers, the mask patterns typically are oriented parallel to the (110) plane provided on the wafer. Both geometries shown in Figure 5.22 provide bias current flow in the [100] direction. That is the optimum bias current direction to minimize the piezoresistive response to mechanical stress. Both geometries have major feature edges aligned to the [110] directions for ease of layout, and both are simple to arrange in dual or quad configurations. Offset voltage cancellation with no loss of magnetic responsivity is accomplished when the bias currents in the dual elements are orthogonal and when similarly named terminals are connected with metal leads. The (100) plane in silicon has fourfold symmetry. Thus, the diagonally located contacts allow orthogonal bias currents with both currents in the optimum [100] direction required to minimize piezoresistive effects.

For accurate dimensional control, the circular contact geometry shown in Figure 5.22(a) normally uses shallow N+ implant/diffusion regions to form the contacts to the Hall element layer. In a bipolar process, that can be the emitter diffusion used in the NPN transistors. In a CMOS process, it can be the source/drain diffusion used in NMOS transistors. As a result, this design has a component of vertical current flow under the contacts. The diagonal contacts shown in Figure 5.22(b) could use either a shallow or a deep N+ contact region. The deep N+ region acts as an equi-potential contact and minimizes the vertical component of current in the Hall element. When used in dual or quad configurations, both geometries provide offset voltage cancellation for uniform sheet resistance gradients, small mask misalignments, and uniform mechanical stress.

Hall elements that are commutated using CMOS switches must have symmetrical contacts, that is, the bias and sense contacts must be identical to achieve maximum offset reduction. Commutation with the diagonal contact arrangement shown in Figure 5.22 switches the bias current between orthogonal $\langle 100 \rangle$ directions. Both single and dual Hall elements may be commutated.

In sensors that do not use commutation, the bias and sense contacts can be made asymmetrical to achieve other performance advantages. A Hall effect simulator developed by Nathan at the University of Alberta [31] has been used to determine the total geometry factor for a variety of bias and sense contact geometries. The simulations showed that the geometry effect of any particular pair of contacts is the same whether used as a bias contact or a sense contact. Thus, the term G_T in (5.17) has the same functional dependence on both $(L/W)_b$ and $(L/W)_s$ and can be represented by the product $G_b G_s$. However, $(W/L)_b$ also appears explicitly in (5.42). The complete geometry term $G_s G_b (W/L)_b$ describes the performance of the Hall element.

Popovic [1] has shown that, when point contacts are used for sense electrodes, G_b approaches 1 for $(L/W)_b > 4$; for $(L/W)_b < 0.25$, G_b approaches $0.742(L/W)_b$. However, in integrated Hall sensors, it is not practical to obtain point contacts, and the geometrical effects of both the bias and sense contacts must be taken into account. For symmetrical Hall elements like those shown in Figure 5.22, the term $G(W/L)$ has a maximum value of ~0.47 for $(L/W) = 1.4$. As (L/W) increases above 1.4, the term (W/L) decreases more rapidly than G increases. As (L/W) decreases below 1.4, the term G decreases more rapidly than (W/L) increases. Symmetrical Hall elements have been used in many commercial integrated Hall sensors.

In general, a decrease in $(L/W)_s$ results in an increase in magnetic responsivity; however, that also causes an increase in the output resistance of the Hall element. Because MOS transistors do not require input current, they can take advantage of the lowest value of $(L/W)_s$ allowed by process geometry, provided commutation is not used on the Hall element. Bipolar transistors require base current, and the base currents of the input differential amplifier may not be perfectly matched. Thus, if $(L/W)_s$ is made too small, the increase in output resistance can introduce a significant source of offset voltage. The output resistance set by $(L/W)_s$ is also important in determining the thermal and $1/f$ noise voltage contributed by the Hall element.

Integrated Hall sensors typically use junction-isolated Hall elements. The N-type Hall element is surrounded by P-type material. If a P-cap is used, the Hall element is surrounded by P-type material on all sides except in the contact areas. When bias is applied, most of the body of the Hall element is reverse biased with respect to the surrounding P-type material. This geometry has a beneficial effect because the surrounding depletion region is continuously sweeping the thermally generated minority carrier holes out of the Hall element. As a result, the integrated Hall element is a true one-carrier device, and no loss of magnetic responsivity due to thermally generated holes occurs at high temperature. The sweep out of the minority carrier holes results in a leakage current distributed over the area of the Hall element. Inhomogeneity in that leakage current can cause increased offset voltage at high temperature.

5.3.7 Vertical Hall Elements

All the Hall elements discussed to this point have been in the plane of the surface of the integrated sensor die. In that configuration, the Hall element is responsive only to the vertical component of magnetic field. Popovic [1] describes a Hall element formed in an N-type substrate that responds to magnetic field in the plane of the surface of the die. The device is called a vertical Hall element. Two vertical Hall elements oriented at 90 degrees allow the magnetic field in the plane of the die surface to be split into orthogonal components.

Biard [32] has developed a vertical Hall element that is compatible with the linear bipolar process used in the fabrication of the more usual horizontal Hall elements that lie in the plane of the die. The device uses the same N-type epitaxial layer used in the conventional Hall element with the bias and sense contacts all on the top surface of an N-type epitaxial resistor, as shown in Figure 5.23. Equation (5.42) is independent of shape and, therefore, applies to both vertical and horizontal Hall elements. When $G_T(L/W)_b$ is the same for vertical and horizontal Hall elements, the magnetic responsivity will be the same for both types. Two of those vertical Hall elements can be integrated in the same linear bipolar chip with a horizontal Hall element with all elements having the same magnetic responsivity. That makes it possible to sense an arbitrary magnetic field as three orthogonal components using a single integrated die. That integrated sensor die can also contain bias and signal conditioning circuitry.

Vertical Hall elements should be oriented such that bias current flows in a [110] direction. That orientation is optimum for minimizing piezoresistive effects in the vertical Hall element. Each vertical Hall element has two currents, which flow in opposite directions (180-degree angles); there is no practical way to achieve orthogonal bias currents. As a result, vertical Hall

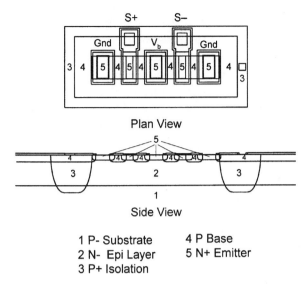

Figure 5.23 Vertical Hall element for linear bipolar process technology.

elements do not lend themselves to offset voltage cancellation by use of multiple elements or commutation. Because of higher offset voltage, the magnetic sensitivity of vertical Hall elements is not as good as for dual or quad horizontal elements. However, for constant voltage bias, the nominal offset voltage of a vertical Hall element is constant with temperature and can be removed by trimming.

The vertical Hall element is biased by connecting the center contact to V_b and the outside contacts to ground. The Hall voltage is sensed between the contacts labeled S+ and S−. In the configuration shown in Figure 5.23, the shape and the location of the N+ contact regions (5) is determined by openings in the P-type base diffusion (4), which also serves as a P-cap. That provides an automatic self-aligned geometry, because all the N-type contacts are determined by features in one mask layer. The vertical Hall element is the N-type epitaxial layer (2).

5.3.8 Packaging for Integrated Hall Sensors

The earliest Hall-effect integrated circuits were packaged using a robust, though costly, approach of flip-chip bonding onto ceramic substrate. The flip-chip approach still enjoys broad application. In that configuration, both the electrical and mechanical connections to the substrate are created by first placing solder bumps on the silicon die. The ceramic is prepared with a pattern of solderable material. When the die is inverted into position over the ceramic and both units are heated, the solder bump on the chip wicks into the surface of the solderable material on the prepared ceramic. The wicking results in self-alignment of the silicon chip. Any shear stresses that arise due to differential thermal expansion of the silicon and ceramic are transmitted only through the bump connection. Therefore, it is relatively easy to manage those stresses, and the piezoresistive packaging effect on Hall element and input amplifier offset voltage is small.

Low-cost packaging of semiconductor dies utilizes the steps of die attach to a lead frame, wire bond, and encapsulation. The lead-frame material typically has a temperature coefficient of expansion (TCE), which differs from that of silicon. Die attachment normally occurs at elevated temperature. Therefore, when the die attach is completed, and the sandwich is cooled to room temperature, a net stress appears in the silicon. This packaging stress contributes to a change in the offset of the Hall element [33] or input amplifier [34], due to piezoresistive effects. A similar effect may occur during plastic encapsulation [35]. Because the piezoresistive effect is the result of mismatched TCE in the sensor package, the resulting offset voltage will be temperature sensitive. Creep or slip between elements of the package can cause hysteresis in offset voltage as a result of temperature cycling.

Various approaches have been applied successfully to reduce the degree of change in the Hall offset during packaging. In the die-attach operation, the properties of the glue layer may be considered. Obviously, the glue layer will also have a TCE that is different from that of silicon. However, the glue layer usually is fairly thin and, therefore, applies little stress to the silicon as it expands and contracts. More important are the mechanical properties of the glue layer. A relatively stiff glue layer will transmit shear stress from the underlying lead frame to the silicon, while a flexible glue layer will not support the transfer of shear stresses.

Some success has been achieved in reducing the effect of encapsulation layers on the offset voltage by the application of thin buffer layers to the surface of the silicon die. Such buffer layers either can be applied as a thin-film overcoat during integrated circuit (IC) fabrication or alternatively may be dispensed onto the surface of the silicon die immediately prior to encapsulation. Examples of thin-film buffer layer materials are polyamide and photoresist. Each material has seen some applicability in the semiconductor industry as a protection layer to reduce mechanical damage and stress during packaging. In dispensed materials, elastomers such as room temperature vulcanizing (RTV) and other silicon compounds have been used.

With regard to the design of the Hall element itself to minimize the packaging effects on offset voltage, both experience and finite element modeling (FEM) have shown that placement of the Hall element symmetrically with respect to the die edges can result in lower offset voltages. In addition, a general trend exists that a larger die (in proportion to the Hall element) will have somewhat lower package-induced offset voltage.

5.3.9 Trimming Methods and Limitations

In most applications, a combination of the Hall and input amplifier offset voltage sets a limit on the ability to detect a magnetic signal. There has long been a need to further improve the magnetic sensitivity by actively adjusting the final circuit to compensate for the inherent offsets. Essentially, a measurement is made of the circuit offset, and a compensating offset is created elsewhere in the circuit. That approach is generally termed *trimming*. Three trimming methods have been widely used.

The oldest approach to trimming is called *diode zap*. The approach is relatively crude but has proved to be reliable. A resistor ladder can be adjusted with this method by the creation of a closed-circuit node, in which a reverse biased diode previously existed. The technique involves application of a high potential across the zap diode junction, with resultant junction breakdown, rapid heating, and migration of aluminum into the underlying silicon. Once cooled, the junction remains as a low-resistance element. While this approach has the least requirements for complex equipment, it tends to take a good deal of layout space on the silicon die. An array of diodes must be defined. In addition, probe pads must be included for each "bit" in the trim ladder. The layout area for the probe pads must be included in calculation of the cost of providing trimming.

A more recent approach is based on laser trimming. In this case, a resistive ladder of thin-film material is included and designed such that certain legs of the ladder can be trimmed open with a laser cut or partially cut to increase the resistance. The former method is generally called *digital trimming*, while the latter is called *analog trimming*. The costs of laser trimming include the costs of depositing and patterning a laser-trimmable thin film, as well as the more obvious costs of both a tester to excite the die magnetically while measuring response and the laser itself. The layout area required for laser trimming depends strongly on the ability to control the positioning of the laser beam on the die. The overall accuracy available for beam positioning has improved significantly in recent years due to advances in similar equipment used in other parts of the semiconductor industry.

Another way to justify the cost of laser trimming is to make the thin-film material perform double duty. A favorite approach is to design the thin film to have the temperature coefficient of resistance (TCR) close to zero.

Analog circuit design is assisted greatly by having a zero-TCR resistor available on chip. Consequently, laser trimming of thin films has become the generally favored approach for Hall effect and most other types of sensors.

Finally, a more recent approach has been to apply both fuse and antifuse approaches to circuit trimming. For processes such as CMOS, in which polysilicon is already available, a fuse can be easily created, essentially using an approach equivalent to that of a diode zap. A resistor network is predefined with trimmable elements. Following measurement of the offset of the individual silicon die, a fuse can be opened by application of an electrical pulse. This approach again requires layout of a probe pad for each "bit" of trim resolution. More exotic processes that enable an antifuse function have been developed, essentially variations of the polysilicon fuse approach. Optimization of cost and reliability of the application determine the approach to be used in each case.

In the case of fuse trimming, there are options to make a digital trim, with completely open elements, or an analog trim, with a controlled increase in the resistance of the equivalent circuit. Various approaches have been applied to analog trimming using electrical pulses applied to the fuse element.

Trimming to reduce or eliminate offset in Hall sensors normally is done in wafer form. As a result, any offset voltage introduced by the mechanical effects of packaging cannot be removed in this way. That limits the applicability of trimming and places more emphasis on the need to improve packaging.

5.3.10 Applications and Trends

The first high-volume application of the Hall element was as a switch used in computer terminal keyboards (1961).

An integrated Hall element switch was the basis for an automotive solid state ignition system. The application used a configuration called a *vane switch*; it was originated in 1965 by Ron Holmes, a Honeywell engineer, and still enjoys broad application today. More recently, however, automotive requirements for increasing fuel economy and reduced emissions are driving the adoption of a more sophisticated engine system. This system makes use of geartooth sensors placed on both the camshaft and the crankshaft of the engine. A custom target is placed on each shaft to enable accurate monitoring of position. The geartooth sensors are required to operate inside the engine and are often immersed in hot motor oil. The ambient temperature of the Hall sensor can be as high as 180°C, which can result in junction temperatures as high as 200°C. Geartooth sensors are also widely used in automotive antilock brake systems to detect wheel speed. Those Hall sensors also must operate correctly over a wide ambient temperature range. Another mass application that requires tens of millions of sensors per year is the brushless dc motors.

Although unit volumes are smaller, many linear Hall sensors are sold. By integrating temperature compensation circuitry with the sensor, the magnetic field can be measured accurately over a broad range of temperatures. Linear sensors have found application in current sensing.

With a wide variety of transducer technologies available, there is a good deal of competition for applications. The integrated Hall sensor has found favor due to its inherent low cost, wide temperature range of operation, insensitivity to dirty environments, and the ruggedness associated with non-contact measurements of the magnetic field.

As microprocessor voltages trend lower, there will be increased pressure to reduce the supply voltage of other circuits to match. Some high-performance applications will require the use of Bi-CMOS technology to combine the desirable characteristics of silicon bipolar with the ideal switch capabilities of CMOS. That combination of technologies will allow offset reduction in the Hall element through commutation and offset reduction in the integrated amplifier by use of chopper stabilization.

Increasingly, some applications are sensitive to the value of overall supply current. That is true mostly of portable or battery-operated applications. In some cases, it is allowable to address that need by reducing the duty cycle of the Hall integrated circuit.

Despite other market drivers, the global marketplace is increasingly demanding that sensors have optimally low costs and high reliability. That requires manufacturers to make robust designs that can be produced anywhere in the world with low field failure rates. The Hall technology in its various forms is particularly well suited to meet that challenge in the area of motion and position sensing and current sensing.

5.4 Nonplatelike Hall Magnetic Sensors

This section surveys a new class of Hall devices that are not platelike but that have three-dimensional structures. This survey covers the vertical Hall device, sensitive to a magnetic field parallel to the chip surface; the cylindrical Hall device, which behaves like a vertical Hall device combined with magnetic flux concentrators; the two-axis vertical Hall device, for the two in-chip-plane components of a magnetic field; and the three-axis Hall device, to measure all three components of a magnetic field. All those Hall magnetic sensors are fabricated using the vertical Hall silicon process, and they feature low supply current, low noise, and high long-term stability. Silicon Hall sensors in general have a lower signal level compared to compound semiconductor Hall sensors, due to the lower electron mobility. However, by achieving a much lower noise level, they still may feature a higher signal-to-noise ratio. On the other hand, a low mobility also reduces the dependence of all other magnetogalvanic effects, for example, the magnetoresistance effect. In such a way, these sensors not only can be employed for the measurement of fields below 1 μT, but they also are particularly adapted to the measurement of very strong fields of up to several tens of teslas. Their applications include high-accuracy magnetometry, long-range position sensing, angular position sensing, current sensing, magnetic scanning of documents, and more. A preliminary version of this section was recently published in a journal [36].

5.4.1 Introduction

A conventional platelike structure of a Hall device is optimal only when we are dealing with a homogeneous magnetic field perpendicular to the chip surface. When the magnetic field to be measured is inhomogeneous over the volume of a Hall device, the structure of the Hall device should be adapted to the shape of the magnetic field lines.

Leaving the solid ground of conventional Hall plates, it is useful for us to formulate a set of general criteria that a good Hall device must meet. In the present context, the following three criteria are essential (in usual notation [1]):

1. The orthogonality of the vectors of the current density and of the magnetic field over the active region of the device, because the Hall electric field is given by

$$\mathbf{E}_H = -R_H [\mathbf{J} \times \mathbf{B}] \qquad (5.43)$$

2. The accessibility of this active region to retrieve the Hall voltage, which is given by

$$V_H = \int_{S_1}^{S_2} \mathbf{E}_H ds \qquad (5.44)$$

(S_1 and S_2 indicate the positions of the sense contacts).

3. Moreover, in the absence of a magnetic field, the voltage difference between points S_1 and S_2 should be zero. That means that along a suitable integration path S_1, S_2 (2), the biasing electrical field $\mathbf{E}$ and the Hall electric field $\mathbf{E}_H$ should be mutually orthogonal, that is,

$$\mathbf{E} \cdot \mathbf{E}_H = 0 \qquad (5.45)$$

Another possible way of thinking about Hall magnetic sensors is as follows. For a given orientation or distribution of a magnetic field, how should we structure a semiconductor device to best fulfill the above three criteria? Can we influence the distribution of a magnetic field in the device to better achieve our objective? Such a way of thinking led to the development of the Hall nonplates presented in this section.

Hall nonplates are not a mere scientific curiosity. They have some features that make them very useful in a number of novel magnetic sensor applications. All the Hall nonplates discussed here are already commercially available [37].

5.4.2 Vertical Hall Devices

The simplest structure for a Hall nonplate is the so-called vertical Hall device [38]. It was devised with the aim of creating a good semiconductor sensor, responding to a magnetic field parallel with the chip surface (Figure 5.24). The symmetry plane of this quasi-plate is vertical with respect to the chip surface, hence the attribute *vertical*.

Figure 5.25 illustrates the technological structure of the vertical Hall device. Its N-type active zone is confined into an approximately platelike structure by a deep P-type ring and the depletion layer between the active zone and the ring. The active region of the device is imbedded into the bulk of a silicon chip. That fact explains in part the long-term stability and robustness of this Hall device: Its active zone is buried in a monocrystal, far away from the chip surface.

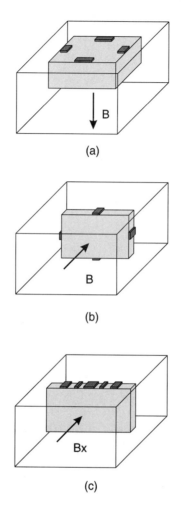

Figure 5.24 Genesis of the vertical Hall device. A conventional Hall plate that is sensitive perpendicular to the chip surface (a) can be rotated so that its active plane comes to lie vertically in the chip (b). A conformal transformation to the plate brings all contacts to the surface (c). The so-obtained vertical Hall sensor is electrically equivalent with the plate (b) and can be manufactured using conventional microelectronic fabrication processes.

Thanks to the quasi-radial form of the current density lines in a vertical Hall sensor, the planar Hall effect in it is much lower than in a Hall sensor of a conventional geometry [39].

All other Hall devices presented in this section are fabricated using the vertical Hall silicon process and share with the vertical Hall device the basic features of low supply current, low noise, and high long-term stability.

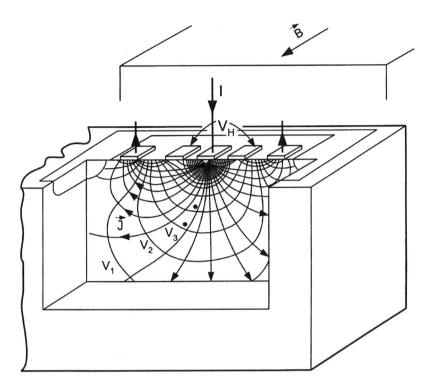

Figure 5.25 A cut through a vertical Hall device, showing the current and equipotential lines (adapted from [1]). The vertical cut plane is a symmetry plane of the sensitive volume (hence the term *vertical*), but the whole sensitive volume is a three-dimensional structure. All three criteria given in Section 5.1 are well met. Notably, the output signal is the Hall voltage V_H, given by the integral (5.19) of the Hall field over a line following an equi-potential line (at $B = 0$) connecting the two sense contacts.

5.4.2.1 Basic Features

The material used for vertical Hall devices is silicon. Because the doping level is low, to achieve a high mobility and a high signal level—see (5.11) and (5.15)—the input resistance is comparatively high, of the order of a few kilo-ohms. For a typical biasing current of 1 mA, a vertical Hall device has a sensitivity of about 400 V/AT.

Unfortunately, silicon is not a perfect material for magnetic field transduction, and the fabrication process is not free of imperfections. That leads to a series of parasitic effects that, if not compensated for, limit the accuracy of a sensor device. How some of the most important effects are generated and how they develop under the influence of temperature and magnetic field are examined next.

5.4.2.2 Offset

The vertical Hall sensor, like any other Hall element, is subject to an error output voltage at zero field, the offset. Offset voltage cannot be distinguished from the output signal resulting from a dc or slowly varying magnetic field. However, according to the resistor-bridge model, offset is directly correlated to the material resistivity, causing it to vary with temperature and with magnetic field in a similar way as the input resistance. It can be compensated by subtracting a part of the biasing voltage from the output voltage. At the same time, this principle also considerably reduces offset drift with temperature.

5.4.2.3 Nonlinearity in the Magnetic Field

The magnetic nonlinearity of the vertical Hall device is determined by three distinct effects related to the three parameters G, R_H, and t in (5.17).

The first effect originates in the modulation of the thickness of the active zone Δt, caused by a variation in the thickness of the depletion layer around the device. In an analogy to the operation principle of a junction-field effect transistor (JFET), the effect is called the junction-field effect. The influence of this parasitic effect depends on the internal potential distribution and for small fields is a linear function of the applied field and the applied current. Typical values are of the order of 1% nonlinearity for a field of 2T and for a biasing current of 1 mA. The effect can be reduced either by using two devices that are matched with opposite sign or by using appropriate biasing conditions for the potential of the surrounding P-ring.

A second reason for magnetic nonlinearity is to be found in the material properties. Up to fields of about 2T, the Hall factor R_H decreases nearly quadratically with the field. That is due to scattering mechanisms resulting from the interaction between charge carriers and crystal lattice. The effect strongly depends on the crystal orientation of the current flow in the device, and the corresponding nonlinearity typically reaches 1% at 2T. This nonlinearity cannot be compensated directly, because it is inherent to the device material.

The third reason for nonlinearity in a silicon Hall sensor is given by the device geometry. The geometrical correction factor G is close to 1 for devices that are long and thin in current direction and that have small contacts. The shorter a device and the longer its contacts, the smaller G becomes. It has been shown that G then increases with the magnetic field in about a quadratic manner [1].

However, because of the opposite sign, mutual compensation between material and geometry-related nonlinearity can be achieved. For an appropriate design for the vertical Hall device, the overall nonlinearity can be as small as 0.1% in magnetic flux density up to 2T [40].

5.4.2.4 Temperature Cross-Sensitivity

The variation of sensitivity due to a variation of temperature is called temperature cross-sensitivity. For strongly extrinsic N-type silicon at room temperature, all impurities have given their electron to the conduction band, so that for small variations of temperature the density of charge carried in the material remains constant. The thermal behavior of the sensitivity depends only on the Hall scattering factor and has been found to be about +0.08%/K for silicon for constant current biasing [41]. That value corresponds well to measured characteristics. Temperature cross-sensitivity can easily be compensated by modulating the current through the device, with the help of an appropriate resistor in parallel to the input contacts. The remaining variation of sensitivity is then usually less than 0.25% over the temperature range from −10°C to 60°C.

5.4.2.5 Planar Hall Effect

It is well known that Hall sensors with conventional plate-shape geometry are sensitive not only to a magnetic field perpendicular to the plate but also to a field in the plane of the plate. The planar Hall effect [42] adds an error voltage to the output Hall voltage. The planar voltage depends on the angle between current and field in the plane and is the strongest at odd multiples of $\pi/4$. It has been shown that the particular crystal orientation and the unique shape of the vertical Hall device reduce the influence of such a planar Hall voltage by about one order of magnitude compared to conventional plate-shape devices [39]. In such a way, vertical Hall sensors are appropriate for the use in precision magnetic measurement instruments used for mapping or control applications.

5.4.2.6 Accurate Magnetic Field Measurement Using Vertical Hall Devices

Silicon vertical Hall devices are commercially available products and are used as key components in a meter for electrical energy and in high-accuracy magnetic field transducers. Their essential features are high current-related sensitivity (400 V/AT), low noise (<10μT from 10^{-4} Hz to 10 Hz), and an unprecedented long-term stability: Magnetic sensitivity does not change more than 0.01% over a year.

Those features are an excellent basis for the application of vertical Hall devices in precision instruments for magnetic field measurement. However, the various parasitic effects have to be corrected to reach high absolute accuracy.

Two matched vertical Hall sensors in the probe head can mutually compensate for offset, junction-field effect, and planar Hall effect. Temperature cross-sensitivity and nonlinear behavior in the magnetic field can be effectively reduced using external compensation. The vertical Hall device follows simple equations with those parameters, so analog electronic compensation functions can be implemented easily. Efficient temperature compensation is achieved by using the Hall sensor itself as a temperature sensor. Operating the Hall devices in the constant current mode allows for implementation of all external compensation through bias-current modulation.

Based on those concepts, single-axis transducers with 0.01% accuracy for a flux density range up to 2T and a temperature range from 15°C to 35°C have been brought to industrial maturity and are now available as commercial products [43].

Adding a temperature stabilization in the probe head and using computer calibration toward a precision nuclear magnetic resonance (NMR) meter, an absolute accuracy of even better than 0.004% for the field range from −6T to 6T was achieved [44]. Thanks to the very small volume of the active zones of the vertical Hall device, even fields with nonlinear gradients of up to 30 T/m were measured with similar precision.

5.4.3 Cylindrical Hall Devices

In some important applications, the magnetic field to be measured naturally has a circular shape, like that around a current-carrying wire or in the narrow gap between two ferromagnetic field concentrators (Figure 5.26). A Hall sensor in a plate shape (conventional or vertical) can in such a case only use a limited portion of the flux lines to generate a Hall voltage between its output contacts.

A geometry that naturally takes advantage of circular field shapes is given with the so-called cylindrical Hall device. How it operates can be illustrated by thinking of it as a vertical Hall device that has been rotated around an axis through the five electric contacts (Figure 5.27). By that operation, the initially thin plate, which is sensitive to a uniform field, unfolds to a half-cylinder, which is sensitive to a circular field.

If such a structure is embedded in a sensor chip and combined with a pair of integrated planar magnetic flux concentrators, it becomes a very sensitive magnetic field sensor also for noncircular fields [45] (Figure 5.28).

Evidently, the structure of a cylindrical Hall device is not even similar to a plate. Nevertheless, this three-dimensional structure is well adapted to a circular magnetic field so that the criteria for a good Hall device, given

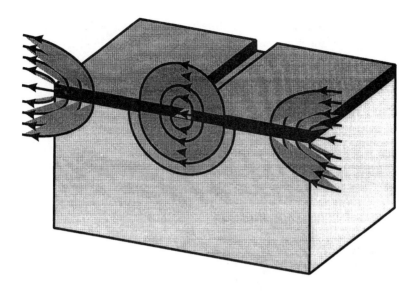

Figure 5.26 The magnetic field around an air gap of ferromagnetic flux concentrators has a circular form.

in Section 5.4.1, are met about as well as (or even better than) in the original vertical Hall device. Compared to traditional platelike Hall devices, the geometry of the cylindrical Hall device allows lower bias voltage and lower offset to be achieved.

The sensitivity of the cylindrical Hall device equipped with ferromagnetic field concentrators is very high for low fields; however, it decreases as the field value reaches the saturation field of the concentrator material at about 20 mT. A cylindrical Hall sensor with integrated magnetic concentrators shows the following characteristics: The mean supply current-related sensitivity S_I is 2,000 V/AT, and the supply voltage-related sensitivity S_V is 0.6 V/VT. The equivalent offset field B_{off} is 0.5 mT with a temperature coefficient α_{voff} of 1%/K. In analogy to the vertical Hall device, a simple analog compensation using the input voltage can reduce the equivalent offset to less than 50 μT between −10°C and 60°C. The relevant low-frequency noise is of $1/f$ type. The noise power at 1 Hz is $4 \cdot 10^{-13}$ V^2/Hz, which corresponds to an equivalent output drift of less then 2.5 μT over 1 hour. The input resistance is 3.8kΩ. Used as a current sensor with 1-mA bias current, the device shows a sensitivity of 1 mV/A.

A cylindrical Hall sensor is applicable in many cases that require high detectivity magnetic sensing, a domain that used to be reserved exclusively for magnetoresistance sensors. For example, a cylindrical Hall sensor is applied

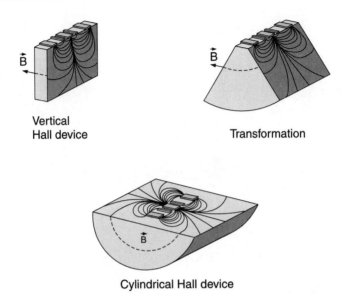

Figure 5.27 Genesis of a cylindrical Hall device by a conformal deformation of a vertical Hall device. The sensitive volume of a cylindrical Hall device is adapted to a circular magnetic field (shown in Figure 5.26). Within the whole active region, the current lines are approximately perpendicular to the magnetic field lines. Moreover, the generated Hall voltage can be retrieved using two sense contacts at the surface of the device.

in a sensitive current sensor [46] (Figure 5.29), for precise gear teeth sensing [47], for contactless angular position sensing (Figure 5.30), for long-range position sensing, for magnetic scanning of documents, and so on.

5.4.4 Two-Axis Vertical Hall Devices

So far, this section has discussed Hall elements that are sensitive to a single component of an applied magnetic field of straight or circular shape. Now we examine structures based on the same technology that are sensitive to two or to all three components of a magnetic field.

A merged combination of two vertical Hall devices gives a magnetic sensor for the simultaneous sensing of the two in-plane components of a magnetic field [48]. By sharing the center current contact, the active zone can be made compact and the two orthogonal field components X and Y are measured in the same spot (Figure 5.31). Although the active regions of the two merged vertical Hall sensors are not electrically isolated, the

Hall-Effect Magnetic Sensors

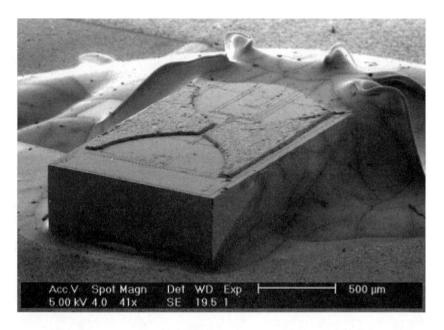

Figure 5.28 Photograph of a cylindrical Hall device with integrated magnetic flux concentrators. Thanks to the concentrators, the sensitivity of this magnetic sensor is as high as 2,000 V/AT. The detectivity threshold for a quasi-static magnetic field (in the bandwidth from 10^{-3} Hz to 10 Hz) is 70 nT, which is a best value for a commercial Hall device.

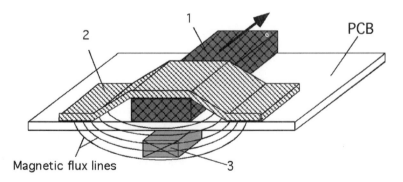

Figure 5.29 Schematic of a current sensor based on a cylindrical Hall magnetic sensor: (1) current lead, (2) ferromagnetic yoke, and (3) magnetic sensor. Items 1 and 2 are batch-processed on a printed circuit board (PCB) (after [46]).

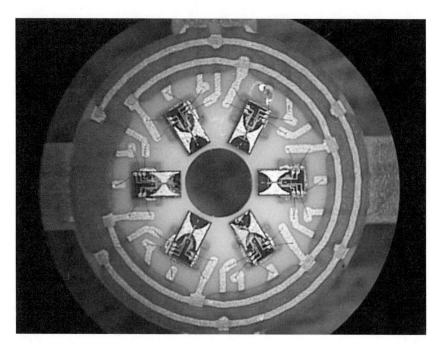

Figure 5.30 A Sentron high-distance magnetic angular position sensor. This figure shows a Hall probe module consisting of six high-sensitivity cylindrical Hall devices mounted on a ceramic substrate.

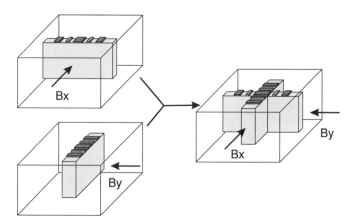

Figure 5.31 A merged combination of two orthogonally placed vertical Hall devices measures simultaneously the two in-plane components of a magnetic field.

orthogonal design decorrelates the magnetogalvanic effects and thus efficiently reduces undesired cross-coupling.

A combination of such a magnetic sensor with a permanent magnet becomes an accurate contactless angular position sensor (Figure 5.32).

The unique feature of such a combination is its great robustness. Relatively large variations in the position of the magnet have little influence on the result of the angle measurement. The information on the angle of

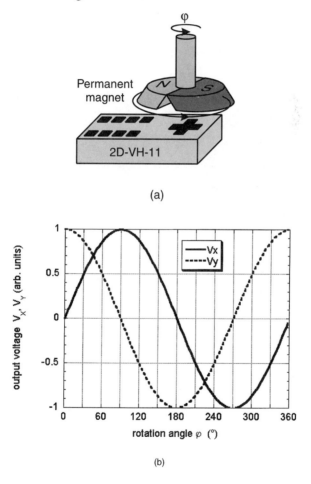

Figure 5.32 (a) In a combination with a permanent magnet, a two-dimensional vertical Hall sensor becomes an accurate angular position sensor. (b) The measured angle depends only on the ratio of the two output voltages and not on their absolute values, so that the sensitivity drift with temperature is directly compensated [49].

the magnetic field vector with respect to the sensitive axes of the sensor can be retrieved from the ratio of the two output signals rather than from their absolute values. That renders the sensor virtually immune to the manufacturing tolerances and temperature fluctuations. An accuracy of ±0.1 degree C can be easily achieved with such a sensor.

Another merged combination of three halves of a vertical Hall device gives a three-phase angular position sensor, which is especially useful for miniaturized brushless motors [49] (Figure 5.33). Because any two half-quasi-plates that complete a Hall device are not in the same plane, sensitivity is reduced, but in return the output signals have an optimal shape and timing and the sensor has a minimum of connections.

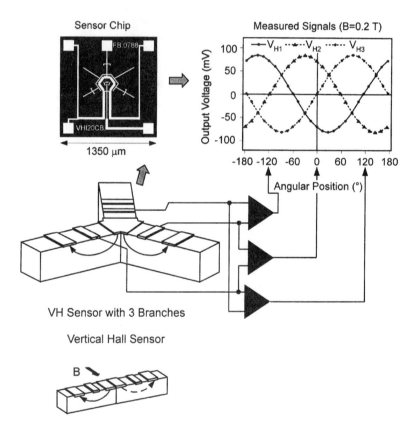

Figure 5.33 A three-phase angular position sensor realized in the vertical Hall technology. The sensor simultaneously gives the three signals phase-shifted for 120 degrees. The three signals can be used directly to control the stator currents of a brushless motor [49].

5.4.5 Three-Axis Hall Devices

Even more challenging than to make a good two-axis single-chip Hall sensor is the realization of a good three-axis single-chip Hall sensor. Such a device fabricated in vertical Hall technology [50] has an active zone in the shape of a square with eight ohmic contacts at its surface (Figure 5.34). Here again the lateral confinements are realized by a P-ring (not visible).

Four contacts in the corners are used alternatingly as current input and current output contacts, and the four midside contacts are Hall sense contacts. The total current flows in such a way through the device that three different Hall voltages corresponding to the three magnetic field components B_x, B_y, and B_z can be measured (Figure 5.35).

From Figure 5.35 it can be noticed that each Hall contact is sensitive to one in-plane magnetic field component and to the perpendicular component. A convenient way to obtain three voltages representing the three magnetic field components independently is illustrated in Figure 5.36.

Applying such an electronic interface, all three magnetic field components are measured independently in the device center. The total active volume can be as small as 0.1 by 0.1 by 0.1 mm^3, allowing for measurements with a high spatial resolution. The sensor axes are perfectly orthogonal because of the photolithographic manufacturing process, and the three voltages are generated simultaneously, without any switching of currents or voltages. Encapsulated in a miniaturized metallic case, the three-axis Hall chip becomes the world's smallest three-axis Hall probe, allowing three-dimensional magnetic field mapping of very small objects in a very convenient way.

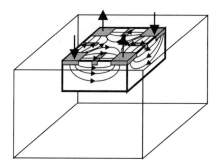

Figure 5.34 The three-axis Hall magnetic sensor. The eight shaded regions at the chip surface represent the contacts; the vertical arrows are the bias currents; and the curved arrows are the current-density lines.

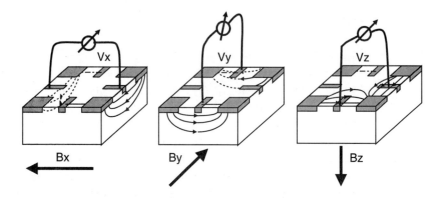

Figure 5.35 The voltmeter symbols indicate how the Hall voltages proportional to various magnetic field components are retrieved. The two in-plane magnetic field components B_x and B_y are measured between opposite Hall contacts. The sensitive regions are reminiscent of that of a vertical Hall device; see Figure 5.24(c). The perpendicular component B_z is measured between adjacent Hall contacts, and the sensitive region is similar to that of a Hall plate with the open bottom [51].

Hall-Effect Magnetic Sensors 239

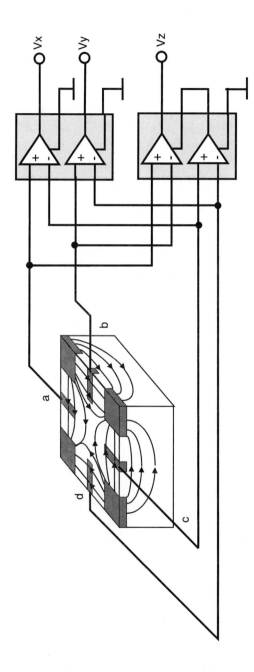

Figure 5.36 The entire signal retrieval of three voltages V_x, V_y, and V_z, representing the three magnetic field components B_x, B_y, and B_z, can be performed by a simple differential amplification stage.

References

[1] Popovic, R. S., *Hall Effect Devices*, in B. E. Jones (ed.), *The Adam Hilger Series on Sensors*, Philadelphia, PA: Adam Hilger, 1991.

[2] Hall, E. H., "On a New Action of the Magnet on Electric Currents," *American Journal of Mathematics*, 2, 1879, p. 287.

[3] Pearson, G. L., "A Magnetic Field Strength Meter Employing the Hall Effect in Germanium," *Review of Scientific Instruments*, Vol. 19, 1948, p. 263.

[4] Popovic, R. S., J. A. Flanagan, and P. A. Besse, "The Future of Magnetic Sensors," *Sensors and Actuators*, A56, 1996, pp. 39–55.

[5] Schott, C., et al., "High Accuracy Analog Hall Probe," *IEEE Trans. Instrum. Meas.*, Vol. 46, No. 2, 1997, pp. 613–616.

[6] Welker, W., "Uber neue halbleitende Verbindungen," *Z. Naturforschung*, 1952, 7a, p. 744.

[7] Guenther, K. G., "Aufdampfschichten aus halbleitenden III-V Verbindungen," *Z. Naturforschung*, 1958, 13a, p. 1081.

[8] Guenther, K. G., "Aufdampfschichten aus halbleitenden III-V Verbindungen," *Naturwissenschaften*, 1958, 45, p. 415.

[9] Sakai, Y., and M. Ohsita, "Preparation of Indium Antimonide Films and Measurement of Hall Characteristics" (in Japanese), *J. IEE Japan*, Vol. 80, 1960, p. 166.

[10] Asahi Kasei Electronics, biographical sketch (in Japanese), 1995, pp. 6–7.

[11] Shibasaki, I., "Properties of Hall Elements by Vacuum Deposition" (in Japanese), in H. Abe (ed.), *The Correction of Advanced Technology of Semiconductor Devices*, Tokyo: Keiei system kennkyu syo, Shinjyuku, 1984, p. 373.

[12] Shibasaki, I., "High Sensitivity InSb Hall Elements and Their Development for Practical Use" (in Japanese), *Monthly Report of Japan Society of Chemical Industries*, Vol. 41, No. 5, 1988, p. 12.

[13] Shibasaki, I., "High Sensitive Hall Element by Vacuum Deposition," *IEE Japan, Tech. Digest on 8th Sensors Symp.*, 1989, p. 211.

[14] Cho, A., and J. R. Arthur, "Molecular Beam Epitaxy," *Prog. Solid-State Chem.*, Vol. 10, 1972, p. 157.

[15] Shibasaki, I., et al., "High Sensitive Thin Film InAs Hall Element by MBE, IEEE," *Digest Tech. Papers on Transducers*, 1991, p. 1069.

[16] Asahi Kasei Electronics, Hall elements catalogs, 1986.

[17] Shibasaki, I., "Mass Production of InAs Hall Elements by MBE," *J. Cryst. Growth*, Vol. 175/176, 1997, p. 13.

[18] Shibasaki, I., "The Practical Hall Elements as Magnetic Sensors by Thin Film Technology," *Proc. IEEE Lasers and Electro-Optics Society 1995 Annual Meeting*, Vol. 1, 1995, p. 85.

[19] Takahasi, K., T. Moriizumi, and Y. Sakai, "Characteristics of InAs Thin Films by Vacuum Deposition" (in Japanese), *Proc. 26th Spring Conf. Japan Society of Applied Physics*, 1965, p. 18.

[20] Iwabuchi, T., et al., "High Sensitivity Hall Elements Made From Si-Doped InAs on GaAs Substrates by Molecular Beam Epitaxy," *J. Cryst. Growth*, Vol. 150, 1995, p. 1302.

[21] Shibasaki, I., T. Kajino, and K. Tajika, Japan Pat. No. P1598818 (application: July 13, 1982).

[22] Nagase, K., et al., "InAs/AlGaAsSb Quantum Well Hall Elements Having High Output Voltage and Good Temperature Characteristics," *Digest of Tech. Papers; Abstract of Late News Papers, 7th Intl. Conf. on Solid-State Sensors and Actuators*, 1993, p. 34.

[23] Kuze, N., et al., "High Sensitivity Hall Elements Made From Si-Doped InAs on GaAs Substrates by Molecular Beam Epitaxy," *J. Cryst. Growth*, Vol. 150, 1995, p. 1307.

[24] Kuze, N., et al., "Molecular Beam Epitaxial Growth of High Electron Mobility InAs/AlGaAsSb Deep Quantum Well Structures," *J. Cryst. Growth*, Vol. 175/176, 1997, p. 868.

[25] Kataoka, S., *Magneto-Electric Transducers* (in Japanese), Tokyo: Nikkan Kogyo Shimbun Press, 1965.

[26] Wiess, H., *Structure and Application of Galvanomagnetic Devices*, London: Pergamon Press, 1969.

[27] Maupin, J. F., and M. L. Geske, "The Hall Effect in Silicon Circuits," in C. L. Chien and C. R. Westgate (eds.), *The Hall Effect and Its Applications*, New York: Plenum Press, 1980, pp. 421–445.

[28] Bilotti, A., G. Monreal, and R. Vig, "Monolitic Magnetic Hall Effect Sensor Using Dynamic Quadrature Offset Cancellation," *IEEE J. Solid State Circuits*, Vol. 32, No. 6, June 1997, pp. 829–836.

[29] Bellekom, A. A., and P. J. A. Munter, "Offset Reduction in Spinning-Current Hall Plates," *Sensors and Materials*, Vol. 5, No. 5, 1994, pp. 253–263.

[30] Kanda, Y., and A. Yasukawa, "Hall-Effect Devices as Strain and Pressure Sensors," *Sensors and Actuators*, Vol. 2, 1982, pp. 283–296.

[31] Nathan, A., H. Baltes, and W. Allegretto, "Review of Physical Models for Numerical Simulation of Semiconductor Microsensors," *IEEE Trans. Computer Aided Design*, Vol. 9, No. 11, Nov. 1990, pp. 1198–1208.

[32] Biard, J. R., Hall effect device formed in an epitaxial layer of silicon for sensing magnetic fields parallel to the epitaxial layer, U.S. Pat. No. 5,572,058, Nov. 5, 1996.

[33] Steiner, R., et al., "Influence of Mechanical Stress on the Offset Voltage of Hall Devices Operated With Spinning Current Method," *J. Micromechanical Systems*, Vol. 28, No. 4, 1999, pp. 466–472.

[34] Gee, S., T. Doan, and K. Gilbert, "Stress Related Offset Voltage Shift in Precision Operational Amplifiers," *Proc. 43rd Electronic Component and Technology Conf.*, New York, 1993, pp. 755–764.

[35] Manic, D., J. Petr, and R. S. Popovic, "Short and Long-Term Stability Problems of Hall Plates in Plastic Packages," *Proc. IEEE International Reliability Physics Symp. (IRPS 2000)*, Apr. 2000, San Jose, CA, pp. 225–230.

[36] Popovic, R. S., "Non-Plate-Like Hall Magnetic Sensors and Their Applications," *Sensors and Actuators A*, Vol. 85, 2000, pp. 9–17.

[37] For details on the commercially available Hall non-plates, see: www.sentron.ch.

[38] Popovic, R. S., "The Vertical Hall-Effect-Device," *IEEE Electronic Dev. Letters*, EDL-5, 1984, pp. 357–358.

[39] Schott, C., P.-A. Besse, and R. S. Popovic, "Planar Hall Effect in the Vertical Hall Sensor," *Proc. EUROSENSORS XIII*, The Hague, Netherlands, Sept. 12–15, 1999, pp. 437–440.

[40] Schott, C., and R. S. Popovic, "Linearizing Integrated Hall Devices," *Proc. Transducers 97, Intl. Conf. on Solid-State Sensors and Actuators*, Chicago, IL, June 1997, Vol. 1, pp. 393–396.

[41] Long, D., "Scattering of Conduction Electrons by Lattice Vibrations in Silicon," *Physics Review*, Vol. 120, No. 6, 1960, pp. 2024–2032.

[42] Goldberg, C., and R. E. Davis, "New Galvanomagnetic Effect," *Physics Review*, Vol. 94, No. 5, 1954, pp. 1121–1125.

[43] Schott, C., et al., "High Accuracy Analog Hall Probe," *IEEE Trans. Instrum. Meas.*, Vol. 46, No. 2, 1997, pp. 613–616.

[44] Schott, C., et al., "High Accuracy Magnetic Field Measurements With a Hall Probe," *Review of Scientific Instruments*, Vol. 70, No. 6, 1999, pp. 2703–2707.

[45] Blanchard, H., et al., "Cylindrical Hall Device," *Proc. IEDM 1996*, San Francisco, CA, 1996, pp. 541–544.

[46] Blanchard, H., J. Hubin, and R. S. Popovic, "Low-Cost Open-Loop Current Sensor in Batch-Process Technology," *Proc. Sensors 99*, Nurnberg, Germany, May 18–20, 1999, 2, P1.32, pp. 421–425.

[47] Dutoit, B., et al., "High Performance Micromachined Sm2Co17 Polymer Bonded Magnets," *Sensors and Actuators A*, Vol. 77, No. 3, Nov. 2, 1999, pp. 178–182.

[48] Burger, F., P. A. Besse, and R. S. Popovic, "New Fully Integrated 3-D Silicon Hall Sensor for Precise Angular Position Measurements," *Sensors and Actuators A*, Vol. 67, 1998, pp. 72–76.

[49] Burger, F., P.-A. Besse, and R. S. Popovic, "New Single Chip Hall Sensor for Three Phases Brushless Motor Control," *Sensors and Actuators A*, Vol. 81, Nos. 1–3, Apr. 1, 2000, pp. 320–323.

[50] Schott, C., J.-M. Waser, and R. S. Popovic, "Single-Chip 3D Silicon Hall Sensor," *Sensors and Actuators A*, Vol. 80, 2000, pp. 167–173.

[51] Randjelovic, Z., et al., "A Non-Plate Like Hall Sensor," *Sensors and Actuators A*, Vol. 76, Nos. 1–3, Aug. 30, 1999, pp. 293–297.

6

Magneto-Optical Sensors
Yuri Didosyan and Hans Hauser

Optics provides rich opportunities for the development of techniques for measurements of a large number of various physical quantities. In optical methods of measurement, parameters of light beams are changed under the influence of the measured quantity. The results of measurements are contained in the optical beams and can be transferred at considerable distances in open air or in optical fibers. Advantages of optical methods include contactlessness, high stability with respect to electromagnetic interference, and (when the measurements are performed in a high-voltage environment) the absence of potential separation problems. In addition, optical methods are often associated with ultimate precision, and the measurements can be performed in very wide frequency and dynamic ranges.

Magneto-optics studies the interaction of light with a medium exerting action of an external magnetic field, including changes in the polarization state and in the propagation direction of light. It greatly contributes to the development of modern optical measurement techniques. Magneto-optics allows measurements of magnetic fields by means of their direct influence on light. It also can be combined with other optical methods of measurements for operation of parameters of the sensing light beams, for example, for their modulation or their scanning.

This chapter briefly covers the basic concepts of the magneto-optical Faraday and Kerr effects and reviews new magneto-optical sensors for measurements of magnetic fields and electric currents. In addition, it briefly discusses new magneto-optical sensors for geometric measurements.

6.1 Faraday and Magneto-Optical Kerr Effects

6.1.1 Faraday Effect

Faraday discovered the following phenomenon: If linearly polarized light passes through a medium placed in the magnetic field and the direction of the magnetic field is parallel to that of light propagation, the plane of polarization of light rotates. Figure 6.1 shows a scheme of the classical Faraday setup: A linearly polarized light passes through magneto-optical material 3 and analyzer 4 and is collected by photoreceiver 5. The rotation of the polarization plane (Faraday rotation) is detected by means of the photoreceiver's signal: According to the Malus law, the intensity of light transmitted by the analyzer is proportional to the cosine squared of the angle between the polarization of the incident light and the main plane of the analyzer. Thus, by measuring the intensity of light for given orientation of the analyzer, we can determine the direction of polarization of light emerging from magneto-optical material 3.

Faraday found that the angle of rotation is proportional to the magnetic field H and to the length of the magneto-optical material l:

$$F = VlH \qquad (6.1)$$

The proportionality constant V is called the Verdet constant. The Faraday effect represents an odd effect with respect to the H direction. A change in the sign of H causes a change in the sense of Faraday rotation. Therefore, when light passes back and forth through the magneto-optical material, the angle of rotation increases two times. That nonreciprocity of the Faraday effect provides the opportunity (successfully being used since experiments conducted by Faraday himself) to substantially increase the angle of the

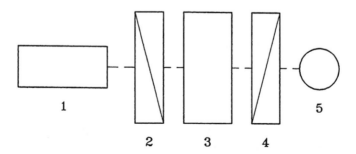

Figure 6.1 Faraday effect scheme.

polarization rotation by passing light many times back and forth through the magneto-optical material.

Equation (6.1) is valid for paramagnets and diamagnets, but it does not apply to magnetically ordered materials, for example, to ferromagnets. In the latter materials, the Faraday rotation is not proportional to the external magnetic field but rather is a function of the magnetization; the sign of the magnetization determines the sense of Faraday rotation.[1]

The Faraday effect is related to the fact that the magnetized state cannot be described by a single refraction index. The magnetic field in the medium makes a difference between refraction indices for circularly polarized light with opposite senses of polarization rotation.

The electric vector of a light wave of unit amplitude polarized along the x-axis is

$$E_x = \exp[i(\omega t - \mathbf{k}z)] \tag{6.2}$$

where ω is the circular frequency, z is the propagation direction, and $\mathbf{k}$ is the wave vector:

$$k = (2\pi/\lambda)n \tag{6.3}$$

λ is the wavelength of light and n is the refraction index. Linearly polarized light can be represented as a sum of two circularly polarized light waves with mutually opposite senses of rotation of the electric vector. Let the indices ' and + correspond to the light with clockwise (cw) direction of rotation of the electric vector and the indices " and − correspond to the counterclockwise (ccw) direction. Then the cw polarized light wave is

$$E'_x = (1/2)\exp[i(\omega t - k_+ z)] \tag{6.4}$$

$$E'_y = (1/2)\exp[i(\omega t + \pi/2 - k_+ z)] \tag{6.5}$$

The ccw polarized wave is

$$E''_x = (1/2)\exp[i(\omega t - k_- z)] \tag{6.6}$$

$$E''_y = (1/2)\exp[i(\omega t - \pi/2 - ik_- z)] \tag{6.7}$$

---

1. In nomagnetic materials, the Faraday rotation is also a function of magnetization. But because the susceptibility of these materials is very low, the function reduces to proportionality to the external field $\mathbf{H}$ in (6.1).

The sum of these two waves makes

$$E_x = [(1/2)\exp(-ik_+z) + \exp(-ik_-z)]\exp(-i\omega t) \qquad (6.8)$$

$$E_y = [(1/2)\exp(-ik_+z) - \exp(-ik_-z)]\exp(-i\omega t) \qquad (6.9)$$

Let us denote

$$k_+ + k_- = 2k; \; k_+ - k_- = 2\chi \qquad (6.10)$$

Then the resulting wave will have the following form:

$$E_x = \cos\chi z * \exp[i(\omega t - kz)] \qquad (6.11)$$

$$E_y = \sin\chi z * \exp[i(\omega t - kz)] \qquad (6.12)$$

That means that at the exit of a magneto-optical material of thickness l the electric vector of the light wave is rotated with respect to the initial polarization direction by the angle F:

$$\tan F = E_y/E_x = \tan(\chi \cdot l) \qquad (6.13)$$

$$F = \frac{k_+ - k_-}{2} \cdot l = \frac{2\pi}{\lambda} \cdot \frac{n_+ - n_-}{2} \times l \qquad (6.14)$$

The difference of the refraction indices $n_+ - n_-$ is called the circular birefringence. It is due to the action of the external magnetic field on the medium and is proportional to the rotation of the polarization plane.

Often the circular birefringence is combined with linear birefringence. Linear birefringence is associated with spatial nonuniformity and means that the refraction indices of the medium depend on the polarization orientation. Linear birefringence originates from the crystallographic structure, from the strains that occurred during the manufacturing process of the sensor material (remanent birefringence), from mechanical stresses taking place during operation, from temperature changes, and so on. Linear birefringence substantially affects the state of polarization of light and distorts the results of measurements.

The formula that describes propagation of light through the medium displaying both the Faraday rotation F and the linear birefringence δ (distributed homogeneously within the medium) has the form [1, 2]:

$$\begin{pmatrix} E_x \\ E_y \end{pmatrix} = \begin{pmatrix} \alpha + i\beta & -\gamma \\ \gamma & \alpha - i\beta \end{pmatrix} \begin{pmatrix} E_{x0} \\ E_{y0} \end{pmatrix} \qquad (6.15)$$

where $\Delta = \left(F^2 + \dfrac{\delta^2}{4} \right)^{1/2}$, $\alpha = \cos\Delta$, $\beta = \dfrac{\delta}{2}\left(\dfrac{\sin\Delta}{\Delta}\right)$, and $\gamma = F\left(\dfrac{\sin\Delta}{\Delta}\right)$.

E_x and E_y are components of the electric vector of the lightwave at the output from the sensor, and E_{x0} and E_{y0} are components of the electric vector of the lightwave incident on the sensor. If the main plane of the polarizer forms an angle θ with the vertical y-axis, then the components are $E_{x0} = \sin\theta$ and $E_{y0} = \cos\theta$.

6.1.2 Magneto-Optical Kerr Effect

The magneto-optical Kerr effect manifests itself in the change of parameters of light reflected from a magnetic sample. That action of the sample on the light strongly depends on the mutual orientations of the plane of incidence of light and of the direction of the sample's magnetization. There are three basic types of the Kerr effect:

- In the *polar Kerr effect*, the magnetization is perpendicular to the reflective surface and parallel to the plane of incidence, as shown in Figure 6.2(a).
- In the *longitudinal (meridional) Kerr effect*, the magnetization is parallel both to the reflective surface and to the plane of incidence, as shown in Figure 6.2(b).
- In the *transverse (equatorial) Kerr effect*, the magnetization is parallel to the reflective surface and perpendicular to the plane of incidence, as shown in Figure 6.2(c).

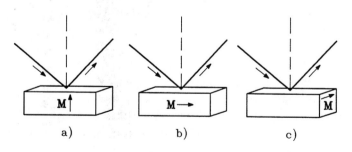

Figure 6.2 Magneto-optical Kerr effect: (a) polar Kerr effect; (b) longitudinal Kerr effect; and (c) transverse Kerr effect.

In the polar Kerr effect, the linearly polarized light becomes elliptically polarized after reflection, with the large axis of the polarization ellipse rotated with respect to the initial polarization direction. Similar changes also are observed in the longitudinal Kerr effect. A common feature of the two effects is that there exists a component of the wave vector of light on the direction of the magnetization vector. That fact is responsible for the analogy of the Kerr effects with the Faraday effect and permits us to consider all those effects as longitudinal effects. The transverse Kerr effect results in the change of the intensity and in the phase shift of the reflected light. The phase shift takes place if the polarization plane of the incident light is not parallel or perpendicular to the plane of incidence.

6.2 Sensors of Magnetic Fields and Electric Currents

Compared to conventional current transformers, the magneto-optical current transformers (MOCTs) based on the Faraday effect have a number of important advantages. They are contactless and the measured signal is transmitted by a light beam propagating in air or in optical fibers. This eliminates costly problems of potential separation. Moreover, the MOCTs are resistant to electromagnetic interference, have a wide frequency bandwidth, and can operate down to dc currents in an extremely high dynamic range.

Measurements of magnetic fields and electric currents are the natural applications of magneto-optics. The main features of the measurement scheme have remained basically unchanged since Faraday's discovery in 1845. The principle of measurements usually is based on determining the polarization rotation of the light in a sensing element subjected to the action of the magnetic field of the current to be measured. The most significant improvements of measurement techniques were made in recent decades and were mainly associated with the materials of the sensing elements.

Paramagnetic glasses have a substantial drawback that complicates their use as magnetic sensors, namely, the strong dependence of the polarization rotation on temperature. The temperature dependence seriously affects the results of measurements, and it is difficult to separate contributions of the magnetic field changes from those due to temperature changes. The rather weak temperature dependence (perhaps less than 10^{-4}%/K) of the Verdet constant characterizes diamagnets. Therefore, diamagnets, despite their much smaller Verdet constants, were chosen for the development of many types of magnetic field (electric current) sensors. Some of these sensors will be described in this chapter.

Revolutionary changes in the sensing elements occured after the synthesis of transparent ferromagnets, which provide uncomparably larger angles of Faraday rotation than paramagnets and diamagnets. Moreover, these materials gave rise to the development of modified schemes of magneto-optical measurements of magnetic fields using ferromagnetic domain structures. In the next paragraphs we will discuss the latest developments in this field.

6.2.1 Polarimetric Measurements

The Faraday rotation depends on the magnetic field. For para- and diamagnets, this dependence is simply proportional to H: relation (6.1). Therefore, by measuring the Faraday rotation angle, the value of the magnetic field is determined. When the conventional Faraday scheme shown in Figure 6.1 is used for measurement of the Faraday rotation, the intensity of light incident on the photoreceiver reads

$$I = I_0 \cos^2(\gamma - F) \qquad (6.16)$$

where I_0 is the intensity of light incident on the magneto-optical material and γ is the angle between the main planes of the polarizer and the analyzer. The sensitivity of the detection with respect to the Faraday rotation is

$$\frac{dI}{dF} = I_0 \sin 2(\gamma - F) \qquad (6.17)$$

In the range of small F, the maximum sensitivity is obtained when $\gamma = 45$ degrees.

The results of measurements are very sensitive to the fluctuations of the output power of the light source. There are several schemes to substantially reduce that drawback. In the so-called dual-quadrature polarimetric configuration, shown in Figure 6.3, a polarization beam splitter is used instead of the analyzer (4) in Figure 6.1 [3].

Light beams polarized in the main plane of the splitter and in the orthogonal plane emerge from the splitter in two different directions. Intensities of the light beams $I_\perp$ and $I_\parallel$ are measured by two separate photoreceivers. If the main planes of the splitter and the polarizer are parallel, then the ratio of the intensities gives the Faraday rotation on the sensing element [4]:

$$\tan^2 F = \frac{I_\perp}{I_\parallel} \qquad (6.18)$$

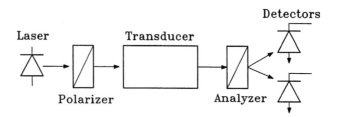

Figure 6.3 Dual-quadrature configuration.

If a magnetic field produced by a pure ac current is measured, then the ac part of the photoreceiver's signal (see Figure 6.1) is independent on the current. Therefore, by dividing the ac part of the photoreceiver's signal by the dc part, we can avoid the influence of the light fluctuations on the measurement results [4].

Another way to eliminate the influence of the intensity fluctuations is to use chromatic modulation [5–7]. The sensing element is illuminated by polychromatic rather then monochromatic light. The light intensity of the infrared part of the light emergent from the polarization filter, used as the analyzer (4) in Figure 6.1, is insensitive to the magnetic field because it is not polarized by the filter. Consequently, this part of the spectrum serves as a reference signal.

The results of measurements are very sensitive to the linear birefringence. Consider the dual-quadrature polarimetric configuration. In this configuration, the result of measurement is expressed in the following form:

$$S = \frac{(I_1 - I_2)}{(I_1 + I_2)} \qquad (6.19)$$

where I_1 and I_2 are the intensities of light measured by each photoreceiver.

If the Faraday rotation is much larger than the linear birefringence ($F \gg \delta$), then we can obtain from (6.15) and (6.19)

$$S = \sin 2F \qquad (6.20)$$

The measured parameter S is directly related to the Faraday rotation F. Moreover, if F itself is small enough, then the dependence between S and the Faraday rotation (and, consequently, the measured current) reduces to the proportionality

$$S = 2F \qquad (6.21)$$

But if the Faraday rotation is much smaller than the linear birefringence $F \ll \delta$, then (6.15) and (6.19) lead to

$$S = 2F \frac{\sin \delta}{\delta} \tag{6.22}$$

We can see that the birefringence may cause a substantial decrease of the measured signal. A decrease of the measured signal S with the linear birefringence occurs in two directions. First, the light emerging from the birefringent medium is elliptically polarized. That apparently decreases the sensitivity of the measurements; in the limiting case of the circular polarization, $I_1 = I_2$ and $S = 0$. Second, due to the linear birefringence, the large axis of the polarization ellipse rotates at the smaller angle than in the case of $\delta = 0$.

The Faraday rotation F and the birefringence delta δ both are linear functions of the length of the sensing element. But while the first multiple in (6.22) grows linearly with the length, the second multiple, $\sin \delta / \delta$, represents the oscillating function of the length. Therefore, for some values of the parameters, the growth of the length of the sensing element causes the decrease rather than the increase of the measured signal. We can also notice that increments of S depend on the actual value of δ, that is, they are different in the different places of the sensing element. That implies that the results of measurements depend on the location of the conductor carrying the measured current.

6.2.2 Magneto-Optical Current Transformers Based on Diamagnets

Verdet constants of diamagnets are very low; and to obtain sufficiently large Faraday rotation angles, the lengths of the sensing elements have to be very large, several meters and more. Large lengths prevent the use of diamagnets for magnetic field measurements (unless the test field is exceptionally uniform). Nevertheless, these materials are successfully used for integral magnetic field measurements of test electric currents. Usually the sensing element surrounds a conductor with the test current. In accordance with the Ampere's law,

$$i = \oint H dl \tag{6.23}$$

the integral Faraday rotation F is a measure of the current i:

$$F = V \oint H dl = NV_i \qquad (6.24)$$

It is necessary to mention, however, that contrary to the Ampere's law, the latter statement is true only in the absence of linear birefringence. Otherwise, as one can see from the previous paragraph, the different parts of the optical path inside the sensing element will have different sensitivities. Respectively, the Faraday rotation angle F will depend on the position of the conductor with the measured current. As long as the sensing element circles the conductor, this relation does not depend on the location of the conductor. N is the number of passes of light around the conductor. We can see that the sensitivity of current measurements is proportional to the product NV and can be enhanced by an increase in N.

Good opportunities for the increase of N provide fiber-type magneto-optical current transformers (MOCTs), in which the fiber is winding the conductor many times. The main problem associated with these sensors is that the linear birefringence in the sensing element seriously affects the results of measurements. The birefringence originates mostly from two sources: thermal effects and mechanical factors. Whereas the Faraday rotation in the closed optical path does not depend on the path length, according to (6.24), the birefringence increases with the increase of the length [8]. To reduce the influence of the birefringence on the results of measurements, various schemes have been developed [2, 9]. Figure 6.4 shows a scheme of a fiber-type MOCT developed at Siemens [9].

The sensor head represents a coil from a standard telecommunication fiber. The coil was previously annealed [10]. This procedure not only reduces the inherent linear birefringence of the fiber but, more important, significantly reduces the bend-induced birefringence (i.e., the birefringence, originating from wrapping of the fiber into the coil). That improves the current sensitivity and reduces the temperature dependence of the output signal. For pure ac current measurements, a further reduction of the temperature sensitivity is done by means of an intelligent signal evaluation. The dc part of the normalized ouput signal is used as a measure of the residual birefringence and, therefore, of the sensor head temperature. The dc part is used to correct the temperature-dependent sensitivity of the ac part of the output signal. The uncorrected output signal drifts with temperature by 8% over about 60°C. The temperature drift compensation scheme reduces the output error by a factor of 10 in a temperature range from 25°C to 80°C.

Because the Verdet constant of the fiber core material is very small, very large light path lengths in the fiber-type MOCTs are needed to provide

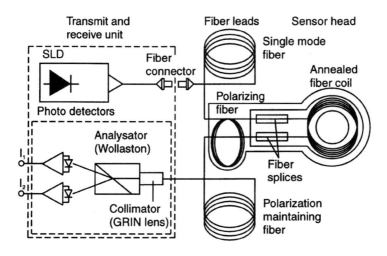

Figure 6.4 Scheme of a fiber-type MOCT.

reasonable sensitivity. That results in a poor time resolution, making the MOCTs unsuitable for measurements of transient currents. Moreover, the large number of turns around the conductor results in an enlarged fiber strain and curvature, thus enhancing problems with the birefringence.

Compared with optical fibers, bulk glasses provide higher values of the Verdet constant, have smaller linear birefringence and better stability, and can be manufactured in different structures [8, 11–13]. High sensitivity of measurements in a bulk-glass MOCT can be achieved if light passes several times through the sensing element. Figure 6.5 [8] shows a geometry of the MOCT.

The sensing element consists of four rectangular prisms and a small 90-degree prism made of SF-6 flint glass. To avoid the reflection-induced phase difference, a dual-quadrature reflection is used [14]. Figure 6.5 illustrates the triple-optical-pass case. By adjustment of the incident spot of light, the number of passes and thus the sensitivity of the MOCT can be changed. A problem of the increased birefringence resulting from the increased optical length can be solved by exploiting the nonreciprocal property of the Faraday effect (similarly to the algorithm described in Section 6.2.1). Moreover, to provide multiple passes, a good collimation of the incident light beam and a sufficiently large extension of the sensing element are necessary.

Another way to increase sensitivity of the MOCTs based on diamagnets is to use ferromagnetic field concentrators [14, 15]. A conductor with the measured current is surrounded by a C-shaped concentrator ring consisting

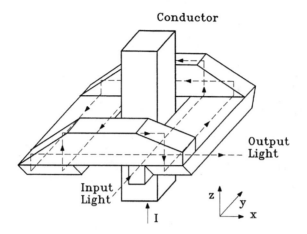

Figure 6.5 Multipass sensing element.

of a ferromagnetic material (see Figure 6.6 [15]). The sensing glass is located in the gap of the concentrator's ring. Because of this geometry, the glass experiences a stronger magnetic field. The light beam propagating through the glass undergoes a larger Faraday rotation. This scheme leads, however, to considerable nonlinearity and hysteresis of measurements caused by the magnetization curve of ferromagnet.

The light experiences multiple reflections on the side walls of the glass. The Faraday rotation in the glass can be considered proportional to the total length of the light path in the glass. The length does not depend on the

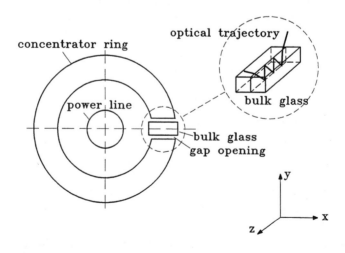

Figure 6.6 Field-concentrator arrangement.

thickness of the glass [15], which is related to the gap size. On the other hand, the magnetic field acting on the glass increases with the narrowing of the gap. That provides opportunity for a further increase in the sensitivity of the MOCT, but it can be obtained only if the reflections on the side walls are strong enough. Otherwise, the increase in the number of reflections, resulting from the narrowing of the glass, will lead to an intensity loss, canceling the sensitivity increase. Neglecting the light absorption in the bulk glass (which for a given length of the optical path is independent of the glass thickness), the transmission coefficient τ of the optical beam is

$$\tau = \gamma^N \qquad (6.25)$$

where γ is a fraction of power of the optical beam remaining in the glass after a single reflection ($\gamma \leq 1$), and N is the number of reflections.

By using reflective coatings, the transmission coefficient can be made rather large, providing a substantial increase in sensitivity with a decrease in gap size. This is illustrated by Figure 6.7 [15]. A family of calculated curves shows the dependence of sensitivity on gap size. The normalization was performed with respect to a gap of 10 mm in the concentrator of the ferromagnetic material with a relative permeability of 1,000; the mean radius of the concentrator was taken to be equal to 28 mm. According to Figure 6.7, if the transmission coefficient is less than 60%, there is little merit in the reduction of the gap size. On the other hand, if τ exceeds 95%, the

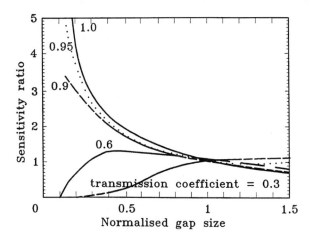

Figure 6.7 Sensitivity ratio as a function of gap size normalized to a reference gap size g_0 at various transmission coefficients.

MOCT's sensitivity can be improved almost twice when the gap size is reduced to the same extent.

In practice, however, multiple reflections can affect the rotation of the polarization plane and consequently reduce sensitivity. That phenomenon originates from the fact that, after a reflection, orthogonal components of the electric vector of the light get different phase shifts, resulting in linear birefringence for the reflected light [14]. The linear birefringence leads to a decrease in the rotation of the polarization plane and to elliptical polarization of the light emerging from the sensing element. But if the initial polarization of light incident on the reflecting surface is parallel or perpendicular to the plane of incidence and—moreover—the Faraday rotation is small, we can (with a good approximation) neglect birefringence and its influence on the sensitivity. According to [15], this distortion is negligibly small for the usual operation rate of ±2° Faraday rotation.

The problem of reducing the birefringence inside the sensing element of the MOCTs with glass ring sensor heads is investigated in [16]. To provide a uniform temperature distribution inside the sensing element under ambient temperature changes, use of a closed-structure shell consisting of a rigid heat-conductive material was proposed. It was also proposed that the sensor be enclosed by buffers having high heat resistance and the same coefficient of thermal expansion as the sensing material—to slow the speed of heat exchange between the sensor and the shell.

These measures provide substantial reduction of the thermal induced birefringence. The sensor, buffers and the shell were combined together without gluing by the well-distributed force which was produced by the shell: Figure 6.8.

Experiments on a MOCT with a glass SF6 sensor and an aluminum shell showed that the transducer provides an accuracy better than 0.3% in the temperature range from −30°C to 70°C for currents from 100 to 2,400 A.

Works are in progress on the development of the measurement systems, including both electric current and voltage sensors [12]. To compensate for temperature effects, the two different sensing materials were used simultataneously. One of them was a terbium gallium garnet (TGG) crystal, the second was FR-5 glass. The outputs of both MOCTs were divided and the ratio was used to determine the sensor temperature and hence to provide the necessary compensatory gain for the main TGG-derived current measurement.

6.2.3 MOCTs Based on Transparent Ferromagnets

Transparent ferromagnets provide a substantial increase in measurement sensitivity (by several orders of magnitude). Transparent ferromagnets are

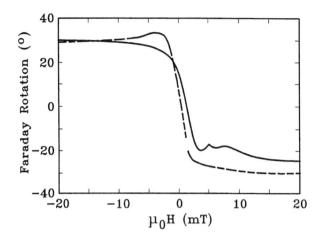

Figure 6.8 Magneto-optic response function in a sample of $Gd_{3-x}Bi_xFe_{5-y}Ga_y$ garnet. [17].

characterized by very high specific Faraday rotation, greatly exceeding Faraday rotation in diamagnets. In ferromagnets the polarization rotation is not always a linear function of the applied magnetic field as it is for para- and diamagnets, since the rotation is affected by a domain structure. For magnetic fields lower than the saturation field, the magneto-optical sensitivity S' is defined as [4]

$$S' = (F_{sat}/H_{sat})N_{eff}/l \qquad (6.26)$$

which is equivalent to the Verdet constant for diamagnets (N_{eff} is the demagnetizing factor). For illustration, we can compare the value of the Verdet constant for a diamagnetic sensor glass SF-57: $V = 2.84 \times 10^{-4}$ deg/A and sensitivity of the yttium iron garnet $Y_3Fe_5O_{12}$ (YIG) and 0.7 deg/A, respectively, for the wavelength $\lambda = 1.3$ μm [4].

The YIG possesses a weak cubic magnetocrystalline anisotropy, and the action of an external magnetic field on the crystal results mainly in the rotation of magnetization. Therefore, its frequency response is exceptionally good: The relative response does not change with frequency up to the gigahertz band. Very good frequency response also characterizes other compositions, for example, the Bi-substituted garnet $Y_{3-x}Bi_xFe_5O_{12}$ in which it is combined with the enhanced sensitivity (the latter, however, exhibits hysteresis behavior at low magnetic fields).

The high sensitivity takes place in the transparent ferromagnets with strong uniaxial magnetic anisotropy possessing very high values of specific

Faraday rotation, and numerous works are being undertaken to develop MOCTs on these crystals. The highest sensitivity has been measured in the $Gd_{3-x}Bi_xFe_{5-y}Ga_y$ garnet: 25 degrees/mT (Figure 6.8) [17]. However, that value was observed only in a very narrow range of magnetic fields: $\mu_0 H < 1$ mT.

An action of the magnetic fields on uniaxial ferromagnets results mostly not in the rotations of magnetization vectors but in motion of domain walls: Domains with spontaneous magnetization parallel to the field grow in width at the expense of oppositely magnetized domains. The motion typically occurs much slower than the rotation of the magnetization vector. Therefore, sensors built from uniaxial ferromagnets have a substantially lower frequency limit than ferromagnets with weak anisotropy. Nevertheless, it is possible to substantially expand the frequency range of magnetic field measurements by using an optical waveguide geometry [18]. In such a geometry, light propagates within the plane of the garnet film with perpendicular uniaxial magnetic anisotropy. A magnetic field applied in the plane of such a film rotates equally the magnetization of both types of domains toward the applied field. Thus, magnetization rotation, as opposed to the motion of domain walls, is the dominant response to in-plane magnetic fields. Experiments have shown that a 100-μm-thick film of the composition $(BiTb)_3(FeGa)_5O_{12}$ in a waveguide geometry exhibits a virtually flat frequency response up to 1 GHz. The waveguide geometry makes it possible to take full advantage of large Faraday rotation values achieved in the epitaxial garnet films. Problems with development of the corresponding magnetic field sensors are associated with low coupling efficiency and birefringence.

The majority of works on epitaxial garnet films for sensor applications make use of the conventional Faraday-effect geometry, in which light is incident normally on the sensor (and parallel to the direction of the magnetic field). In the case of uniaxial ferromagnets, the character of the Faraday rotation differs from that for nonmagnetic materials. Consider the Faraday rotation by a multidomain structure when domain walls are aligned perpendicular to the surface. The oppositely magnetized domains rotate polarization of the transmitted light at the same angles but in opposite directions. Because in the absence of resulting magnetization the widths of the oppositely magnetized domains are equal to each other, the resulting rotation of the polarization plane of transmitted light equals zero. When some external magnetic field changes the sizes of the domains, the resulting rotation of the polarization plane differs from zero and grows with the field up to the value corresponding to the Faraday rotation of the monodomain state—when all domains with a magnetization antiparallel to the field direction disappear, and the whole ferromagnet becomes magnetized in the direction of the field. Contrary to all other media, absolute values of Faraday rotation in any point of the uniaxial ferromagnet are the same, and the action of the external magnetic field consists in redistribution of the sizes of areas with different senses of Faraday rotation, causing a resultant rotation of the polarization plane of the transmitted light beam. When the widths of the domains are comparable to the wavelength of light (as it is in the case of garnets), a magneto-optical diffraction of light on the domain stripe structure takes place. The polarization of the diffracted light beams is perpendicular to the initial polarization (i.e., polarization of the incident light) and does not depend on the applied magnetic field. The polarization of the zeroth-order beam, however, is a function of the applied magnetic field. It coincides with the initial polarization when the resulting magnetization is zero and rotates when the widths of the oppositely magnetized domains become unequal; the maximum rotation angle corresponds to the monodomain sample. Thus, by mounting a diaphragm transmitting only the zeroth-order beam and measuring the polarization of the transmitted light, we obtain a value of the test magnetic field. However, the problem of the temperature dependence of the Faraday rotation still remains [19–21].

Furthermore, using the conventional Faraday setup even in the case of ferromagnets does not provide sensitivity (ratio of Faraday rotation to magnetic field) sufficient for many applications. A number of investigations have been devoted to the increase of the sensitivity of conventional MOCTs. One of the best values of sensitivity obtained in the U.S. National Institute of Standards and Technology and Deltronic Crystal Industries, Inc. equals 25 degrees/mT. But this value took place only in a very limited range of magnetic fields, namely in the range < 1 mT, and sharply decreased outside this range.

The drawbacks of all the methods described here are associated with the fact that the value of the magnetic field is obtained from the measurements of polarization rotation. Results of those measurements are very sensitive to the temperature of the magneto-optical material, to the birefringence, and to the variations of the output power and wavelength of the light source. To transmit light from the sensor to the photoreceiver, optical fibers are used. Because the main parameter is the polarization of light emerging from the sensor, the results of measurements are very sensitive to polarization distortions caused by the optical fibers. Last but not least, in all existing methods, the measured signal—usually the intensity of light transmitted through the analyzer—is a substantially nonlinear function of the magnetic field. The measured light intensity is a nonlinear (cosine) function of the

Faraday rotation. In addition, in ferromagnets the Faraday rotation itself is a nonlinear function of the magnetic field (due to nonlinearities of magnetization curves and hysteresis).

6.2.4 MOCTs With Direct Registration of the Domain Wall Positions

New ways of solving these problems consist in the direct measurements of the domain walls displacements caused by the applied magnetic field. The domain wall motion in so-called canted ferromagnets—orthoferrites—is described in [22]. In a wide frequency range, this motion is independent of the temperature of the orthoferrite crystal providing opportunity for elimination of corresponding uncertainties. The results of measurements are also insensitive to a large number of factors affecting conventional MOCTs. Variations of the Faraday rotation, of the sensor's transparency, and of the output power and wavelength of the light, and polarization distortions in the optical fibers can cause changes in the sensitivity of the measurements. However, they do not affect the measured parameter, the position of the domain wall image.

Figure 6.9 is a photo of the two-domain structure in the plate of yttrium orthoferrite. This domain structure is obtained by mounting two small magnets with opposite (with respect to the plate's surface) magnetization directions near the top and the bottom of the plate. The structure is visualized by means of conventional Faraday effect techniques, the plate is illuminated by linearly polarized light. Behind the plate, an analyzer and a camera are

Figure 6.9 Two-domain structure in the plate of yttrium orthoferrite.

mounted. The analyzer is rotated to a position in which it extinguishes light passing through one (lower) domain. Because the senses of Faraday rotation by the adjacent domains (characterized by mutually opposite magnetization directions) are different, the analyzer transmits light passing through the other (upper) domain.[2]

When a magnetic field of the measured current is acting on the plate, the position of the domain wall, that is, of the boundary between "light" and "dark" domains, changes in accordance with the field, while the Faraday rotation in the domains and thus the domains' contrast remains unchanged. By monitoring the position of the domain wall image, we obtain the value of the magnetic field of the test current. The principle of the measurement is illustrated in Figure 6.10. The photo in the figure was obtained when the magnetic field of the measured current acted on the domain structure shown in Figure 6.9. The measured current represented rectangular pulses of 300 kHz frequency and of different amplitudes. The domain wall moves up and down synchronously with the change of the direction of the current.

In Figure 6.10, the zone of the domain wall motion is seen as a zone of intermediary brightness. The lower, black and the upper, white parts of

Figure 6.10 Domain wall oscillations in the magnetic field of 400 A/m. Frequency: 300 kHz.

---

2. It should be mentioned that high contrast of the domain structure is obtained only in orthoferrite plates cut normally to their optical axes. If light propagates along another direction, the character of the Faraday rotation in the plate substantially changes with subsequent reduction of the contrast of the domain structure's image.

the photo represent parts of the sample permanently occupied by "dark" and "light" domains, respectively. The mean part of the photo, characterized by the intermediary brightness, represents the part of the sample alternately occupied by the "light" and "dark" domains. In other words, this is the range of the domain wall motion. We can see that the contrast of the domain structure does not depend on the value of the acting magnetic field.

The positions of the upper and lower boundaries of the range of the domain wall motion are proportional to the amplitudes of the current pulses. By measuring the size of the intermediary contrast zone, we can obtain the value of the current amplitude. With an ordinary ruler, we can measure currents from dc to hundreds of kilohertz! (In the case of dc current measurements, the position of the boundary between the images of two domains is measured.)

The scheme of the MOCT based on the direct registration of the domain wall position (DRDWP) is shown in Figure 6.11. A polarized light beam from the source (1) illuminates the sensor, the orthoferrite plate (2). By means of two small magnets (not shown), a two-domain structure is formed in the plate. The plate is located in the vicinity of the conductor (3), which carries the test current. Light emerging from the sensor propagates through the analyzer (4) and is incident on the position-sensitive photoreceiver (5). In the plane of the photoreceiver, the image of the domain structure is similar to that shown in Figure 6.9. The task of the photoreceiver is to determine the boundary between the images of the two parts of the domain structure of different brightness—of the domain wall position.

The maximum difference between the intensities of light transmitted through the oppositely magnetized domains and the analyzer is obtained in

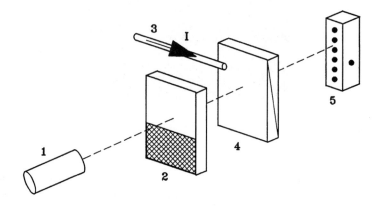

Figure 6.11 Scheme of the MOCT based on the DRDWP.

the light modulator mode when the analyzer (4) makes an angle of 45 degrees with the polarization of the incident light. Light from the source (1) to the sensor and from the analyzer (4) to the photoreceiver (5) can be transmitted via optical fibers. The results of measurements are not affected by the polarization distortions caused by the fibers.

The velocity of the domain wall motion depends on the applied magnetic field; in yttrium orthoferrite, it can reach cosmic values, up to 20 km/s. In the range of small magnetic fields, up to 100 mT, the dependence of the domain wall velocity on the magnetic field is close to linear. For slowly changing electric currents, up to the kilohertz range, the instantaneous positions of the domain wall (and thus the signal from the photoreceiver) are linear functions of the electric current. At higher frequencies, the domain wall's instantaneous positions lag from the acting field, and the corrections given in [22] have to be used. Another way of measuring fast-changing magnetic fields is to measure the domain wall velocity [23] which can also be performed by the DRDWP method.

Up to now, only sensors based on the Faraday effect have been described. Application of the Kerr effect provides much smaller angles of polarization rotation. Therefore, results of magnetic field measurements by means of the Kerr effect usually are characterized by a rather low signal-to-noise ratio. Nevertheless, there are works on the improvement of Kerr effect techniques for magnetic field and electric current measurements. In [24] a sensitivity of 0.3 A/m is reported, with a sampling time of 2 seconds. In [25] investigations on cerium monopnicides are described. Those crystals possess very high figures of merit and could be used as Kerr effect sensors for magnetic fields. Currently used materials can operate only at a very low temperature (1.5 K).

6.3 Geometric Measurements

Small displacements of the sensing element between two magnets down to 1 μm, corresponding to the changes of the Faraday rotation of 0.01 degree, can be measured by use of a technique for the high sensitive measurements of the Faraday rotation [26]. The technique employs a rotating polarizer and a sensitive electronic system that detects the phase difference between the main optical beam that passes through the sensing element and a reference beam. That method, however, substantially restricts sampling rates of measurements.

In [27], a scanning beam collimator method combined with a centerposition detection technique was used to measure the direction of the light beam. Other examples of measurements in which precise, high-speed determination of light spot positions is crucial include measurements of positions of robot manipulators, motion capture, and range-imaging sensors [28]. Existing photoreceivers, however, do not provide spatial resolution and speed of measurements necessary for many applications [29].

In magneto-optical measurements of light spot positions [30–32], the moving knifelike diaphragm is formed by the domain structure in a uniaxial ferromagnet and an analyzer. That method has been used for measurements of the atmospheric refraction angle of light beams at long distances, where the limiting factor is the signal-to-noise ratio of the photoreceiver signal [33].

References

[1] Tabor, W. J., and F. S. Chen, "Electromagnetic Propagation Through Materials Possessing Both Faraday Rotation and Birefringence: Experiments With Ytterbium Orthoferrite," *J. Appl. Phys.*, Vol. 40, 1968, pp. 2760–2765.

[2] Rogers, A. J., J. Xu, and J. Yao, "Vibration Immunity for Optical-Fiber Current Measurements," *J. Lightwave Technol.*, Vol. 13, 1995, pp. 1371–1377.

[3] Rochford, K. B., A. H. Rose, and G. W. Day, "Magneto-Optic Sensors Based on Iron Garnets," *IEEE Trans. Magn.*, Vol. 32, 1996, pp. 4113–4117.

[4] Wagreich, R. B, and C. C. Davis, "Accurate Magneto-Optic Sensitivity Measurements of Some Diamagnetic Glass and Ferrimagnetic Bulk Crystals Using Small Applied AC Magnetic Fields," *IEEE Trans. Magn.*, Vol. 33, 1997, pp. 2356–2361.

[5] Chu, B. C. B., Y. N. Ning, and D. A. Jackson, "Optical Current Comparator for Absolute Current Measurement," *Sensors Actuators A*, Vols. 37–38, 1993, pp. 571–576.

[6] Jones, G. R., et al., "Faraday Current Sensing Employing Chromatic Modulation," *Optics Commun.*, Vol. 145, 1998, pp. 203–212.

[7] Aspey, R. A., et al., "Elliptical Polarization Effects in a Chromatically Addressed Faraday Current Sensor," *Meas. Sci. Technol.*, Vol. 10, 1999, pp. 25–30.

[8] Yi, B., et al., "A Novel Bulk-Glass Optical Transducer Having an Adjustable Multiring Closed-Optical-Path," *IEEE Trans. Instrum. Meas.*, Vol. 47, 1998, pp. 240–243.

[9] Wilsch, M., P. Menke, and T. Bosselmann, "Magneto-Optic Current Transformers for Applications in Power Industry," *Proc. 2nd Cong. Optical Sensor Technology OPTO'96*, Leipzig, Germany, Sept. 1996, pp. 237–242.

[10] Tang, D., et al., "Annealing of Linear Birefringence in Single-Mode Fiber Coils: Application to Optical Fiber Current Sensors," *J. Lightwave Technol.*, Vol. 9, 1991, pp. 1031–1037.

[11] Ning, Y. N., et al., "Recent Progress in Optical Current Sensing," *Rev. Sci. Instrum.*, Vol. 66, 1995, pp. 3097–3111.

[12] Cruden, A., et al., "Optical Crystal Based Devices for Current and Voltage Measurements," *IEEE Trans. Power Delivery*, Vol. 10, 1995, pp. 1217–1223.

[13] Cruden, A., et al., "Compact 132 kV Combined Optical Voltage and Current Measurement System," *IEEE Trans. Instrum. Meas.*, Vol. 47, 1998, pp. 219–223.

[14] Song, J., et al., "A Prototype Clamp-On Magneto-Optical Current Transducer for Power System Metering and Relaying," *IEEE Trans. Power Delivery*, Vol. 10, 1995, pp. 1764–1770.

[15] Li, G., et al., "Sensitivity Improvement of an Optical Current Sensor With Enhanced Faraday Rotation," *J. Lightwave Technol.*, Vol. 15, 1997, pp. 2246–2252.

[16] Ma, X., and C. Luo, "A Method to Eliminate Birefringence of a Magneto-Optic Current Transducer With Glass Ring Sensor Head," *IEEE Trans. Power Delivery*, Vol. 13, 1998, pp. 1015–1019.

[17] Deeter, M. N., et al., "Sensitivity Limits to Ferrimagnetic Faraday Effect Magnetic Field Sensors," *J. Appl. Phys.*, Vol. 70, 1992, pp. 6407–6409.

[18] Deeter, M. N., et al., "Novel Bunk Iron Garnets for Magneto-Optic Magnetic Field Sensing," *IEEE Trans. Magn.*, Vol. 30, 1994, pp. 1464–1468.

[19] Kamada, O., "Magneto-Optical Properties of (BiGdY) Iron Garnets for Optical Magnetic Field Sensors," *J. Appl. Phys.*, Vol. 79, 1996, pp. 5976–5978.

[20] Shirai, K., "Magnetic garnet single crystal for measuring magnetic field intensity and optical type magnetic field intensity measuring apparatus," European Patent 0 521 527 A2, 1996.

[21] Imamura, M., M. Nakahara, and S. Tamura, "Output Characteristics of Field Sensor Used for Optical Current Transformers Applied to the Flat-Shape Three-Phase Bus-Bar," *IEEE Trans. Magn.*, Vol. 33, 1997, pp. 3403–3405.

[22] Didosyan, Y. S., et al., "Magnetic Field Sensor by Orthoferrites," *Sensors Actuators*, Vol. A59, 1997, pp. 56–60.

[23] Didosyan, Y. S., "Measurements of Domain Wall Velocity by the Dark Field Method," *J. Magn. Magn. Mater.*, Vol. 133, 1994, pp. 425–428.

[24] Oliver, S., et al., "Magnetic Field Measurements Using Magneto-Optic Kerr Sensors," *Opt. Eng.*, Vol. 33, 1994, pp. 3718–3722.

[25] Pittin, R., and P. Wachter, "Cerium Compounds: The New Generation Magneto-Optical Kerr Rotators With Unprecedented Large Figure of Merit," *J. Magn. Magn. Mater.*, Vol. 186, 1998, pp. 306–312.

[26] Villaverde, A. B., E. Munin, and C. B. Pedroso, "Linear Displacements Sensor Based on the Magneto-Optical Faraday Effect," *Sensors and Actuators A*, Vol. 70, 1998, pp. 211–218.

[27] Zeng, L., H. Matsumoto, and K. Kawachi, "Scanning Beam Collimator Method for Measuring Dynamic Angle Variations Using an Acousto-Optic Deflector," *Opt. Eng.*, Vol. 35, 1996, pp. 1662–1667.

[28] Bouhacina, T., et al., "Oscillation of the Cantilever in Atomic Force Microscopy: Probing the Sample Response at the Microsecond Scale." *J. Appl. Phys.*, Vol. 82, 1997, pp. 3652–3660.

[29] Schaefer, P., et al., "Accuracy of Position Detection Using a Position-Sensitive Detector," *IEEE Trans. Instrum. Meas.,* Vol. 47, 1998, pp. 914–919.

[30] Gaugitsch, M., et al., "Investigations of a Position Resolving Light Modulator for Refraction Studies," *J. Appl. Electromagnet. Mech.,* Vol. 7, 1996, pp. 11–20.

[31] Didosyan, Y. S., and H. Hauser, "Anwendungen Transparenter Magnete," *Elektrotechnik und Informationstechnik,* Vol. 115, 1998, pp. 378–381.

[32] Didosyan, Y. S., et al., "Application of Orthoferrites for Angular Measurements," *ISSE'99 Proc.,* Dresden, Germany, May 1999, pp. 204–207.

[33] Gaugitsch, M., and H. Hauser, "Optimization of a Magneto-Optical Light Modulator, Part I: Modeling of Birefringence and Faraday Effect," *J. Lightwave Technol.,* Vol. 17, 1999, pp. 2633–2644.

7

Resonance Magnetometers
Fritz Primdahl

7.1 Magnetic Resonance

Nuclear magnetic resonance (NMR) is a common term for a wide range of phenomena used for imaging diagnostics in medicine, spectroscopic investigations of materials and compounds in chemistry, and absolute measurement of weak magnetic fields.

A coherent ac-magnetic NMR signal can be picked up by a coil surrounding a sample of atomic nuclei after suitable excitation. The various signal decay times in a constant large magnetic field are used for imaging diagnostics of different types of biological tissue, and the enhancements of specific ac frequencies, while the sample is exposed to a changing magnetic field, are used for chemical spectroscopy.

For precise measurement of the magnitude of a weak magnetic field, the ac frequency emitted by a suitably excited and well-defined sample of atomic nuclei is determined in the nuclear resonance magnetometers. The resonance frequency is proportional to the magnetic field, and of all the stable nuclei, the proton has the largest proportionality constant or gyromagnetic ratio. The magnetic resonance of protons (in an aqueous sample of spherical shape) constitutes an atomic reference for the SI units of current (ampere) and magnetic field (tesla).

Electron spin resonance (ESR) is the analogous phenomenon for electrons. It also finds wide application in medicine, material sciences, and chemistry. The much higher spin frequency (about 600 times that of protons) offers the possibility of constructing highly sensitive and fast-responding scalar magnetometers. Where the classical proton magnetometer is excited by a dc-magnetic field followed by measurement of the frequency of the decaying nuclear spin signal, the optically "pumped" electron spin magnetometers use light in resonance with an optical spectral line of the sample and produce a continuous ESR signal.

The Overhauser effect proton magnetometer combines the two phenomena. It uses a radio frequency (RF) magnetic signal for ESR excitation of the electrons, which through collisions transfer their excitation to the protons. Continuous excitation of the protons is thereby established, which allows for a continuous proton resonance signal.

7.1.1 Historical Overview

The use of NMR was initiated by the publication of the now classical paper by Block [1]. Between 1948 and 1954, the proton-free precession principle was developed by Varian Associates Research Laboratory [2] into an instrument offering up to one measurement per second of the scalar value of a weak magnetic field (see [3]).

Soon thereafter, the proton magnetometer was introduced at geomagnetic observatories for precise measurement of the Earth's scalar magnetic field, as a welcome absolute standard that replaced the older methods, which used induction coils, balanced magnetic needles. and calibrated permanent magnets [2, 4–7].

Very early on, proton magnetometers were carried to an altitude of about 33 km by a balloon and later launched into the Earth's ionosphere onboard a "rockoon," a small sounding rocket fired from a high-altitude balloon [8]. Other sounding rocket experiments carrying proton magnetometers abounded, spanning from Burrows [9] to Olesen [10]. The popularity of the proton magnetometer for space experiments was, to a large part, due to the fact that no attitude information was needed to interpret the data, because only the scalar value (the magnitude) of the field was measured.

Soon after the first Soviet Sputnik, in 1957, Sputnik-3 was launched on May 15, 1958, and it placed the first (fluxgate) magnetometer in space. In 1961, the U.S. Vanguard III satellite carried a proton scalar magnetometer. Later, many U.S. and Russian satellites used proton magnetometers and cesium magnetometers for monitoring the Earth's field (Cosmos-26/49/321 and OGO-2/4/6) [11].

In 1953, Overhauser published his discovery of the proton spin alignment effect by coupled RF-ESR polarization [12]. For geophysical prospecting and at magnetic observatories, the continuously oscillating and less-power-consuming Overhauser proton magnetometer was extensively used (this chapter's author had the opportunity to see an operating Overhauser magnetometer for airborne survey at the Geological Survey of Canada in 1966), and the development of a field instrument [13] for prospectors made this scalar magnetometer even more popular.

An Overhauser magnetometer made by LETI, France [14], was launched on the Danish geomagnetic mapping satellite Ørsted in February 1999, and a second LETI Overhauser instrument was launched in July 2000 on the German CHAMP fields and potentials satellite.

Following Kastler's description of using optical techniques to produce magnetic polarization of electrons in a vaporous sample [15], the development of optically pumped scalar magnetometers progressed from 1957. Where protons have the largest nuclear gyromagnetic ratio converting field to signal frequency of about 42.5 MHz/T, optical magnetometers using ESR have much larger conversion constants of between 3.5 and 7.0 GHz/T for the alkali metals and up to 28 GHz/T for the metastable He^4 optical magnetometer.

Besides being continuously oscillating, optical magnetometers thus have the advantages of a higher frequency response and less sensitivity to platform rotation compared to the proton magnetometers [16].

Optically pumped scalar magnetometers are used for land, sea, and airborne surveys for prospecting and to some extent at geomagnetic observatories for measuring the field magnitude. They have also flown on many satellites. The absolute magnetic field standard onboard the NASA Magsat Earth's field mapping satellite (1979–1980) was a Cs^{133} optically pumped scalar magnetometer from Varian Associates [17], and a scalar metastable He^4 magnetometer from Jet Propulsion Laboratory is planned to fly onboard the Argentine-U.S. SAC-C Earth observation satellite as part of the Danish magnetic mapping payload (MMP).

Vector measurements using scalar magnetometers have been developed for geomagnetic observatories by sequentially adding stable bias fields to the background field in two or three orthogonal directions. Proton magnetometers have been used, but because the field biasing sequence must be completed in a short time (to confidently follow the changes of the Earth's field), the faster responding optical magnetometers seem to be preferred [3, 18].

The Rice University Group has used rocket launches of optical vector magnetometers with bias coils extensively to observe magnetic field-aligned electric currents in the Earth's ionosphere [19, 20]. A metastable He^4 vector magnetometer from the Jet Propulsion Laboratory is one of the two magnetometers onboard the Ulysses Mission in polar orbit about the Sun at 1AU [21]. In deep space, the magnetic fields are too low to be measured directly using a resonance magnetometer, but the addition of bias fields brings the combined magnitude into the instrument's measuring range.

7.1.2 Absolute Reproducibility of Magnetic Field Measurements

The Système International d'Unités (SI) is the internationally adopted unit system for measurements based on the equations of physics and on an international agreement on actual basic physical samples (like the kilogram) and on adopted exact values of specific constants of nature (such as the velocity of light).

It is an interesting and at times intriguing experience to study the development and follow the discussions in this field of physics and technology. At the turn of the last century, an abundance of "practical" units existed, based on reference values (e.g., the Weston cell voltage, now replaced by the Josephson effect) and agreed upon at large international conferences. However, such reference values proved to conflict with other internationally agreed-upon reference values and constants in the application of the laws of physics [22].

The need is evident for universally and reliably reproducible experiments providing time-invariant sample values (e.g., a definite voltage for calibration of voltmeters) and being maintained at nationally accessible laboratories. But as technology advances and the measurement accuracies in all branches of physics progress, the industrial and scientific communities have to accept slight changes of the reproducible reference values at irregular time intervals to keep the SI and avoid confusion like the one that existed about one hundred years ago.

Such a reference value used for establishing the proper SI measure of magnetic field (the tesla) and also used for establishing the SI unit for electrical current (the ampere) is the proton gyromagnetic ratio, γ_p'. With that atomic constant, a measurement of the magnitude of a magnetic field is converted to the much easier and precise determination of the frequency of an induced ac signal. The prime (') mark is there to remind us that γ_p' is the proton gyromagnetic ratio in pure water and in a spherical container. The chemical liquid has a slight influence on the frequency and so has the shape of the vessel [23, 24]. That is why γ_p' is different from the single proton γ_p.

The currently accepted value is

$$\gamma_p' = 2.675\ 152\ 55 \times 10^8 \text{ radians}/(s \cdot T) \tag{7.1}$$

as given in Cohen and Taylor [25]. The published uncertainty of that value is 81 radians/(s · T), or about 3×10^{-7}, corresponding to 15 pT in the Earth's field of 50,000 nT.

The accepted value in 1960 was $2.675\,13 \cdot 10^8$ and before that, $2.675\,23 \cdot 10^8$ was widely used [26]. At some time, different values for γ'_p in high fields (1T) and low fields (< 1 mT) were in use, but this has been reconciled by a slight change of the reference for the ampere [27].

The accuracy of the reference value of γ'_p is one thing, but the final measurement accuracy of a carefully constructed proton precession magnetometer is quite another. Primarily, the time standard for the frequency determination has to be sufficiently reliable and accurate, but that is fairly inexpensive to establish. Other sources of errors exist. If kerosene or hexane is used as the proton-rich liquid sample, the deviant chemical shifts of those liquids will change the instrument calibration (the difference between γ'_p and γ_p amounts to about 1.3 nT in the Earth's field). If the sample shape is nonspherical, then another shift at the 0.1-nT level may be introduced. If magnetic impurities (e.g., from the machining tools) are accidentally enclosed in the construction elements of the sensing head, a head orientation-dependent error may be introduced, similar to the effect of a nonspherical shape. The important conclusion here is that even a supposedly absolute instrument should be verified against a standard instrument or field.

7.2 Proton Precession Magnetometers

7.2.1 Mechanical Gyroscopes

To understand the basic principle of nuclear resonance magnetometers (and of ESR magnetometers), let us briefly consider the mechanical gyroscope. Figure 7.1 shows a toy gyroscope consisting of a fast spinning wheel with the axle supported at one end by a bearing, free to rotate about the support. When the spinning wheel (spin rate ω_s) is let loose, it starts to revolve about the vertical support in a slow precession with the rate Ω.

The key to understanding the motion is the inertial force acting on all parts of the spinning wheel. A differential volume at the edge of the wheel is kept in a circular orbit about the axis by the stress in the material, which delivers the centripetal force, just as the gravity of the Sun keeps a planet in orbit. The volume continues revolving as long as the breakdown limit of the material is not exceeded. If that happens, a catastrophic explosion

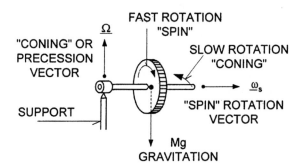

Figure 7.1 The mechanical gyroscope. The force of gravitation and the reaction from the support (not shown) constitute a mechanical torque, which makes the spinning wheel revolve slowly about the support.

of the wheel may result, demonstrating that we are dealing with real and very strong forces.

From the point of view of the wheel (i.e., in a coordinate system rotating with the wheel), it "feels" the inertial force, called the *centrifugal force*, tending to pull the wheel apart in all directions perpendicular to the axis of rotation. (Throw a stone, and your hand feels an inertial force: the stone's reaction against being accelerated.) Figure 7.2 shows the situation when the wheel rotates about the body axis and perfect balance exists. The centrifugal forces are directed symmetrically outward at right angles to the axis, and all the parts of the wheel are kept in place by internal stresses in the material.

A quite different situation exists if the wheel rotates about a skew axis not aligned with the body symmetry axis. Then the centrifugal forces no

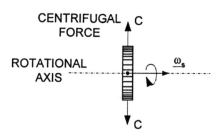

Figure 7.2 A spinning wheel in perfect balance while rotating about the body symmetry axis. The centrifugal forces **C** pass through the body center of gravity, and the parts of the wheel are forced to revolve about the axis by the stress forces in the material.

longer pass through the center of gravity, and the balance of forces is upset. Figure 7.3 shows this situation; clearly the centrifugal forces tend to rotate the wheel into a position symmetrical about the axis of rotation. The inertial centrifugal forces result in a mechanical torque on the wheel trying to twist the body axis parallel to the rotation axis.

If the spinning wheel is rigidly mounted on a skew axle, severe vibrations result, which ultimately may destroy the system. That underlines the importance of balancing a high-speed revolving machine element to prevent destruction. However, a symmetrically mounted spinning wheel can be made to perform a stable rotation about an axis slightly different from the body axis. This follows from the fact that rotation vectors can be added vectorially to form a resultant instantaneous rotation vector. Figure 7.4 shows how that acts on the toy gyroscope.

The wheel is spinning at the rate ω_s, and at the same time it is slowly revolving about the vertical axis of the support with the coning rate Ω (here shown displaced to the center of gravity). The combination of these two rotations is a resulting instantaneous rotation vector ω, which is twisted upward a small angle relative to the symmetry axis of the wheel. The tangent to this small angle is Ω/ω_s.

The coning rate Ω adjusts to a magnitude so that the centrifugal forces exactly balance the torque from the gravitation. A formal description uses the moment of inertia, I, about the axis:

$$I = \Sigma \Delta M r^2 \qquad (7.2)$$

Here ΔM is the mass of a differential volume of the wheel, r is the perpendicular distance of the volume from the axis, and the summing is extended over the total mass of the wheel. The angular momentum **L** is then given by:

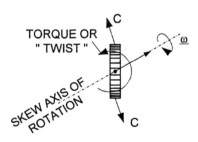

Figure 7.3 Rotation about a skew axis. The centrifugal forces miss the center of gravity and combine to form a mechanical torque on the wheel.

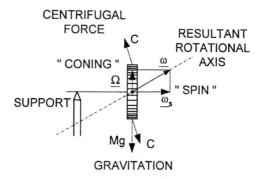

Figure 7.4 The spinning (ω_s) and coning (Ω) gyroscope supported at one end of the axle (free to spin and rotate in all directions) and acted on by the mechanical torque from gravity (*M***g**) and the reaction of the support. The resulting rotation vector ω deviates from the symmetry axis and creates a balancing torque by the centrifugal forces **C**.

$$\mathbf{L} = I \cdot \boldsymbol{\omega}_s \qquad (7.3)$$

The mechanical torque **T** acting on the wheel from the force of gravity is

$$\mathbf{T} = \mathbf{l} \times M\mathbf{g} \qquad (7.4)$$

where **l** is the vector from the supporting point to the center of gravity and $M\mathbf{g}$ is the gravitational force on the wheel (M is the total mass and **g** is the acceleration of gravity). From the second Newton's law follows the equation of motion:

$$\dot{\mathbf{L}} = \mathbf{T} \qquad (7.5)$$

stating that the time change of the angular momentum equals the mechanical torque vector. For example, if we know the angular momentum (or "spin") **L** of the wheel and observe the rate of precession (or coning), Ω, then (knowing the total mass M and the length l of the axle), the acceleration of gravity **g** can be calculated.

7.2.2 Classic Proton-Free Precession Magnetometer

The first and simplest of the nuclear precession magnetometers exploits the fact that protons possess a magnetic moment as well as an angular momentum (spin). Quantum mechanics tells us that the spin **L** and the magnetic dipole moment μ_p of the proton are atomic constants; they are parallel and related by a fixed scalar constant called the proton gyromagnetic ratio, γ_p:

$$\mu_p = \gamma_p \cdot \mathbf{L} \tag{7.6}$$

Apart from that, nothing is said about the internal structure of the spinning proton, but a mental picture illustrating the collective behavior in the classical limit of a large sample of protons may be like the one shown in Figure 7.5. The proton behaves like a small mechanical gyroscope with the axle made of a bar magnet having north and south poles of magnitude $\pm m$ separated by the vector **l**. If the proton is placed in an external field **B**, the poles feel the forces $\mathbf{F} = \pm m \cdot \mathbf{B}$ constituting a mechanical torque $\mathbf{T} = m\mathbf{l} \times \mathbf{B} = \mu_p \times \mathbf{B}$.

The proton magnetic moment is $m\mathbf{l} = \mu_p$, and from a table of quantum mechanical constants, we have $\mu_p \cong 1.41 \times 10^{-26}$ Am². Using the value of $\gamma_p \cong 2.675 \times 10^8$ radians/s · T, then the proton angular momentum (or spin) is calculated to be

$$L \cong 5.27 \times 10^{-35} \text{ kg} \cdot \text{m}^2/\text{s} \tag{7.7}$$

which is equal to $\frac{1}{2}\hbar$, as the proton spin should be. (Planck's constant $\hbar \cong 1.0546 \cdot 10^{-34}$ J · s.)

In the coordinate system of Figure 7.5, the angular momentum **L** is precessing about the $\hat{z}$-axis:

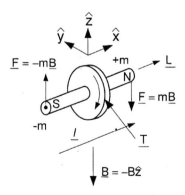

Figure 7.5 Protons have angular momentum or spin **L** and magnetic moment $\mu_p = m\mathbf{l}$, much like a gyroscope with a bar magnet as axle. An external field *B* gives torque **T** on the moment.

$$\mathbf{L} = L(\cos(\omega t)\hat{\mathbf{x}} + \sin(\omega t)\hat{\mathbf{y}}) \tag{7.8}$$

$$\boldsymbol{\mu}_p = \gamma_p L(\cos(\omega t)\hat{\mathbf{x}} + \sin(\omega t)\hat{\mathbf{y}}) \tag{7.9}$$

$$\dot{\mathbf{L}} = \omega L(-\sin(\omega t)\hat{\mathbf{x}} + \cos(\omega t)\hat{\mathbf{y}}) \tag{7.10}$$

$$\mathbf{T} = \boldsymbol{\mu}_p \times \mathbf{B} = \begin{bmatrix} \hat{\mathbf{x}} & \hat{\mathbf{y}} & \hat{\mathbf{z}} \\ \cos(\omega t) & \sin(\omega t) & 0 \\ 0 & 0 & -B \end{bmatrix} = \gamma_p LB(-\sin(\omega t)\hat{\mathbf{x}} + \cos(\omega t)\hat{\mathbf{y}}) \tag{7.11}$$

Using $\dot{\mathbf{L}} = \mathbf{T}$, we get

$$\omega L = \gamma_p LB, \text{ and } \omega = \gamma_p B \tag{7.12}$$

The proton magnetic moment rotates clockwise about the magnetic field B with the frequency

$$f = (\gamma'_p/2\pi)B \tag{7.13}$$

where γ'_p is a modified gyromagnetic ratio.

In a spherical sample of water, the value $\gamma'_p = 2.67515255 \times 10^8$ rad/s · T should be used, as explained above ($\gamma'_p/2\pi$ = 42.576375 MHz/T). The modification of the proportionality constant is not caused by any real change of the proton gyromagnetic ratio but because the external magnetic field B is slightly modified by the diamagnetism of water combined with the demagnetizing factor of the spherical shape of the sample. However, rather than computing the internal field B_i in the water sample and then correcting for shape and diamagnetism, the external field is obtained directly by using γ'_p.

When a volume of a proton-rich liquid is exposed to a strong dc-polarizing magnetic field $\mathbf{B}_p$, the proton magnetic moments tend to be aligned along $\mathbf{B}_p$ (Figure 7.6). That, however, is counteracted by thermal

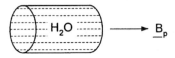

Figure 7.6 A sample of water is exposed to a strong polarizing dc-magnetic field $\mathbf{B}_p$. The field tends to align the magnetic moments of the protons.

agitation, resulting in the net alignment of only a small fraction of the protons. The fraction of aligned (or polarized) protons equals for $\mu_p B_p \ll kT$ the ratio of magnetic energy to the average thermal energy:

$$\text{Fraction of aligned protons} \approx \frac{\text{Magnetic Energy}}{\text{Thermal Energy}} \approx \frac{\mu_p B_p}{kT} \quad (7.14)$$

In a polarizing field $B_p = 14.4$ mT, the magnetic energy of a proton is $\mu_p B_p \cong 2.03 \times 10^{-28}$ J. The average thermal energy is $kT = 4.14 \cdot 10^{-21}$ J, where $k \cong 1.381 \cdot 10^{-23}$ J/K is Boltzmann's constant, and $T = 300$K is the absolute temperature. The magnetic energy is much smaller than the thermal energy, and so the fraction of aligned protons is

$$\frac{\delta n}{n} \approx \frac{\mu_p B_p}{kT} \cong 4.83 \cdot 10^{-8} \quad (7.15)$$

We can now calculate the magnetization of water exposed to a polarization field of 14.4 mT. One cubic meter of water weighs 1,000 kg, and one H_2O molecule weighs 18 atomic mass units (AMU), or $18 \cdot 1.67 \cdot 10^{-27}$ kg $= 3.01 \cdot 10^{-26}$ kg. Each water molecule has two protons, so the density of protons in water is

$$n \cong 2 \times 10^3 / 3.01 \times 10^{-26} \cong 6.64 \times 10^{28} \text{ protons/m}^3 \quad (7.16)$$

The resulting magnetization (or the magnetic moment per unit volume) for a polarization field of $B_p = 14.4$ mT is then

$$\begin{aligned} M_0 &= n \times (\delta n/n) \times \mu_p \\ &= 6.64 \times 10^{28} \times 4.83 \times 10^{-8} \times 1.41 \times 10^{-26} \text{ A/m} \quad (7.17) \\ &= 4.52 \times 10^{-5} \text{ A/m} \end{aligned}$$

M_0 is the saturation magnetization approached exponentially with a time constant of a few seconds in pure water after application of the polarization field $\mathbf{B}_p$. Suddenly removing the polarization field leaves the magnetization of the water $\mathbf{M}$ (approximately) in the direction of $\mathbf{B}_p$, and $\mathbf{M}$ then starts to rotate clockwise about the external field $\mathbf{B}$ with the frequency $f = (\gamma_p'/2\pi)B$. The changing magnetic field from the rotating magnetization then induces an ac signal in a pickup coil wound around the water sample.

The magnetization of the water decreases exponentially with time, so the decaying signal can be picked up for only a couple of seconds (Figure 7.7).

The largest signal is obtained if the pickup coil axis is oriented at right angles to the external **B**-field because **M** rotates about **B**. The polarization field should also be applied at right angles to the external field **B**, because that leaves the largest component of **M** perpendicular to **B**. The same coil can be used for polarization and for signal pickup, but if the coil is a cylindrical solenoid, a dead zone for **B** will exist close to the coil axis. First, because **M** will be left mostly along **B** with a very small component rotating about **B**, and second, as **M** is rotating about an axis close to the coil axis, the induced signal will be still smaller and disappear in noise.

Serson [6] introduced the omnidirectional toroid sensor, and the author used that shape (Figures 7.8 and 7.9) for sounding rocket experiments in the 1970s [10].

Regardless of the orientation, some parts of the coil are always perpendicular to the external **B**-field. The largest signal is obtained when the field is along the symmetry axis, with a factor-of-2 decrease in signal amplitude, when the field is in the plane of the ring. The signal strength goes as $(2 - \sin^2\alpha)$, where α is the angle between the toroid axis and the external **B**-field [28].

Because of its symmetry, the toroid sensor attenuates homogeneous ac noise fields from the environment; still, very strong noise signals may impair the proper functioning of the sensor. During a 1977 sounding rocket launch

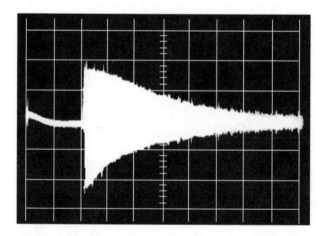

Figure 7.7 A decaying proton precession signal of about 2 kHz in the Earth's field. (Vertical scale: 1μV/div; horizontal scale: 1 s/div.)

Figure 7.8 Omnidirectional toroid proton magnetometer sensor. A hollow acrylic ring with kerosene and 1,000 turns of 1-mm-diameter aluminum wire. An external electrostatic screen improves the noise immunity of the toroid. (Overall dimension: 16 cm.)

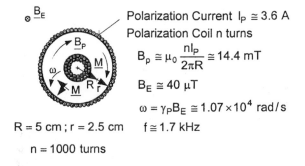

Polarization Current $I_P \cong 3.6$ A
Polarization Coil n turns

$$B_P \cong \mu_0 \frac{nI_P}{2\pi R} \cong 14.4 \text{ mT}$$

$B_E \cong 40 \ \mu T$

$\omega = \gamma_P B_E \cong 1.07 \times 10^4$ rad/s

$R = 5$ cm ; $r = 2.5$ cm $f \cong 1.7$ kHz

$n = 1000$ turns

Figure 7.9 Cross-sectional view of a toroid sensor after polarization by the field B_P. The magnetization **M** rotates about the direction of the Earth's field B_E with the angular velocity ω.

into the ionospheric E-region (95–120 km altitude) from Andenes, Norway, this chapter's author experienced a high level of auroral hiss, completely wiping out the proton signal from a toroid sensor. That phenomenon has been observed frequently onboard sounding rockets, particularly when the sensor is made of a single solenoid [L. J. Cahill, Jr., private communication, 1977]. Sensors made of two antiparallel solenoids are also relatively noise immune, but in contrast to the toroid sensor they still exhibit a null zone close to the common axis.

Immediately after the polarization field $\mathbf{B}_P$ is removed from the toroid proton-rich liquid sample by rapidly switching off the polarization current I_P, the sample magnetization $\mathbf{M}_0$ starts to precess about the Earth's field $\mathbf{B}_P$. The magnetization along the center circle of the ring is

$$M_{\text{axis}} = M_0 \cos \omega t \qquad (7.18)$$

The magnetic field from that magnetization follows concentric circles along the ring, and the magnitude is given by

$$B_M = \mu_0 (1 - D) M_{\text{axis}} \qquad (7.19)$$

where the demagnetizing factor is $D \cong 0$ along the ring. We then have

$$B_M \cong \mu_0 M_0 \cos \omega t \qquad (7.20)$$

The induced voltage in the toroid coil immediately after polarization is

$$A = \pi \cdot r^2 = 2.0 \cdot 10^{-3} \qquad (7.21)$$

$$n = 1{,}000 \qquad (7.22)$$

$$e(t) = nA\mu_0 \omega M \sin(\omega t) = 1.3 \ \mu\text{V} \qquad (7.23)$$

The signal-to-noise ratio depends on the toroid coil and on the amplifier input circuit. A representative example of the circuit is shown in Figure 7.10.

The toroid coil inductance is series-tuned to a frequency corresponding to the average value of the expected B-field magnitudes. For most applications, the coil losses are sufficient to give a suitable bandwidth for the range of fields to be measured. Assuming a total bandwidth of $\Delta f = 500$ Hz (corresponding to about $\pm 6 \ \mu$T), the noise in the $R_{\text{coil}} = 6.1 \Omega$ resistor is

$$e_{\text{noise}} = (4kT\Delta f R_{\text{coil}})^{1/2} = 7.1 \ \text{nV}_{\text{rms}} \qquad (7.24)$$

where k is Boltzmann's constant and T is the absolute temperature in Kelvin.

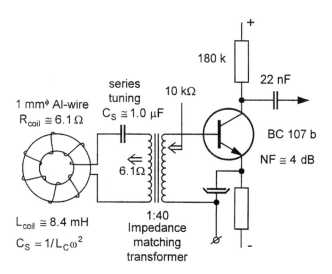

Figure 7.10 Example of an (old) coil impedance-matching input stage for the toroid sensor coil. The angular frequency ω corresponds to the average field to be measured.

The input transistor increases the noise by about 4 dB (the noise factor for the old BC107), resulting in e_{noise} = 11.3 nV$_{rms}$. The theoretical initial signal-to-noise ratio is then

$$e/e_{noise} = 1.3\ \mu V/11.3\ nV_{rms} = 115 \tag{7.25}$$

That number is clearly an overestimate, first because amplitude value is compared to rms noise, and second, because the sample magnetization decreases exponentially with time, and it might not even have reached the equilibrium value before switch-off of the polarization current. Koehler [29] gives the more realistic estimate of

$$S/N = e/e_{noise} \cong 40 - 50 \tag{7.26}$$

The switching off of the polarization current requires some special attention. If the current is removed too slowly, the sample polarization rotates into alignment with the Earth's field and no precession signal is observed. A too rapid switch-off (by just opening a relay contact) probably results in arcing because a large inductor is involved, and the energy in the magnetic polarization field has to be dissipated somewhere. For reliable operation, a controlled removal of the field (and the current) must be established.

During polarization, the sample is exposed to the $\mathbf{B}_{res}$, the combination of the polarization field $\mathbf{B}_P$ and the Earth's field $\mathbf{B}_E$ at right angles to each other (Figure 7.11).

From $\mathbf{B}_P(t) = \mathbf{B}_{P\max}$ to $\mathbf{B}_P(t) = 0$, the change of the direction φ of the resultant field $\mathbf{B}_{res}$ occurs with rapidly increasing speed. At first, the magnetization $\mathbf{M}$ follows $\mathbf{B}_{res}$, but at some point $\mathbf{B}_{res}$ starts to "run away" from the magnetization. That happens when the rate of change of the angle φ becomes larger than the natural angular velocity ω of the protons, if precessing in the instantaneous field $\mathbf{B}_{res}$.

$$\frac{d\varphi}{dt} \geq \omega_p = \gamma_p \cdot B_{res}; \; B_{res} = \sqrt{B_P^2 + B_E^2} \qquad (7.27)$$

$$\cot\varphi = \frac{B_P}{B_E}; \; \frac{d\cot\varphi}{dt} = -\frac{1}{\sin^2\varphi}\frac{d\varphi}{dt} = \frac{1}{B_E}\frac{dB_P}{dt} \qquad (7.28)$$

If $B_P(t)$ decreases linearly with time toward zero during the switching time T_S, then $dB_P/dt = -B_{P\max}/T_S$ is constant. Counting time t from the beginning of the switch-off operation, we have

$$B_P(t) = B_{P\max} \cdot \left(1 - \frac{t}{T_s}\right) \qquad (7.29)$$

Further, we can write (using $d\varphi/dt = \omega_P$)

$$\sin^2\varphi = \frac{B_E^2}{B_{res}^2}; \; \frac{B_E^2}{B_{res}^2} \cdot \frac{B_{P\max}}{B_E \cdot T_S} = \gamma_p B_{res} \qquad (7.30)$$

The time for switching off the polarizing field follows from the switch circuit parameters, and a reasonable value is $T_S \cong 1$ ms. That means that at

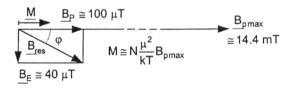

Figure 7.11 The polarizing field $\mathbf{B}_P$ decreases rapidly, and at some point the resultant field $\mathbf{B}_{res}$ starts to rotate by the angle φ away from the polarizing direction and toward the direction of the Earth's field $\mathbf{B}_E$.

the point where $\mathbf{B}_{res}$ runs away from $\mathbf{M}$, we have (see Figure 7.11 for the remaining numerical values)

$$B_{res} = \sqrt[3]{\frac{B_E B_{P\max}}{T_S \gamma_p}} = 129 \ \mu\text{T} \tag{7.31}$$

The magnitude of the polarizing field is reduced from $B_{P\max}$ to

$$B_{P0} = \sqrt{B_{res}^2 - B_E^2} = 123 \ \mu\text{T} \tag{7.32}$$

The time t_0 from the beginning of the switching sequence until runaway starts is

$$t_0 = T_S \cdot \left(1 - \frac{B_{P0}}{B_{P\max}}\right) = 0.991 \text{ ms} \tag{7.33}$$

At runaway, the rotation angle φ_0 of M away from being perpendicular to the Earth's field B_E is

$$\varphi_0 = A\tan\left(\frac{B_E}{B_{P0}}\right) = 18 \text{ degrees} \tag{7.34}$$

Because of the cubic root in the expression for B_{res}, B_{P0} and φ_0 are relatively insensitive to changes in T_S. A doubling of T_S means an increase in the start-runaway angle of a factor of only about 1.28, to 23 degrees.

An example of an electronic switching circuit is shown in Figure 7.12. A large number of relay or electronic switching circuits and combinations of mechanical and electronic systems have been used over the years. The switch shown in Figure 7.12 has been used successfully for six sounding rocket launches of proton magnetometers into the polar ionosphere in 1974 to 1977. In addition to the circuit shown, a small space-qualified relay disconnected the coil from the electronic switch and connected it to the amplifier input when the polarizing current was quenched, and the precession signal was available over the coil terminals. That was a protection of the amplifier input and extra security against having residual currents from the switch flowing in the coil during the measuring cycle. Such currents might perturb the external magnetic field.

At the start of the switching cycle, about 3.6A flow from the 22.4V battery via the transistor switch through the coil. The 2.2-kΩ shunt resistor

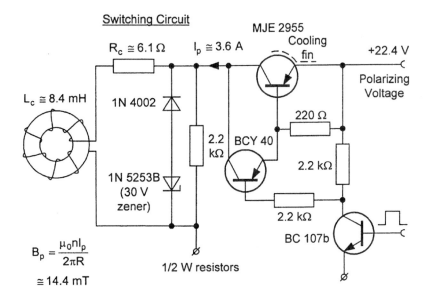

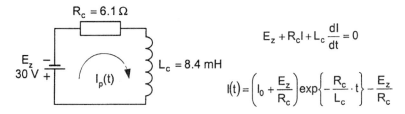

Figure 7.12 Electronic polarizing switch and equivalent circuit.

provides ground potential for the collectors of the switch transistors, and the diode in series with the 30V Zener is reverse polarized and open (disconnected). Immediately after opening the transistor switch at $t = 0$, the coil voltage jumps from +22.4V to about −30.7V, and the diode-Zener combination conducts, providing an alternative return path for the coil current. Save for the voltage over the coil resistance, the diode-Zener combination maintains a large but controlled voltage drop over the coil, preventing any damage to the switch transistors and ensuring a rapid dissipation of the coil magnetic energy into heat in the coil resistance and in the diode-Zener combination.

From the circuit parameters and using the equivalent circuit from Figure 7.12, the switching time t_0 is calculated. The time rate of change of the polarization current $I_P(t)$ is not strictly constant, but because of R_C it follows an exponential decrease from I_0 = 3.6A and approaching $I = -E_Z/R_C$ = −4.9A for $t \to \infty$ (see the equation in Figure 7.12). The zero crossing time t_0 for I is calculated as follows:

$$I = 0 \text{ for } t_0 = \frac{L_C}{R_C} \ln\left(1 + \frac{I_0 R_C}{E_Z}\right) \cong 0.756 \text{ ms} \quad (7.35)$$

That time is equal to the previously introduced switching time T_S. That means the assumption of a constant time rate of decrease of the polarization current is only a fair approximation because of the coil ohmic resistance, and the total switching time is, for the same reason, slightly less than the 1 ms used above. The time from when the $\mathbf{B}_{res}$ vector starts to run away from the magnetization $\mathbf{M}$ to the polarization current reaches zero is, of course, much smaller than T_S. It takes less than 10 μs from the runaway point until the polarizing field has completely disappeared.

The coil has some self-capacitance. When the polarizing current reaches zero and the series resistance of the diode-Zener combination becomes very high, the self-capacitance is charged to approximately the Zener voltage, and the self-resonance of the coil may be excited by the discharge of this voltage. That dampened oscillation may interfere with the proton precession signal if the coil Q is too large. The 2.2-kΩ shunt resistor (see Figure 7.12) also serves to lower the coil Q and increase the dampening of the self-resonance oscillation, which easily can be orders of magnitude larger than the precession signal.

The growth of the magnetization of the proton-rich liquid sample follows an exponential curve toward the saturation value after application of the constant polarization field. Similarly, an exponential decay of the magnetization toward the equilibrium value in the Earth's field follows after removal of the polarization. $M(t)$ grows as

$$M(t) = M_0 \cdot \left(1 - e^{-\frac{t}{T_1}}\right) \quad (7.36)$$

where T_1 is the "spin-lattice" relaxation time constant. Figure 7.13 shows the growth of magnetization in oxygen-free water. Dissolved oxygen in the water leads to a rapid decay of the proton signal. Different proton-rich chemical liquids have, as is to be expected, different relaxation time constants.

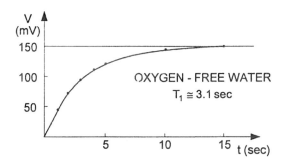

Figure 7.13 Amplified precession signal strength dependence of the polarization time. Growth of the magnetization with polarization time follows the same curve (after [30]).

$$\text{Distilled Water } T_1 \cong 2\text{-}3\text{s} \tag{7.37}$$

$$\text{Kerosene } T_1 \cong 0.5\text{s} \tag{7.38}$$

$$\text{Oxygen-free Water } T_1 \cong 3.1\text{s} \tag{7.39}$$

The decay of the magnetization after switch-off of the polarization goes exponentially toward the very small Earth's field equilibrium value as:

$$M(t) = M_1 \cdot e^{-\frac{t}{T_2}} \tag{7.40}$$

where M_1 is the start magnetization and T_2 is the "spin-spin" relaxation time. In general, $T_2 \leq T_1$, and the time constants are selected according to the application of the proton magnetometer. For ground-based use, the time constants often are chosen to be as long as possible to get as much time as possible to measure the precession frequency. For sounding rockets, a short time constant is needed to increase the sampling rate to once per second or more.

A magnetic field gradient over the sensor decreases the time constant of the precession signal decay. That is because the proton precession frequency will vary over the sample, and the resulting interference will lead to a faster signal decay (Figure 7.14).

After the polarization, the protons start rotating from the same direction. However, because of the gradient, the fields at the ends of the sample are different and the precession frequencies are also different:

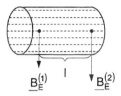

Figure 7.14 The effect of a field gradient: Different fields lead to different precession frequencies, and signal interference shortens the signal decay time.

$$\omega_1 = \gamma'_P \cdot B_E^{(1)} \quad \omega_2 = \gamma'_P \cdot B_E^{(2)} \tag{7.41}$$

After the time t, a phase difference between the protons at the two positions evolves (they will have rotated through slightly different angles):

$$\Delta\varphi = (\omega_1 - \omega_2) \cdot t = \gamma'_P \cdot (B_E^{(1)} - B_E^{(2)}) \cdot t = \gamma'_P \cdot \frac{\partial B}{\partial r} \cdot \lambda \cdot t \tag{7.42}$$

When $\Delta\varphi = \pi$, the magnetizations at the two ends of the sensor are antiparallel, and the signal is very much reduced. For $\lambda = 10$ cm and the gradient $\partial B/\partial r = 400$ nT/m, the time t_0 for signal disappearance is

$$t_0 = \pi/(\gamma'_P \cdot (\partial B/\partial r) \cdot \lambda) \cong 0.3\text{s} \tag{7.43}$$

which is much shorter than the relaxation time for oxygen-free water. That effect can be used to measure weak field-gradients.

Magnetic material close to the sensor will expose it to a large field gradient and destroy the signal. That can be used to verify that a precession signal is seen: The signal should disappear when a magnet is brought near the sensor.

Previously, the proton precession frequency measurement was done by counting the number of pulses from an absolute reference frequency generator during a predetermined number of precession periods or multiplying the precession frequency (using a PLL) by a suitably high number N and then determining the number of pulses of the N-times-precession frequency for 1-sec or 0.1-sec periods derived from an absolute frequency standard. Both methods are sensitive to time jitter in the start and stop edges of the counting period determination, and to the ±1 count truncation error in the number of received impulses. The system can, of course, be optimized rendering the error sources negligible, and, in particular, the filtering properties of the PLL should be fully exploited.

However, with the availability of powerful digital signal processors (DSPs), more efficient and better noise suppressing digital algorithms offer superior signal analysis methods compared to the classical frequency determination. Controlled by a stable and accurate time base crystal, the decaying precession signal is digitized after suitable amplification to match the ADC's input voltage range. The digital time series representing the precession signal is then analyzed in the DSP, most simply by fitting to the signal the initial amplitude a_0, the angular frequency ω, and the time decay constant δ in the following expression:

$$a(t) = a_0 \cdot e^{-\delta t} \cdot \sin(\omega t) \qquad (7.44)$$

That immediately provides the precession frequency and the measurement quality parameters a_0 and the decay constant, as the result of an averaging process using all signal data (R. L. Snare, IGPP, UCLA, private communication, 1994). About 1,000 points of the signal can be digitized in 100 ms and analyzed in real time.

More advanced NMR spectral analysis routines also exist. They are based on fast Fourier transforms (FFTs), which are well suited for DSPs. A predetermined number of spectral peaks come out as eigenvalues of an analysis matrix, and the algorithm may resolve double peaks caused by field gradients with picotesla resolution ([31]; Jan Henrik Ardenkjaer-Larsen, private communication, 1995).

The classical proton magnetometer can be turned into a continuously oscillating instrument. Reimann [32] and Sigurgeirsson [33] describe a system whose basic principle is shown in Figure 7.15. This chapter's author had the opportunity to see the operating Icelandic instrument during a visit to Reykjavik in 1968.

Water is polarized in the strong field of a permanent magnet and the water is pumped to stream laminarly to a signal pickup coil at a safe distance from the disturbance of the magnet. The polarization is directed along the Earth's field, and it can stay in that direction long enough to reach the signal coil. The weak field from a so-called α-coil turns the polarization perpendicular to the Earth's field. On leaving that coil and entering the signal coil chamber, the protons start precessing about the Earth's field with a continuous signal, because freshly polarized water is steadily supplied by the pump. The principle is neat, but because of the permanent magnet and

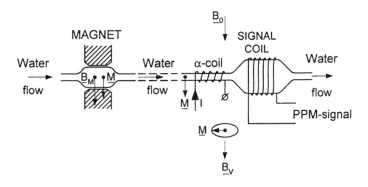

Figure 7.15 Continuously oscillating classical proton precession magnetometer. The polarization is done by a permanent magnet, and the water is pumped to a signal pickup coil at a safe distance. The α-coil field rotates the magnetization perpendicular to the Earth's field.

the "plumbing," the system is heavy, bulky, and suited for ground observatory use only.

7.2.3 Overhauser-Effect Proton Magnetometers

In 1953, Alfred W. Overhauser (see Overhauser, [12]) suggested that the large polarization of the electron magnetic moments would be transferred to the protons of the same sample by the various couplings between the electrons and the protons. That offers an alternative method for proton spin alignment [34] requiring far less power than the classic Block method [1], which uses large dc polarization fields. The upper energy level of the free electrons in an external dc magnetic field can be continuously saturated using an RF signal in resonance with the corresponding ESR spectral line, which is determined by the environmental magnetic field. In some cases, the ESR peak is narrow, meaning that the RF power needed for polarization saturation of the electrons is modest, even if continuous RF pumping of the electrons is maintained. This is the dynamic nuclear polarization (DNP), and the theoretical enhancement of the proton polarization by DNP is about 330 times their natural polarization in the same magnetic field [13].

The trick is to have a proton-rich liquid sample and at the same time to have some "free" electrons available for RF ESR. In most stable chemical compounds, the electrons are paired and unavailable for ESR, but in the free radicals a single, unpaired electron exists. That, in passing, also makes the free radical somewhat aggressive and with a tendency to be chemically unstable.

The Tempone® nitroxide free radical (see Figure 7.16) is basically a carbon ring that includes one oxidized nitrogen atom, and it is used almost universally in Overhauser-effect magnetometers. The free electron is associated with the nitrogen atom, and Tempone is chemically stable in the long term in a range of solvents and is not excessively aggressive. The Tempone compound also has other important advantages for the use in scalar magnetometers. The RF-resonance frequency of electrons is 28 GHz/T, which corresponds to a spectral peak at 1.4 MHz in the Earth's field of about 50,000 nT. Generally, to maintain resonance, the pumping signal frequency must (within the width of the spectral line) follow the large-scale changes of the field. However, the free electron of Tempone resides in the large nuclear magnetic field from the nitrogen atom, which moves the spectral line up to about 60 MHz, corresponding to an effective average field of about 2.1 mT. This is the so-called "zero field splitting" of the Tempone electron energy levels, and it is useful for two reasons: (1) the changes of the Earth's field are negligible compared to the large nuclear field, which means that the pumping frequency does not need to be tuned to follow the changes in the Earth's field, and (2) the large nuclear magnetic field enhances the electron polarization, resulting in an increase of the proton polarization of more than 1,000 times over the natural polarization in the much weaker Earth's field [13]. The various couplings between the electrons and the protons transfer the polarization to the protons, leaving the magnetization either parallel or antiparallel to the environmental magnetic field and at a level 1,000 times or more over the natural polarization in the same background field.

The ESR line for nitroxide is fairly broad, up to about 100 μT in the Earth's field. By using perdeuterated Tempone (i.e., substituting deuterium

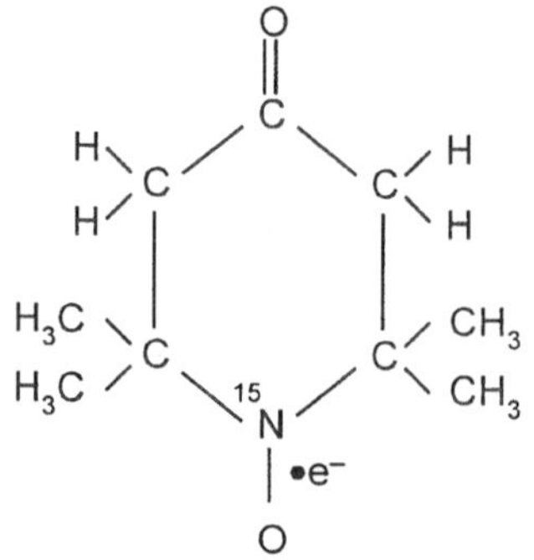

Figure 7.16 Tempone® nitroxide free radical. The unpaired electron indicated by the dot near the nitrogen atom is available for ESR polarization (after [35]).

for the hydrogen atoms), the ESR line width is reduced to 20 μT to 30 μT, with a resulting substantial reduction of the RF power needed to saturate the electron polarization [35]. In contrast, the proton NMR spectral line can be less than 2 nT in solvents of reasonably long relaxation-time constants, and the inaccuracies can be limited to a small fraction of a nanotesla [13].

The Overhauser scalar magnetometer sensor has the liquid proton sample (with an optimized amount of Tempone dissolved) inside a 60-MHz cavity resonator and with the cavity connected to a continuously operating 60-MHz RF oscillator via a coaxial cable. A sustained forced-precession oscillation of the protons at the precession frequency $f = \gamma'_p \cdot |\mathbf{B}_{Earth}|$ may then be obtained by letting the proton sample be the part of a feedback circuit that determines frequency. Intermittent proton free precession also can be established by first RF-polarizing the electrons of the sample for a (short) time long enough to establish the proton polarization. After switching off the RF, a short dc-current impulse in the pickup coil turns the proton moments perpendicular to the Earth's field, and the decaying proton precession signal can be observed, just as in the classic proton magnetometer.

As for the dc-polarized free-precession magnetometer, a null line exists for **B**-vector directions along the pickup coil axis, close to which the proton signal disappears into the noise. Because of the large dynamic proton polarization by the Overhauser effect, the half-opening angle of the theoretical "null cone" is quite small. For the commercial GEM-19 magnetometer [13], the null cone is barely noticeable as a slight increase in the measurement noise from typically 0.1 nT for **B** perpendicular to the null line to about 1.0 nT for **B** closer than about 5 degrees. It is of little concern for most applications, because the instrument still produces valid, albeit slightly noisier, measurements. The LETI Overhauser magnetometer sensor [14] for the Oersted and CHAMP satellite missions was constructed to be omnidirectional by the use of highly inhomogeneous ac-excitation fields from the proton precession forcing coils. Regardless of the direction of the external field, some parts of the liquid always have the precession forcing field perpendicular to the external field and to the sensor coil axis.

For low fields below about 15,000 nT, the proton free precession rate becomes proportionally smaller and the amplitude of the induced signal correspondingly lower. However, the signal-to-noise ratio can be maintained by narrowing the frequency band pass of the analog signal-handling circuits at the expense, of course, of a longer sampling time. At low fields, the Overhauser effect based on Tempone presents a slight disadvantage because of a double positive/negative response of the DNP factor as a function of the ESR RF-signal frequency. The double peak response is particular to Tempone and caused by a complicated ESR spectral line structure.

Figure 7.17 shows the Tempone double structure in two different solvents. That is elegantly used by LETI in their magnetometer to obtain proton polarization along the environmental field in one cell and antiparallel to the field in a second cell using the same RF signal. However, the ESR frequency separation between the positive and the negative peaks depends on the external field. At low fields (below about 16,000 nT), the peaks are so close together that a substantial reduction in the DNP factor results. The consequence is that the signal-to-noise ratio approaches zero more rapidly than does the proton precession frequency, when the **B**-field goes to zero, an effect that is of little concern for Earth's field measurements. However, for interplanetary applications the deep-space low-fields performance of a scalar instrument is of importance.

For satellite instruments, the weight and the power consumption are driving parameters; for deep-space missions away from the large planetary magnetic fields, the low-field performance is an additional concern. The

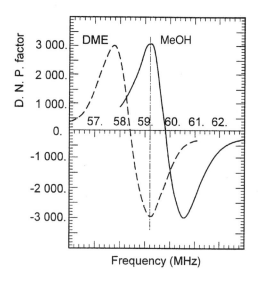

Figure 7.17 The double positive/negative proton polarization structure of the Tempone nitroxide free radical (after [36]). The DNP factor is shown against the ESR RF-pumping frequency for Tempone in two solvents: methanol (MeOH) (solid line) and dimethoxyethane (DME) (dashed line). The "chemical shift" in the two solvents is demonstrated by the frequency displacement of the two curves.

ESR RF power decreases dramatically with a narrowing ESR line, and a free radical not having the double positive/negative peak structure of Tempone presents a superior low-field performance. An alternative free radical of the Trityl-group has a single ESR line of only 2.5 μT width (see [37]), compared to 20 μT or more for the nitroxide. The RF-saturation power of the ESR is proportional to (at least) the square of the line width; thus, the power is milliwatts compared to watts for the nitroxide free radical. Figure 7.18 shows a comparison between the nuclear polarization degree against RF power for the nitroxide and for Trityl.

Trityl does not have the zero field splitting, so the free electron sees only the external field; no effective additional nuclear field exists, as in the case of the nitroxide. Full ESR saturation is obtained at a modest RF-power level, and the free radical concentration can be optimized for maximum polarization coupling between the electrons and the protons, because there is room for the consequential slight broadening of the line and associated increase in saturation power. DNP either can take place in the Earth's field requiring the RF to track the field, or a homogeneous dc-field of about 10 mT can be applied, thereby further enhancing the proton polarization and reducing the need for RF tracking. A laboratory instrument with a 7-ml sample and using the dc and RF double polarization yields a signal-to-noise ratio of 55 dB/Hz$^{1/2}$ in 50,000 nT for an average dc and RF polarization power of less than 700 mW. With 200 ms for polarization and 150 ms for

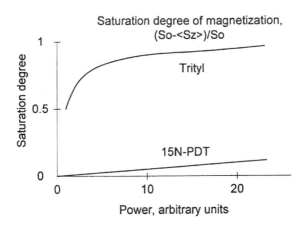

Figure 7.18 Polarization saturation degree versus ESR RF power for Trityl and Tempone. Trityl reaches saturation polarization for much less power than does perdeuterated Tempone nitroxide 15N-PDT (Ardenkjaer-Larsen, private communication, 1997).

readout, the instrument takes three samples per second. About 1,000 points of the proton precession signal can be digitized in 100 ms and subsequently analyzed in a DSP, retrieving the proton precession frequency and the signal decay time constant. The principle seems promising for the development of a low-power space instrument; the low-field performance is particularly attractive for a deep-space scalar magnetometer [37].

7.3 Optically Pumped Magnetometers

7.3.1 Metastable He^4 Magnetometers

The most common He^4 isotope has two protons and two neutrons in the nucleus with no net nuclear magnetic moment. Furthermore, the two electrons fill up the first shell, also without any uncompensated electron spin or magnetic moment. For observing the electron Zeeman splitting in a gaseous sample of He^4, the atom is lifted from the 1^1S_0 ground state into the 2^3S_1 metastable level by a sustained high-frequency (HF) glow discharge excitation. The He^4 atom in that condition can then be regarded as the ground state of a new atom, with both electrons available for ESR.

In passing, the He^3 isotope (a decay product of tritium) has a net nuclear magnetic moment with a gyromagnetic ratio of about 60% that of the proton. The He^3 nuclear magnetic moment can be polarized by interactions with optically pumped electrons in the RF-excited metastable state, as for He^4; a nuclear resonance magnetometer similar to the Overhauser effect proton magnetometer can then be constructed.

A sealed glass vessel with He^4 gas of suitable partial pressure is exposed to an electrodeless HF glow discharge, producing the metastable 2^3S_1 ground state. The external magnetic field B_0 splits the energy into three Zeeman levels, designated $m = +1$, $m = -1$, and $m = 0$, where $\Delta E = h\nu_0$ (Planck's constant $h = 6.626075 \times 10^{-34}$ J · s) is the energy difference between $m = 0$ and $m = \pm 1$. From a macroscopic point of view, the +1 level has the magnetic moment of the electrons antiparallel to the external magnetic field and the −1 level has the moment parallel to the field. The transition from ± 1 to 0 can be induced by a resonance ac-magnetic field $B(t) = B_1 \cos(2\pi\nu_0)$, where ν_0 is the electron Larmor frequency or the precession frequency of the electrons about the external field B_0. The two electrons in the metastable 2^3S_1 state are decoupled almost entirely from the atom and behave as if in vacuum; then the energy difference ΔE equals twice the free electron magnetic moment ($\mu_e = 9.2847701 \times 10^{-24}$ J/T) times the external field. The determination of the Larmor resonance frequency ν_0 then yields the scalar magnitude of the external field according to

$$\Delta E = h\nu_0 = 2 \cdot \mu_e \cdot B_0 \text{ or } \nu_0 = 2 \cdot \gamma_e \cdot B_0 \qquad (7.45)$$

where the free-electron gyromagnetic ratio $\gamma_e = (2 \cdot \mu_e/h) = 28.0$ GHz/T, which is the largest conversion factor of any optically pumped magnetometer, larger than those of the alkali metal vapor instruments [38, 39].

The determination of the frequency ν_0 is established in the following way. The electrons in the triplet metastable 2^3S_1 state can be optically excited into the higher 2^3P_0 energy level by infrared light of about 1,083 nm wavelength. Of the three spectral lines D_0, D_1, and D_2, corresponding to the transitions from the 2^3S_1, $m = 0, \pm 1$ levels to the 2^3P_0 level, the D_0-line is well separated from D_1 and D_2 and can be picked out with a suitably narrow bandwidth infra-red (IR) light beam [40, 41]. The D_0 ($m = 0$) transition is excited by unpolarized or circularly/linearly polarized light with the light propagating along (a component of) the external magnetic field $\mathbf{B}_0$, which is termed the M_z-mode, because the z-axis is along $\mathbf{B}_0$. Originally, the three states $m = 0, \pm 1$ are equally populated at normal temperatures, and light is absorbed as long as $m = 0$ electrons are available for transition into the 2^3P_0 state. Relaxation occurs with equal probabilities into the three $m = 0, \pm 1$ states, so after some time the $m = 0$ has a much reduced population, and absorption of the light cannot take place. This is line saturation, when (almost) all the $m = 0$ electrons are removed, and the cell then recovers the full transparency. However, by applying the ac-magnetic resonance field $B_1 \cos(2\pi\nu_0)$ at right angles to the external field, electrons are induced to go from the $m = \pm 1$ states into the $m = 0$ state, and $m = 0$ electrons again become available for light absorption [39].

The general principle of an M_z-mode magnetometer is shown in Figure 7.19. For the He^4 magnetometer, the circular polarizer shown can be replaced by a linear polarizing filter, and the need for an interference filter depends on the actual light source used. A laser can be locked to the single D_0 line, whereas an He-lamp transmits the D_1 and D_2 lines as well and requires some filtering to minimize absorption transitions at those lines [42].

Figure 7.20 shows the light absorption versus the ac-magnetic resonance field frequency. By frequency modulating the resonance field RF-oscillator with a suitable low frequency (LF) Ω and an RF swing $\pm\Delta f$, then close to the Larmor frequency, the amplitude and phase of the 1Ω output signal from the photo cell is a direct measure of the deviation of the RF-oscillator center frequency from the Larmor frequency ν_0. That deviation signal can

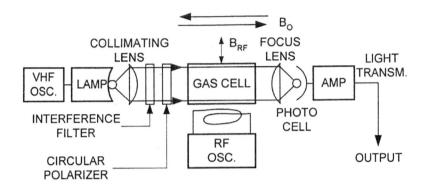

Figure 7.19 M_z-mode optically pumped magnetometer. The optical absorption is measured for a light-transmission path along the external magnetic field B_0, the z-axis. The cell becomes less transparent when the frequency of the ac-magnetic field B_{RF} (or B_1) perpendicular to B_0 approaches the Larmor frequency $\nu_0 = 2 \cdot \gamma_e \cdot B_0$ of the electrons (after [43]).

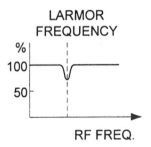

Figure 7.20 Absorption of the transmitted light through the cell against the RF of the transverse ac-magnetic field. Sweeping the frequency periodically $\pm \Delta f$ about the center frequency results in a pure sweep second harmonic output of the photo cell when the center frequency equals the Larmor frequency ν_0 (after [43]).

be used in a feedback loop to control and maintain the center frequency at ν_0. In lock, the 1Ω signal goes to zero and the second harmonic 2Ω signal is maximum. The presence of a large 2Ω signal is a safe indication of proper RF-oscillator lock to the Larmor frequency ν_0 [44]. The RF-oscillator center frequency is subsequently determined and transmitted as the measure of the external magnetic field B_0.

The sweep frequency is 200–300 Hz, the modulation sweep range is ±1,400 Hz corresponding to ±50 nT, and the Larmor precession line width is larger than 1 kHz, so the system responds relatively fast [42, 44]. He^4 is a gas in the relevant temperature range, so the He^4 cell needs no heating or any tight temperature control, and the magnetometer starts immediately after switch-on. The instrument operates for B_0-field directions out to about 60 degrees from the optical path [45]; combining two sensors at right angles reduces the null zone to a single narrow cone, which is acceptable for most magnetic mapping missions (Figure 7.21). The sensor has no heading error outside the null cone; because only one spectral line is involved (D_0), the instrument is absolute in the sense that the conversion factor depends on an atomic constant. The overall sensitivity is relatively low, so a fairly intense light source is needed, which introduces a small shift of the order of ±0.5 nT, depending on the light intensity. To obtain and maintain that low shift, a precise and stable alignment of the RF coil and the optical system has to be established. An absolute comparison between the Jet Propulsion Laboratory (JPL) scalar helium magnetometer and a proton magnetometer showed a scatter of ±0.5 nT, including the errors from both instruments [44].

A highly stable and very low-noise vector-field-measuring version of the JPL helium magnetometer with added bias coils flies onboard the Ulysses solar polar mission [21]. On the coming Cassini mission, JPL has a single-

Figure 7.21 JPL's two-cell He^4 scalar sensors for the SAC-C satellite mission. The combined sensor has only one null line along which the field cannot be measured [45].

cell vector/scalar helium magnetometer [46]. In the scalar mode, the magnetometer will measure fields of intensities between 256 nT and 16,000 nT (upward limited by the RF range) and with directions less than 45 degrees from the optical axis. The absolute accuracy is <1 nT. In vector mode, all field directions and intensities down to zero can be measured; in the absolute scalar mode, however, the angular dependence of the measurement capability presents a challenge to mission planning.

An omnidirectional scalar He^4 magnetometer was recently presented by LETI, CEA Advanced Technologies, Grenoble, France [39]. As for the JPL scalar helium magnetometer, the He^4 atoms are brought into the 2^3S_1 metastable state. The RF-ESR coil and an optical linear polarizer are mechanically coupled and can be rotated simultaneously by the piezoelectric motor controlled by the 1Ω modulation signal output from the photosensitive detector. For any orientation of the external $\mathbf{B}_0$-field, both the $\mathbf{E}_0$ light polarization direction and the resonance ac-magnetic field $\mathbf{B}_1$ are maintained at right angles to the external $\mathbf{B}_0$-field, as required for isotropic operation. Rotation of the sensor in two planes proved the isotropy to be better than 20 pT. The instrument resolution is 1 pT/$\sqrt{Hz}$ in a bandwidth of (dc-) 300 Hz and the band noise is 17 pT$_{rms}$. The linewidth of the RF resonance corresponds to about 70 nT, which, combined with the modest cell dimensions (4 cm by 6 cm), makes it tolerant to gradients up to 1 μT/m.

7.3.2 Alkali Metal Vapor Self-Oscillating Magnetometers

A single electron exists in the outer shell of the alkali metal atoms with an unpaired spin and thus with a magnetic moment available for ESR. Small amounts of the solid metal are contained in a glass vessel with a buffer gas, and the cell is heated and maintained at a suitable temperature for obtaining the optimum alkali vapor partial pressure in a buffer gas, minimizing the effects of interatomic collisions. Coating the cell walls with a material having argon electron structure reduces the wall collision effects. The temperature for a vapor pressure of 10^{-6} torr is 126°C for Na^{23}, 63°C for K^{39}, 34°C for both Rb isotopes, and 23°C for Cs^{133}; of all the alkali atoms, cesium needs the least heating power [38]. The spectral line structure of the alkali atoms is more complex than that of the metastable helium atom, and (except for potassium) the single lines cannot be resolved by ordinary laboratory techniques. An optically pumped/swept RF-oscillator M_z-mode magnetometer like the He^4 light absorption instrument can be constructed, but the unresolved line structure means that the resultant RF-oscillator frequency is a weighted average over several lines. While the short-term noise is low, of the order of 0.2 nT, the absolute level may change because of the spectral line averaging, up to 182 nT for Rb^{85}, 82 nT for Rb^{87}, and 6 nT for Cs^{135} [43, 47]. The cesium magnetometer seems to be the preferred type of the common optical magnetometers because of its high accuracy and low heating demands. In the alkali metals, the outer electron is closely coupled to the atom and not quasi-"free," as for the He^4 atom, which means that the effective gyromagnetic ratios are smaller than that of He^4: about 7.00 Hz/nT for Na^{23}, K^{39}, and Rb^{87}; about 4.66 Hz/nT for Rb^{85}; and 3.50 Hz/nT for Cs^{135} [38]. The gyromagnetic ratio is still larger than that of the proton magnetometer by a factor of about 100, and the optically pumped magnetometers are thus proportionally less sensitive to the platform rotation [16].

Varian Associates developed the Larmor frequency self-oscillating optical magnetometer in great detail [48]. Figure 7.22 shows the principle of the simplest type [43, 38].

In Figure 7.22, the very high frequency (VHF) oscillator excites the alkali metal vapor in the lamp, and the light is collected by the collimating lens (CL) and sent through the absorption gas cell containing the same alkali vapor as the lamp. The interference filter (IF) attenuates the unwanted D_2 line and permits the D_1 line after circular polarization (CP) to pass through the absorption cell and to be focused by the field lens (FL) on the photo cell (PC) [43, 49]. The electrical photo cell output depends on the gas cell transmission coefficient. The optical pumping polarizes the electrons, and the RF-coil B-field (oscillating at the Larmor frequency) causes the electrons to precess in-phase about the external B_0-field. Whenever the coherently rotating electron magnetic moments are closest to the propagation direction

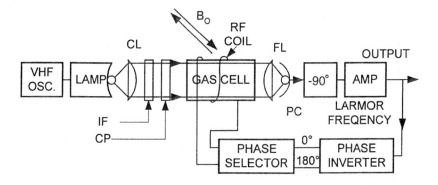

Figure 7.22 The principle of Varian Associates' self-oscillating optically pumped alkali metal vapor magnetometer (after [43]).

of the optical path, the cell transmission increases. Half a Larmor period later, when the electrons are most antiparallel to the optical path, the transmission drops and the photo cell output decreases. A simple explanation is that when all the electron magnetic moments (and spins) are (most) parallel to the optical path, very few electrons with spins in other directions are available for light absorption. Similarly, when the electron spins are rotated away from the direction of the optical path, a large number of electrons are available for absorption of the light. That is an M_x-mode operation because light absorption with the optical path (having some component) across the B_0-field is observed.

When the system is running, the photo cell output is modulated at the Larmor frequency, and after 90-degree phase shift, that signal is used to drive the RF-B coil in a self-oscillating system. That operates only when the external B_0-field is directed toward +135 degrees/−45 degrees relative to the optical path. The angle can be changed to +45 degrees/−135 degrees if the feedback signal phase is inverted 180 degrees. Depending on the feedback phase, a practical system will operate when the external field is inside an angular range close to one of those two directions. The magnetometer does not operate when B_0 is closer than 15 degrees to the optical axis or closer than 10 degrees to being perpendicular to the optical path. Within the operating angular ranges, the magnetometer output shifts slightly in frequency when the B_0-aspect angle changes because of a shift in the weights of the multiple-line average.

Varian Associates has developed a symmetrical dual-cell system [50, 51] that operates in both ±45-degree/±135-degree sectors and has much reduced angular dependence of the photo cell output frequency. The Magsat scalar magnetometer was a Cs^{133} twin dual-cell system mounted with the cell optical axes at 55 degrees to each other, reducing the null zones to two thin cones at right angles to the optical paths. The system measured fields in the range of 15,000 to 64,000 nT with errors from 0.5 to 1.5 nT. Misalignment of the RF-feedback coil with the optical axis will cause large frequency shifts, particularly when the field is close to the null zones and the winding of the coil and the positioning of the optics must be tightly controlled [49, 52]. For optimum operation, the cells have to be heated and maintained at a temperature between 25°C and 30°C. Modern dual-cell Cs-magnetometers are available from the Scintrex company (0.25-nT heading error and 2 pT_{p-p} noise in 0.01–1 Hz bw) and from Geometrics Inc. (20 pT noise at 10 samples per second).

Potassium has a fully resolved optical spectrum throughout the entire range of the Earth's magnetic field. A very high-resolution absolute accuracy has been developed based on a mixture of K^{41} and K^{39}. Sensitivities of 1 $pT/\sqrt{Hz}$ and no systematic errors exceeding 10 pT are reported [47]. The instrument development is ongoing in collaboration with GEM Systems Inc., Toronto, Canada.

References

[1] Block, F., "Nuclear Induction," *Phys. Rev.*, Vol. 70, 1946, pp. 460–485.

[2] Varian, R. H., "Methods and Means for Correlating Nuclear Properties of Atoms and Magnetic Fields," U.S. Pat. Off., Re. 23,767. Original No. 2,561,490, dated July 24, 1951, Serial No. 55,667, Oct. 21, 1948 (reissued Jan. 12, 1954).

[3] Alldredge, L. R., and I. Saldukas, "The Automatic Standard Magnetic Observatory," Technical Bulletin No. 31, U.S. Department of Commerce, Environmental Science Services Administration, Coast and Geodetic Survey, Washington, D.C.: GPO, 1966.

[4] Packard, M. E., and R. H. Varian, "Free Nuclear Induction in the Earth's Magnetic Field" (abstract only), *Phys. Rev.*, Vol. 93, 1954, p. 941.

[5] Waters, G. S., and G. Phillips, "A New Method for Measuring the Earth's Magnetic Field," *Geophys. Prospecting*, Vol. 4, 1956, pp. 1–9.

[6] Serson, P. H., "A Simple Proton Precession Magnetometer," Report, Ottawa: Dominion Observatory, May 1962a.

[7] Serson, P. H., Proton precession magnetometer, U.S. Pat. No. 3,070,745, 1962b.

[8] Cahill, L. J., Jr., "Investigation of the Equatorial Electrojet by Rocket Magnetometer," *J. Geophys. Res.*, Vol. 64, 1959, pp. 489–503.

[9] Burrows, K., "A Rocket Borne Magnetometer," *J. Brit. IRE*, Vol. 19, 1959, pp. 767–776.

[10] Olesen, J. K., et al., "Rocket-Borne Wave, Field and Plasma Observations in Unstable Polar Cap E-Region," *Geophys. Res. Lett.*, Vol. 3, 1976, p. 711.

[11] Dolginov, Sh. Sh., "The First Magnetometer in Space," in G. Haerendel, et al. (eds.), *40 Years of COSPAR*, ESA Publications Divisions, ESTEC: Noordwijk, The Netherlands, 1998, pp. 55–63.

[12] Overhauser, A. W., "Dynamic Nuclear Polarisation," in D. M. Grant and R. K. Harris (eds.), *Encyclopedia of Nuclear Magnetic Resonance*, Vol. 1, New York: Wiley, 1966, pp. 513–516.

[13] Hrvoic, I., "Proton Magnetometers for Measurement of the Earth's Magnetic Field," GEM Systems Inc., Toronto, *Proc. Internat. Workshop on Geomagn. Observatory Data Acquisition and Processing*, Nurmijärvi Geophysical Observatory, Finland, May 1989, Helsinki, Finland: Finnish Meteorological Institute, 1990, Section 5.8, pp. 103–109.

[14] Duret, D. N., et al., "Overhauser Magnetometer for the Danish Oersted Satellite," *IEEE Trans. Magn.*, Vol. 31, 1995, pp. 3197–3199.

[15] Kastler, A., "Quelques Suggestions Concernant la Production Optique et la Détection Optique d'une Inégalité de Population des Niveaux de Quantifications Spatiale des Atomes. Application de l'Experiénce de Stern et Gerlach et la Résonance Magnetique," *J. de Phys. et le Radium*, Vol. 11, 1950, p. 255.

[16] Alexandrov, E. B., and F. Primdahl, "On Gyro-Errors of the Proton Magnetometer," *Meas. Sci. & Technol.*, Vol. 4, 1993, pp. 737–739.

[17] Langel, R. A., et al., "The Magsat Mission," *Geophys. Res. Lett.*, Vol. 9, 1982, pp. 243–245.

[18] De Vuyst, A. P., "Magnétomètre Theodolite a Protons," Institut Royal Météorolique de Belgique, Miscellanea—SERIE C, No/r 2, 23 pp., Report to the International Association of Geomagnetism and Aeronomy, General Assembly, Moscow, Russia, 1971.

[19] Park, R. J., and P. A. Cloutier, "Rocket-Based Measurement of Birkeland Currents Related to an Auroral Arc and Electrojet," *J. Geophys. Res.*, Vol. 76, 1971, pp. 7714–7733.

[20] Sesiano, J., and P. A. Cloutier, "Measurement of a Field-Aligned Current System in a Multiple Auroral Arc System," *J. Geophys. Res.*, Vol. 81, 1976, pp. 116–122.

[21] Balogh, A., et al., "The Magnetic Field Investigation on the Ulysses Mission: Instrumentation and Scientific Results," *Astronomy & Astrophysics*, Suppl. Series, Vol. 92, 1992, pp. 221–236.

[22] Giacomo, P., "News From the BIPM Concerning the Meter, the Kilogram, the Second and the Ampere," *IEEE Trans. Instr. Meas.*, Vol. IM-36, 1987, pp. 158–160.

[23] Belorisky, E. W., et al., "Sample-Shape Dependence of the Inhomogeneous NMR Line Broadening and Line Shift in Diamagnetic Liquids," *Chemical Phys. Lett.*, Vol. 175, 1990, pp. 579–584.

[24] Belorisky, E., et al., "Demagnetizing Field Effect on High Resolution NMR Spectra in Solutions With Paramagnetic Impurities," *J. Phys. II*, Vol. 1, 1991, pp. 527–541.

[25] Cohen, E. R., and B. N. Taylor, "The 1986 Adjustment of the Fundamental Physical Constants," *Rev. Modern Phys.*, Vol. 59, 1987, pp. 1121–1148.

[26] Driscol, R. L., and P. L. Bender, "Proton Gyromagnetic Ratio," *Phys. Rev. Lett.*, Vol. 1, 1958, pp. 413–414.

[27] Schlesok, W., and J. Forkert, "A New Determination of the Ampere and the Gyromagnetic Ratio γ_p' in a Low and High Magnetic Field," *IEEE Trans. Instrum. Meas.*, Vol. 34, 1985, pp. 173–175.

[28] Acker, F. E., "Calculation of the Signal Voltage Induced in a Toroid Proton Precession Magnetometer Sensor," *IEEE Trans. Geosci. Electronics*, Vol. 9, 1971, pp. 98–103.

[29] Koehler, J. A., "Proton Precession Magnetometer," Manuscript, about 55 pages plus figures, 1999, Adobe.pdf file at http://www.diamondjim.bc.ca.

[30] Faini, G., A. Fuortes, and O. Svelto, "A Nuclear Magnetometer," *Energia Nucleare*, Vol. 7, 1960, pp. 705–716.

[31] Pijnappel, W. F., et al., "SVD-Based Quantification of Magnetic Resonance Signals," *J. Mag. Resonance*, Vol. 97, 1992, pp. 122–134.

[32] Reimann, R., "Messung von Swachen Magnetfeldern Mittels Kernresonanz," *Nuclear Instruments and Methods*, Vol. 45, 1968, pp. 328–330.

[33] Sigurgeirsson, T, "A Continuously Operating Proton Precession Magnetometer for Geomagnetic Measurements," *Scientia Islandica* (Science in Iceland), Vol. 2, 1970, pp. 67–77.

[34] Hartmann, F., "Resonance Magnetometers," *IEEE Trans. Magn.*, Vol. 8, 1972, pp. 66–75.

[35] Kernevez, N., et al., "Weak Field NMR and ESR Spectrometers and Magnetometers," *IEEE Trans. Magn.*, Vol. 28, 1992, pp. 3054–3059.

[36] Kernevez, N., and H. Glenat, "Description of a High-Sensitivity CW Scalar DNP-NMR Magnetometer," *IEEE Trans. Magn.*, Vol. 27, 1991, pp. 5402–5404.

[37] Primdahl, F., "Scalar Magnetometers for Space Applications," in R. F. Pfaff, J. E. Borovsky, and D. T Young, *Measurement Techniques in Space Plasmas: Fields*, Geophys. Monograph 103, Washington, D.C.: American Geophysics Union, 1998, pp. 85–99.

[38] Lokken, J. E., "Instrumentation for Receiving Electromagnetic Noise Below 3,000 CPS," in D. F. Bleil (ed.), *Natural Electromagnetic Phenomena Below 30 kc/s*, New York: Plenum Press, 1964.

[39] Alcouffe, F., et al., "An Isotropic Helium 4 Scalar Magnetometer With Extended Frequency Capabilities," *Proc. EMMS Conf.*, Brest, France, July 5–9, 1999.

[40] Smith, E. J., et al., "The Helium Magnetometer: Potential Reductions in Physical Requirements Without a Loss in Performance," in B. T. Tsurutani (ed.), *Small Instruments for Space Physics; Proc. Small Instruments Workshop*, Pasadena, CA, Mar. 20–23, 1993, pp. 2–2 to 2–9, Space Phys. Div., NASA, Nov. 1993.

[41] McGregor, D. D., "High-Sensitivity Helium Resonance Magnetometers," *Rev. Sci. Instrum.*, Vol. 58, 1987, pp. 1067–1076.

[42] Slocum, R. E., and D. D. McGregor, "Measurement of the Geomagnetic Field Using Parametric Resonance in Optically Pumped He^4," *IEEE Trans. Magn.*, Vol. 10, 1974, pp. 532–535.

[43] Ness, N. F., "Magnetometers for Space Research," *Space Sci. Rev.*, Vol. 11, 1970, pp. 459–554.

[44] Slavin, J. A., "JPL Scalar Helium Magnetometer, Report," Tech. Note, Pasadena, CA: Jet Propulsion Laboratory, 1984.

[45] Smith, E. J., et al., "ARISTOTELES Magnetometer System," *Proc. Workshop on Solid-Earth Mission ARISTOTELES*, Anacapri, Italy, Sept. 23–24, 1991, ESA SP-329, Dec. 1991, pp. 83–89.

[46] Dunlop, M. W., et al., "Operation of the Dual Magnetometer on Cassini: Science Performance," *Planet. Space Sci.*, Vol. 47, 1999, pp. 1389–1405.

[47] Alexandrov, E. B., and V. A. Bonch-Bruevich, "Optically Pumped Magnetometers After Three Decades," *Optical Engineering*, Vol. 31, 1992, pp. 711–717.

[48] Bloom, A. L., "Principles of Operation of the Rubidium Vapor Magnetometer," *Applied Optics*, Vol. 1, 1962, pp. 61–68.

[49] Fathing, W. H., and W. C. Folz, "Rubidium Vapour Magnetometer for Near Earth Orbiting Spacecraft," *Rev. Sci. Instrum.*, Vol. 38, 1967, pp. 1023–1030.

[50] Ruddock, K. A., "Optically Pumped Rubidium Vapour Magnetometer for Space Experiments," *Space Research*, Vol. II, 1961, pp. 692–700.

[51] Morris, R., and L. Langan, "Varian Associates' Space Magnetometers 1956–1961," Geophysical Tech. Memo. No. 8, Palo Alto, CA: Varian Associates, Quantum Electronics Div., 1961.

[52] Mobley, F. F., et al., "MAGSAT—A New Satellite to Survey the Earth's Magnetic Field," *IEEE Trans. Magn.*, Vol. 16, 1980, pp. 758–760.

8

Superconducting Quantum Interference Devices (SQUIDs)
Robert L. Fagaly

8.1 Introduction

8.1.1 Superconductivity

At temperatures approaching absolute zero, certain materials undergo a transition to what is known as the superconducting state. In 1911, Kamerlingh-Onnes [1] discovered that the resistance of mercury when cooled below 4.2K dropped to an immeasurably small value [Figure 8.1(a)].

This transition from normal resistance to resistanceless behavior takes place over a narrow temperature range: about 0.001K for pure, strain-free metals and a degree or more for alloys and ceramics. Below that temperature, known as the transition temperature (T_c), the material is characterized by a complete lack of electrical resistance. Subsequent investigations have indicated that a large number of materials undergo a similar superconducting transition. According to the Bardeen-Cooper-Schriefer (BCS) theory [2], the mechanism that permits superconductivity is the phonon exchange between paired electrons (Cooper pairs). The average distance between the electron pairs is the coherence length ξ. Electrons can exchange partners, and they can be depaired by thermal (critical temperature), kinetic (critical current density), or magnetic (critical field) interactions. The temperature, current density, and magnetic field under which the particular material is superconducting form the so-called phase space. In 1986 Bednorz and Müller [3]

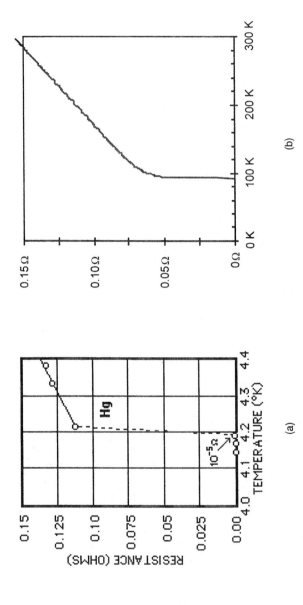

Figure 8.1 Resistance of (a) mercury (after [1]) and (b) YBa$_2$Cu$_3$O$_{7-\delta}$ versus absolute temperature.

discovered a new class of ceramic oxides that became superconducting near 30K, significantly hotter than any previously known superconductor. Since then, newer materials have been developed with the superconducting transition temperatures above 130K, well above the boiling point of liquid nitrogen. $YBa_2Cu_3O_{7-\delta}$ (often referred to as YBCO), as shown in Figure 8.1(b), with $T_c > 90K$ is the most commonly used superconducting ceramic oxide. To distinguish between the types of materials used in making SQUID sensors, we denote the traditional metallic superconductors that typically operate at liquid-helium temperatures (4.2K) as low-temperature superconductors (LTS) and the newer materials that can operate at liquid nitrogen temperatures (77K) as high-temperature superconductors (HTS).

8.1.2 Meissner Effect

An interesting property of the superconducting state is observed if a superconductor is put in a magnetic field and then cooled below its transition temperature [4]. In the normal state, magnetic flux lines can penetrate through the material, as shown in Figure 8.2(a). As the material becomes superconducting, the magnetic flux is expelled, as can be seen in Figure 8.2(b). That is a consequence of Maxwell's equations, in which an electric field gradient cannot exist inside a superconductor. If the superconducting material forms a ring, the flux interior to the ring is trapped when the ring becomes superconducting, as shown in Figure 8.2(c).

If the magnetic field is turned off, a current is induced that circulates around the ring, keeping the magnetic flux ($\Phi = \int BdA$) inside constant, as shown in Figure 8.2(d). Because of the electrical resistance, the current in

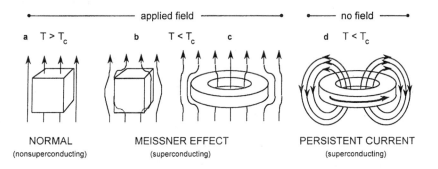

Figure 8.2 Meissner effect in a superconducting ring cooled in an externally applied magnetic field: (a) normal (nonsuperconducting); (b) and (c) Meissner effect (superconducting); and (d) persistent current (superconducting).

a ring made of a normal (nonsuperconducting) metal will quickly decay. The current decay is exponential with a time constant that is related to the resistance (R) and inductance (L) of the ring, $I(t) = I_o \, e^{-tR/L}$. For a superconducting ring $R = 0$, and a persistent current is established. The current continues to circulate as long as the ring is kept cold (below T_c). It should be noted that superconducting rings of tin have been made to carry circulating dc currents for a period much greater than a year, disconnected from any power source, without any measurable decrease in current [5]. That is equivalent to saying that the resistivity of superconducting tin is at least 17 orders of magnitude less than that of room-temperature copper.

8.1.3 Flux Quantization

As long as the current persists, the magnetic flux remains "trapped." The trapped flux has some very unusual properties. First, we cannot change the level of magnetic flux in the ring in a continuous manner. We can only trap discrete levels of the magnetic flux (Figure 8.3). In other words, the magnetic flux is quantized and exists only in multiples of a very small fundamental quantity called a flux quantum (Φ_o) whose magnitude is $2.068 \cdot 10^{-15}$ Wb (1 Wb/m^2 = 1T). Equivalently, for a ring with an area of 1 cm^2, the field inside the ring can exist only in discrete steps of $2.068 \cdot 10^{-11}$ T.

8.1.4 Josephson Effect

For a loop of superconducting wire interrupted by a normal, resistive region, we would expect it to behave the same as a continuous loop of normal metal, that is, a current flowing in the loop would quickly decay. In 1964, Josephson [6] predicted the possibility of electrons tunneling from one superconducting

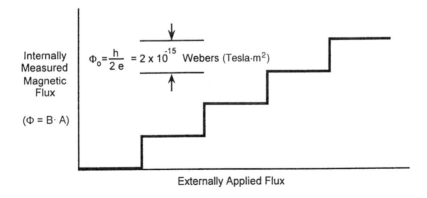

Figure 8.3 Flux quantization.

region to another that had been separated by a resistive (insulating) barrier (usually referred to as a "weak link"). For distances less than the coherence length and currents less than a critical current I_c that is characteristic of the weak link, a current can penetrate the resistive barrier with no voltage drop (Figure 8.4).

There are many ways to make a weak link. The junction can be an insulating (superconducting-insulator-superconducting, or SIS) barrier such as a point contact, as shown in Figure 8.5(a) or a normal metal (superconducting-normal-superconducting, or SNS) or a microbridge, as shown in Figure 8.5(b). Present-day LTS devices use tunnel junction weak links like that shown in Figure 8.5(c). HTS devices use either intrinsic [bicrystal, as shown in Figure 8.5(d), or step-edge grain boundary, as shown in Figure 8.5(e)] junctions or extrinsic [step-edge SNS, as shown in Figure 8.5(f), or ramp-edge, as shown in Figure 8.5(g)] structures.

The weak link can be a region in which the current flowing is higher than the current needed to drive the superconductor normal (I_c). A typical weak link might have a critical current of 10 μA. If the loop has a diameter of 2 mm, that is equivalent to several flux quanta. For a loop of superconductor interrupted by a weak link Josephson junction, magnetic flux threading through a superconducting loop sets up a current in the loop. As long as the current is below the critical current, the complete loop behaves as if it were superconducting. Any changes in the magnetic flux threading through the loop induce a shielding current that generates a small magnetic field to oppose the change in magnetic flux.

8.1.5 SQUIDs

Superconducting quantum interference devices (SQUIDs) use Josephson effect phenomena to measure extremely small variations in magnetic flux.

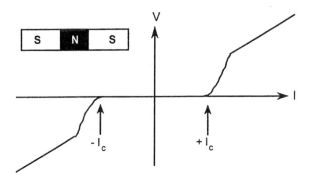

Figure 8.4 Current versus voltage curve of a Josephson tunnel junction.

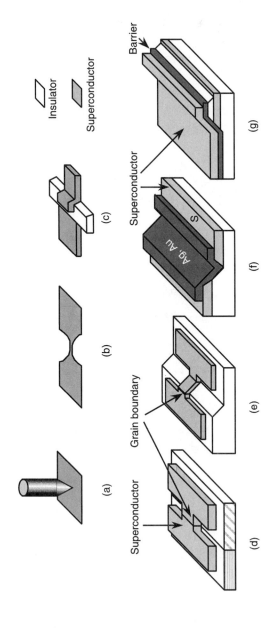

Figure 8.5 Different types of Josephson junctions: (a) point contact; (b) microbridge, also known as a Dayem bridge; (c) thin-film tunnel junction; (d) bicrystal; (e) step-edge grain boundary; (f) step-edge SNS; (g) ramp-edge SNS with a $PrBa_2Cu_3O_{7-\delta}$ barrier.

The theory of different types of SQUIDs is described in detail in the literature [7–9]. SQUIDs are operated as either RF or dc SQUIDs. The prefix RF or dc refers to whether the Josephson junction(s) is biased with an alternating current (RF) or a dc current. Flux is normally (inductively) coupled into the SQUID loop via an input coil, which connects the SQUID to the experiment. Because the input coil is superconducting, its impedance is purely inductive.

8.2 SQUID Sensors

There are fundamental differences between low- and high-temperature superconductors (LTS and HTS). LTS materials are metallic (although some nonmetallic and organic compounds have been found to be superconducting) and isotropic, and have coherence lengths (ξ, the average distance between the Cooper pairs) that are tens to hundreds of interatomic distances. HTS materials are ceramics, brittle, and anisotropic (essentially planar), and have coherence lengths in the c direction (perpendicular to the a-b plane) that are significantly smaller (Table 8.1). Not only is there a temperature limitation to superconductivity, there is a field dependence. The material remains in the superconducting state below a critical field $H_c(T) = H_{c0}[1 - (T/T_c)^2]$, where H_{c0} is the critical field at $T = 0$.

Table 8.1
Characteristics of LTS and HTS Materials

LTS	T_c (K)	$\mu_0 H_{c0}$ (T)	ξ_0 (nm)
Lead	7.19	0.08	90
Mercury	4.15	0.04	–
Niobium	9.25	0.19	40
NbTi	9.5	13	4
Nb$_3$Sn	18	23	3

HTS	T_c (K)	ξ_{ab} (nm)	ξ_c (nm)
YBa$_2$Cu$_3$O$_{7-\delta}$	95	4	0.7
Bi$_2$Sr$_2$Ca$_2$Cu$_2$O$_{10}$	110	4.5	0.2
Tl$_2$Bi$_2$Ca$_2$Cu$_3$O$_{10}$	125	–	–
HgBa$_2$Ca$_2$Cu$_3$O$_{8+\delta}$	134	–	–

8.2.1 Materials

Materials (typically pure metallic elements such as Hg and Pb) that totally exclude flux inside a superconductor (up to a well-defined transition temperature T_c) are referred to as type I superconductors, as shown in Figure 8.6(a). Materials that exhibit a partial Meissner effect are referred to as type II superconductors, as shown in Figure 8.6(b).

Below a lower critical field H_{c1}, type II superconductors act as type I superconductors. Above H_{c1}, the material is (incompletely) threaded by flux lines, and the material can be considered to be in a vortex state [8]. The higher the applied field, the greater number of allowed flux lines until the upper critical field H_{c2} is reached, where the entire material transitions to the normal state. Type II superconductors (e.g., NbTi and Nb_3Sn) tend to be alloys or transition metals with high electrical resistivity in the normal state. All HTS materials are type II superconductors. Superconducting magnets are fabricated from type II materials.

LTS devices have significant advantages and one disadvantage—operating temperature—over HTS devices. Because LTS materials are isotropic and have long coherence lengths (see Table 8.1) relative to their interatomic distances, it is possible to fabricate devices with three-dimensional structures. That allows crossovers and multilayer structures that permit higher sensitivity than single-turn devices. HTS crossovers (needed for multiturn coils) require larger dimensions than the coherence length of YBCO (see Table 8.1) in the c-direction. The effect is that an HTS crossover acts as a Josephson or insulating junction with the addition of significant $1/f$ noise (Section 8.3.5). The associated flux creep that can occur (particularly in HTS materials) by operating in the mixed (vortex) state can lead to nonlinearity or

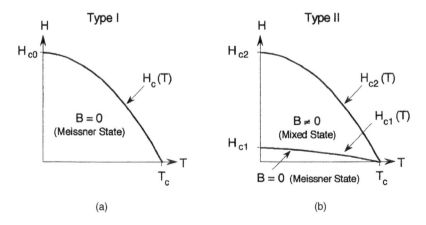

Figure 8.6 The H-T phase space for (a) type I and (b) type II superconductors.

hysteretic effects. While their pinning energies are somewhat lower than that of YBCO, bismuth and thallium compounds seem to have much lower densities of pinning sites [10]. As a result, the flux creep (and its associated resistivity) is considerably higher. By operating at lower temperatures (20K to 30K), it is possible to freeze out the hysteretic effects seen in bismuth and thallium devices at 77K. Because of that, most HTS SQUIDs are fabricated from YBCO because of its sufficiently strong (and intrinsic) flux pinning at 77K.

Another difference is that LTS materials (e.g., NbTi) are ductile and—in wire form—can be made into complex three-dimensional structures such as axial gradiometers (Section 8.4.2). Additionally, using NbTi (or Nb_3Sn) allows detection coils to be in high-field regions, while the actual LTS SQUID sensor can be placed in a low-field environment. Because of the inability to make a truly superconducting flexible three-dimensional structure, axial HTS gradiometers are not possible (although thin-film planar gradiometers are). Even if it were possible to make separate HTS coils, the inability to make superconducting joints (or joints with contact resistances at the subpicowatt level) prevents dc response in discrete element HTS circuits [10].

Another advantage of LTS materials is that they are stable in air, whereas moisture degrades the HTS structure. Thus, passivation layers or overcoatings are required and add to the complexity of manufacture. Although LTS materials have superior properties, the ability to operate a device at liquid nitrogen temperatures rather than at liquid helium temperatures gives HTS devices significant operational advantages and cannot be discounted.

The barrier used for fabricating the weak link is critical. Early point-contact devices used an oxide (NbO) layer. The first commercial dc SQUID used an amorphous SiO_2 barrier that allowed junctions with critical currents within a few percentage points to be easily fabricated. An unfortunate side effect was that the amorphous barrier had significant temperature-dependent resistivity that caused frequency-dependent noise (Section 8.3.5). Present-day LTS SQUIDs utilize AlO_x barriers with temperature independent critical currents. Bicrystal [Figure 8.5(d)] and ramp-edge [Figure 8.5(g)] junctions have become popular in commercial HTS devices. Reference [11] gives an excellent overview on HTS SQUIDs.

8.3 SQUID Operation

The major difference between RF and dc SQUIDs is that the dc SQUID may offer lower noise. The cost of that increase in sensitivity can be the complexity of the electronics needed to operate a dc SQUID and the difficulty in fabricating two nearly identical Josephson junctions in a single device. From a historical viewpoint, although the LTS dc SQUID was the first type of SQUID magnetometer made, early LTS development was with RF SQUIDs. With modern thin-film fabrication techniques and improvements in control electronics design, the dc SQUID offers clear advantages over the RF SQUID for many applications.

8.3.1 RF SQUIDs

The RF SQUID utilizes a single Josephson junction, and flux is normally (inductively) coupled into the SQUID loop via an input coil that connects the SQUID to the experiment and an "RF" coil that is part of a high-Q resonant circuit to read out the current changes in the SQUID loop (Figure 8.7).

This tuned circuit (typically operated at 19 MHz) is driven by a constant current RF oscillator that is weakly coupled to the SQUID loop. As the oscillator drive amplitude is increased, the (peak) detected output of the RF amplifier increases until a critical level is reached. That is, an increase in drive amplitude produces a very small increase in output. The RF drive is set so the SQUID operates on that plateau, where the detected output depends on the current flowing in the input coil. Any changes in the input coil current induces a change in the current flowing in the SQUID ring. The shielding current causes the total flux linking the SQUID ring to remain constant as long as the ring remains superconducting. Another contribution to the total flux in the SQUID (and to the shielding current) comes from the RF current in the (19 MHz) tuned circuit. Thus, we can consider the shielding current to consist of an ac component and a dc component, which biases the junction. When the amplitude of the ac component increases, the critical current of the weak link will be reached and a transition will occur, changing the flux state of the SQUID by a single flux quantum. That transition temporarily reduces the level of oscillation in the RF coil, which then builds up again to its maximum value, and the process repeats itself. Just after the transition, the weak link again becomes a superconductor, and the shielding bias current due to the current in the input coil reestablishes itself to quantize the flux in the SQUID. When the RF oscillations have been reduced sufficiently, the Josephson junction will again be superconducting and the amplitude of the RF oscillations will begin to increase again.

If the dc current in the input coil is changed, the dc bias of the shielding current in the SQUID is changed so that the RF-induced transition occurs

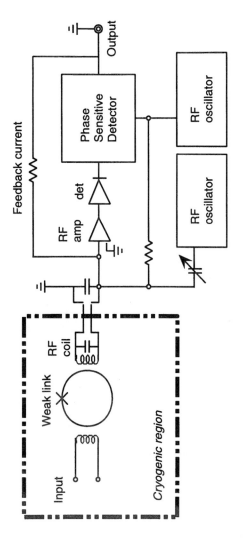

Figure 8.7 Block diagram of SQUID input and electronics for locked-loop operation of an RF SQUID.

at a different level of oscillation in the RF coil. The detected RF output is found to be the periodic function shown in Figure 8.8.

One way to measure the change in input coil current is simply to count the number of periods it produces in the detected RF output. A more commonly used mode of operation is a feedback scheme (see Figure 8.7), which locks in on either a peak or a valley in the triangle pattern output from the RF peak detector. A feedback flux is applied to the SQUID through the RF coil that just cancels the change in flux from the input coil. That allows flux resolution to $\mu\Phi_0$ levels.

An audio frequency oscillator provides a reference for a phase-sensitive detector and also modulates the flux linking the SQUID due to the RF coil by an amount $\Phi_0/2$ (peak to peak). That low-frequency flux modulation, typically several tens of kilohertz, modulates the detected RF output. The amplitude of the RF modulation is zero when the total flux in the SQUID corresponds to a peak or a valley (see Figure 8.8) and increases linearly to a maximum as the flux departs from an extremum one-fourth of the period ($\mu\Phi_0/4$). A change in flux in the SQUID results in an output from the PSD, which is fed back through a resistor to the RF coil. The feedback current through the RF coil counters the flux change from the input coil and maintains the flux in the SQUID locked at a value corresponding to an extremum in the RF response. This manner of operation is called a flux-locked loop.

It should be noted that the RF and input coils (see Figure 8.7) are not wound around the SQUID loop but inductively coupled to the SQUID loop. A minor advantage of the RF SQUID is that its cryogenic connections are simple—only a single coax or a single twisted pair of leads.

8.3.2 dc SQUIDs

The dc SQUID differs from the RF SQUID in the manner of biasing the Josephson junction and the number of junctions. Because there are two

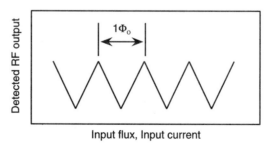

Figure 8.8 Triangle pattern showing detected output (RF) voltage versus flux coupled into the SQUID.

junctions, they need to be matched, as do the shunt resistors. The ideal shunt resistor should have a temperature-independent resistivity. The use of noble metal resistors such as Pd is preferred over amorphous materials or materials with superconducting transitions, which would prevent operation of LTS devices below 4.2K.

Figure 8.9 is a schematic of a typical dc SQUID. Like the RF SQUID (see Figure 8.7), the input, feedback, and modulation coils are not wound around the SQUID loop but inductively coupled to it. It is biased with a dc current approximately equal to twice I_c and develops a dc voltage across the junctions (and shunt resistors). A change in the magnetic flux applied through the SQUID loop induces a wave function phase change that enhances the current through one Josephson junction and reduces the current through the other. That asymmetry, which is periodic in Φ_0, is used to provide a feedback current that nulls the flux penetrating the SQUID loop. Like the RF SQUID, this feedback current (presented as a voltage at the output) is a direct measure of changes in flux applied to the SQUID. Additional details on the operation of dc SQUIDs can be found in [12]. The dc SQUID typically requires at least four pairs of leads (bias, modulation, signal, feedback) between each SQUID sensor and its electronics.

8.3.3 Noise and Sensitivity

SQUID noise is often presented as the spectral density of the equivalent flux noise $S_\Phi(f)$ as a function of frequency or noise energy per unit bandwidth $E_N(f) = S_\Phi(f)/2L$, where L is the inductance of the input coil. To

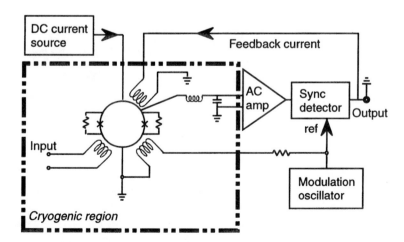

Figure 8.9 Block diagram of a typical dc SQUID.

allow devices with differing input inductances to be directly compared, the sensitivity of SQUID devices is best discussed in terms of the energy sensitivity:

$$E_N = L_{input} I_N^2 = \frac{\Phi_N^2}{L_{input}} \qquad (8.1)$$

where L_{input} is the input inductance of the device, I_N is the current noise, and Φ_N is the flux sensitivity. E_N is often expressed in terms of Planck's constant, $h = 6.6 \cdot 10^{-34}$ J/Hz.

Figure 8.10 shows typical energy sensitivities for LTS and HTS SQUIDs. As can be seen, the noise can be described as the sum of frequency-independent (white) and frequency-dependent ($1/f$) terms.

Magnetometers are often discussed in terms of field sensitivity. However, because the field sensitivity is as much dependent on the geometry of the detection coil (area, number of turns, etc.) as the SQUID itself, energy

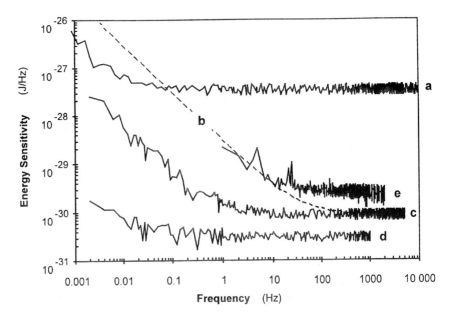

Figure 8.10 Energy sensitivity versus frequency for a number of different SQUID devices. (a) is an LTS RF SQUID operated at a bias frequency of 19 MHz; (b) is a dc-biased LTS dc SQUID with amorphous silicon barriers; (c) is (b) using ac biasing; (d) is a dc-biased LTS dc SQUID with AlO_x barriers; and (e) is an ac-biased HTS dc SQUID utilizing a ramp-edge junction [see Figure 8.5(g)].

sensitivity allows comparison of devices independent of coil geometry or input impedance. Calculation of magnetic field sensitivity is discussed in Section 8.4.

8.3.3.1 White Noise

RF SQUIDs

Typical white noise for an LTS RF SQUID operating at 19 MHz is 10^{-28} J/Hz [see Figure 8.10(a)]. The major limiting factor in the white noise of an RF SQUID is the bias frequency (f_B) used to excite the tank circuit. Noise in RF SQUIDs is proportional to $1/\sqrt{f_B}$ [13]. Increasing the bias frequency from 19 MHz [14] to 180 MHz [15] has been shown to reduce noise by a factor of 3. As f_B increases, the complexity of the electronics also tends to increase. With the discovery of high-temperature superconductivity, the first HTS SQUIDs were single-junction RF SQUIDs. Because it is not known (at least theoretically) if there is a fundamental limit to the white noise of RF SQUIDs, at 1 GHz, an RF HTS SQUID may be quieter than a dc HTS SQUID.

dc SQUIDs

The minimum noise energy for a dc SQUID is given by [16]

$$E_N = k_B T \sqrt{L_{\text{loop}} C} \tag{8.2}$$

where k_B is Boltzmann's constant, L_{loop} is the inductance of the SQUID loop, and C is the capacitance of the junction. Substituting appropriate numbers indicates that the minimum noise energy (E_N) for a dc SQUID is of the order of $h/2$. Devices with sensitivities of $\sim h$ have been constructed. The extremely low-noise levels are achieved by limiting dynamic range and avoiding feedback. The need for practical (useful) devices require that feedback be used and that the SQUID electronics have a reasonable dynamic range. Commercially available dc SQUIDs have noise levels approaching 10^{-31} J/Hz.

8.3.3.2 Temperature Dependence

Due to the strong temperature dependence of superconducting properties, such as energy gap, $\Delta(T)$; critical current, I_c; surface resistivity, R_S; and so on, especially near T_c, we find that basic SQUID parameters such as bias and modulation drive experience variations as a function of temperature.

The best noise performance is obtained when the SQUID is operated at or below $1/2 T_c$. A niobium LTS SQUID operating at 4.2K has

T/T_c = 0.44, while a YBCO SQUID at 77K has T/T_c = 0.83. Operating at lower temperatures also means potentially lower noise (Section 8.3.4). A significant temperature variation may require retuning for optimum performance. A typical LTS SQUID, for example, Figures 8.10(c) and (d), will have a temperature variation of ~0.1 Φ_0/K, an HTS device, for example, Figure 8.5(f), has $d\Phi/dT \approx 0.015\ \Phi_0$/K. Because liquid cryogens are normally used to cool SQUID devices, ambient pressure and pressure variations can change the cryogen temperature with a commensurate change in SQUID operating parameters. For example, geophysical applications may require operation in deep mines or underwater. A 1-torr pressure variation at 760 torrs is equivalent to a 1.5-mK change in the temperature of a liquid helium bath and a 11-mK change in the temperature of a liquid nitrogen bath.

8.3.3.3 Field Dependence

In normal operation, LTS SQUIDs are operated in the Earth's magnetic field or lower (<60 μT) environments. LTS experiments requiring operation in high fields usually employ coils that transport the measured flux or currents to the LTS SQUID sensor located in a low field region of the cryostat. Because HTS coils are located on the same substrate as the SQUID itself, HTS SQUIDs are subjected to the environment being measured. Some HTS devices simply will not function in fields exceeding tens of microteslas. In addition, the decreased flux pinning in HTS devices at high fields results in reduced slew rates. The development of ramp edge junctions [17] has allowed HTS devices to operate in fields up to 0.1T with white noise scaling as $\sqrt{H}$.

8.3.3.4 1/f Noise

Along with white noise, there exists a frequency-dependent contribution that increases as the frequency decreases (see Figure 8.10). The onset of this $1/f$ noise can be dependent on the ambient magnetic field when the SQUID sensor is cooled. Cooling the SQUID sensor in low ambient magnetic fields (< 1 μT) may significantly improve the $1/f$ performance, particularly with HTS SQUIDs using grain boundary junctions. It should be noted that measurements of $1/f$ noise, usually taken at frequencies well below 1 Hz, are difficult. The SQUID sensor should be placed in a superconducting can that is itself inside Mumetal® shielding. Care should be taken to eliminate any potential mechanical motion. Because vibration often appears as a $1/f^2$ contribution, excessive low frequency noise may be identified by its spectral content.

Sources of 1/f Noise

Thermally activated critical current fluctuations due to trapping and release of electrons in the barrier produce fluctuations in the (Josephson) barrier height $\Delta(T)$ with variations in I_c. A large contribution to the noise in some dc SQUIDs can arise from the presence of the dc current bias. By chopping the dc bias in combination with the conventional flux modulation techniques, it is possible to reduce the added $1/f$ noise. This ac bias reversal approach [18] separates the original signal waveform from the noise associated with the dc bias and can reduce $1/f$ noise at very low frequencies; see Figures 8.10(b) and 8.10(c). Bias reversal limits the maximum bandwidth to less than half the bias reversal frequency. Flux noise probably arises from motions of flux lines trapped in the body of the SQUID and is thought to be thermally activated. Flux noise fluctuations have not been able to be reduced by any known modulation scheme.

8.3.4 Control Electronics

The system output voltage is the voltage drop across the feedback resistor in a negative feedback loop controlled by the SQUID electronics. The feedback signal is generated in response to changes in the output signal of the SQUID sensor. The output of the SQUID sensor is periodic in the field coupled into the SQUID loop. Negative feedback (similar to a PLL technique) is used to maintain the system operating point at a particular (and arbitrary) flux quantum. When operated in this mode, the system is in a flux-locked loop.

One important factor of SQUID design is for the feedback electronics to be able to follow changes in the shielding currents. If the shielding current changes so fast that the flux in the SQUID loop changes by more than $\Phi_0/2$, it is possible that the feedback electronics will lag behind the rapidly changing flux. When the electronics finally catch up, they can lock on an operating point (see Figure 8.6) different from the original. In that case, the SQUID has "lost lock" because the SQUID has exceeded the maximum slew rate of the electronics. That places an upper limit on the bandwidth of the system. The typical bandwidth of commercially available SQUID systems is dc to 50+ kHz. Custom electronics have been built extending bandwidths above 5 MHz. Typical slew rates for SQUIDs are in the range of $10^5 \sim 10^6$ Φ_0/sec.

Even though we may not need or want to observe rapidly changing signals, situations may arise when ambient noise (e.g., 60 Hz) may determine the slew rate requirements of the system. To recover a signal from such interference, it is necessary that the system be able to track all signals present at the input, including the noise. When system response is sped up to handle very fast signals, sensitivity to RF interference and spurious transients is also increased. Because ability to remain locked while subjected to strong electrical transients is greatest when the maximum slew rate is limited (slow), while ability to track rapidly varying signals is greatest when the maximum slew rate is greatest (fast), it is desirable to be able to match the maximum slew rate capability to the measuring situation. As a matter of convenience, many commercial SQUID systems offer user-selectable slew rates along with high-pass and low-pass filters for noise reduction.

8.3.5 Limitations on SQUID Technology

When we utilize SQUID-based measurement systems and data reduction algorithms, it is important to bear in mind several fundamental limitations:

- SQUIDs are sensitive to relative (field or current) changes only. That is a consequence of the fact that the output voltage of a SQUID is a periodic function (see Figure 8.6) of the flux penetrating the SQUID loop. The SQUID is "flux-locked" on an arbitrary maximum (or minimum) on the V-Φ curve, and the SQUID output is sensitive to flux changes relative to the lock point.
- Although the SQUID has an intrinsic bandwidth of several gigahertz, when operated with standard flux-locked loop electronics using ac flux modulation, the maximum bandwidth is typically 50 to 100 kHz. Another limitation is the presence of $1/f$ noise. The use of ac biasing limits the maximum bandwidth to less than half the bias reversal frequency. If the bias reversal frequency is too high, noise can be induced due to voltage spikes in the transformer coupled preamplifier input circuit. Because of that, the maximum bandwidth of present-day HTS SQUIDs is limited to ~30 kHz. If MHz bandwidths are required, the ac bias is not used; however, there will be excess noise below 1 kHz.
- SQUID magnetometers are vector magnetometers. For a pure magnetometer operating in the Earth's magnetic field, a 180-degree rotation will sweep out a total field change of ~100 μT. If the magnetometer has a sensitivity of 10 fT/$\sqrt{\text{Hz}}$, tracking the total field change requires a dynamic range of 100 μT/10 fT = 200 dB, well beyond the capabilities of current electronics. In addition, the rotational speed must not cause the current flowing through the

SQUID sensor to exceed its slew rate limitations. Gradiometers (Section 8.4.4) are insensitive to uniform fields and do not suffer this dynamic range limitation.

8.4 Input Circuits

8.4.1 Packaging

Although it is possible to couple magnetic flux directly into the SQUID loop, environmental noise considerations make that difficult, if not impossible, in an unshielded environment. In addition, the area (A) of a typical SQUID loop is small (<0.1 mm^2), and its resulting sensitivity to external flux changes ($\Delta\Phi = A \cdot \Delta B$) small. Although a larger loop diameter would increase the SQUID's sensitivity to external flux, it also would make it much more susceptible to environmental noise. For that reason, external flux is normally inductively coupled to LTS SQUID loops by a flux transformer. Because HTS devices have the detection coil grown on the same substrate, the Josephson loop is exposed to the environment being measured. That limits the ability of HTS devices to operate in high fields (Section 8.3.3).

The packaging of the sensor should be sufficiently rugged to allow use under adverse conditions. A niobium can will shield LTS devices from external fields greater than 20 mT. If an experiment requires high fields, it is desirable to use a magnet with a compensation coil that generates a null field region. The SQUID sensor is placed there and connected to the experiment by twisted pair(s) of NbTi leads for fields less than 10T. Nb$_3$Sn leads may permit superconducting connections in fields above 20T. HTS devices require isolation from humid environments to prevent degradation of the YBCO film. Normally a surface passivation layer over the YBCO is combined with encapsulating the HTS SQUID in a gas-filled, sealed G-10 or plastic enclosure.

Today, SQUIDs are fabricated as planar devices. In such a configuration, the superconducting loop, Josephson junctions, and coils (input, feedback, and modulation) are patterned on the same device. Multilayer deposition techniques are used (primarily in LTS devices), and coils normally are in the form of a square washer. The planar configuration leads to quite small devices, occupying only a few cubic millimeters compared to 5+ cm^3 (1.2 cm diam. by 5 cm) for older toroidal RF SQUIDs [19]. Another advantage of the planar device is that it is possible to have the detection coils as part of the SQUID sensor, eliminating the need for separate (three-dimensional) detection coils. Such an integrated sensor has the potential to significantly reduce the complexity of multichannel systems.

8.4.2 The SQUID as a Black Box

Whether an RF or a dc SQUID, a SQUID system can be considered as a black box that acts like a current (or flux)-to-voltage amplifier with extremely high gain. In addition, it offers extremely low noise, high dynamic range (>140 dB), excellent linearity (>1:10^7), and a wide bandwidth that can extend down to dc. Conceptually, the easiest input circuit to consider for detecting changes in magnetic fields is that of a SQUID sensor connected to a simple superconducting coil (Figure 8.11).

Because the total flux in a superconducting loop is conserved, any change in the external field through the signal coil will induce a current in the flux transformer, which must satisfy

$$\Delta\Phi = NA\Delta B = (L_{coil} + L_{input})\Delta I \qquad (8.3)$$

where ΔB is the change in applied field; N, A, and L_{coil} are the number of turns, area, and inductance of the detection coil, respectively; L_{input} is the inductance of the SQUID input coil; and ΔI is the change in current in the superconducting circuit. If the lead inductance is not negligible, it must be added to L_{coil} and L_{input}.

8.4.3 Sensitivity

Maximum sensitivity is almost never the optimum sensitivity. Nevertheless, an understanding of the techniques used to maximize sensitivity is essential to any discussion of optimum sensitivity. Because the SQUID system has an output that is proportional to the input current, maximum sensitivity is

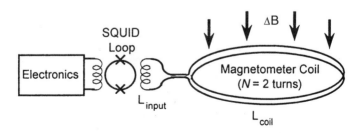

Figure 8.11 Schematic diagram of typical SQUID input circuit.

obtained by using the input circuit that provides the maximum current into the SQUID and satisfies all other constraints of the experimental apparatus.

A common constraint is the physical size of the detection coil. As seen in (8.3) and (8.4), it is clear that maximum sensitivity to uniform fields is obtained with an infinitely large coil. Infinitely large coils fit quite nicely into dewars with infinitely large necks, but they have the serious disadvantage of boiling off liquid cryogens at an infinitely fast rate. Therefore, it is common practice to build the largest-diameter detection coil that will fit in a physically realistic dewar neck.

Another constraint on coil design is spatial resolution, which is dependent on the nature of the source, the geometry of the detection coil, and the distance between the coil and the source. If nearby objects are to be measured (e.g., biomagnetism or magnetic microscopy), spatial resolution may be more important than absolute sensitivity. One rule of thumb is not to have the coil diameter significantly less than the distance between the coil and the source. That distance includes the tail spacing (gap) of the dewar used to provide the cryogenic environment.

The object(s) being measured determine the tradeoff between sensitivity and spatial resolution. For example, from the $1/z^3$ field dependence of a single magnetic dipole source less than one coil diameter from the detection coil, it is easily shown that spatial resolution can be better than one-tenth the coil diameter. Multiple or higher order (quadruple, etc.) sources may have spatial resolution for the order of the coil diameter. For sources many diameters distant from the coil, resolution may be multiples of the coil diameter. In that situation, larger coils are recommended. However, it makes no sense to design coils for significantly higher sensitivity than environmental constraints (noise) permit.

To calculate the sensitivity and noise level of a simple detection coil system, the inductance of the detection coil must be known. The inductance of a flat, tightly wound, circular multiturn loop of superconducting wire is given by [20]

$$L = 4 \cdot 10^{-7} N^2 \pi r \left[\ln\left(\frac{8r}{r_{\text{wire}}}\right) - 2 \right] \quad (8.4)$$

where r is the radius of the detection coil and r_{wire} is the radius of the (superconducting) wire. Knowing the coil inductance L_{coil}, we can rewrite (8.4) as

$$\Delta B = (L_{\text{coil}} + L_{\text{input}}) \Delta I / NA \quad (8.5)$$

Because the SQUID system has an output proportional to the input current, maximum sensitivity is obtained by using the input circuit that provides the maximum current into the SQUID and satisfies all other constraints of the experimental apparatus. For a pure magnetometer of a given diameter, the maximum sensitivity occurs when the impedance of the detection coil matches that of the SQUID sensor ($L_{coil} = L_{input}$).

8.4.4 Detection Coils

Several factors affect the design of the detection coils [21], including the desired sensitivity of the system, the size and location of the magnetic field source, and the need to match the inductance of the detection coil to that of the SQUID. The ability to separate field patterns caused by sources at different locations and strengths requires a good signal-to-noise ratio. At the same time, we have to find the coil configuration that gives the best spatial resolution. Unfortunately, the two tasks are not independent. For example, increasing the signal coil diameter improves field sensitivity but sacrifices spatial resolution. In practice, system design is restricted by several constraints: the impedance and noise of the SQUID sensors, the size of the dewar, and the number of channels, along with the distribution and strength of noise sources.

It is extremely important for dc response that the detection coil(s) be superconducting. Resistance in the detection circuit has two effects: (1) attenuating the signal and (2) adding Nyquist noise. Resistive attenuation is important only below a frequency f_0, such that the resistive impedance is equal to the sum of the inductive impedances in the loop (e.g., $f_0 \approx R/L_{tot}$, where L_{tot} is the total inductive impedance of the loop). Resistive noise is important only if it becomes comparable to other noise sources or the signal ($<10^{-30}$ J/Hz for biomagnetism, $<10^{-26}$ J/Hz for geophysics). For a SQUID with $E_N \sim 10^{-30}$ J/Hz, the total resistance of the circuit, including any joints, must be less than 10^{-13} Ω [10]. Thus, it is important that all solder joints, press-fits, or connections have as low a joint resistance as possible.

Figure 8.12 displays a variety of detection coils. The magnetometer in Figure 8.12(a) responds to the changes in the field penetrating the coil. More complicated coil configurations provide the advantage of discriminating against unwanted background fields from distant sources while retaining sensitivity to nearby sources.

Because of our current inability to make flexible wire or true superconducting joints in HTS materials, three-dimensional HTS coil structures, for

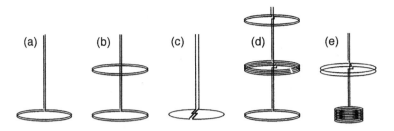

Figure 8.12 (a) Magnetometer; (b) first derivative gradiometer; (c) planar gradiometer; (d) second derivative gradiometer; and (e) first derivative asymmetric gradiometer.

example, those in Figures 8.12(b), 8.12(d), and 8.12(e), are not possible. Present-day HTS magnetometers are fabricated as planar devices and are available only as pure magnetometers [Figure 8.12(a)] and planar gradiometers [Figure 8.12(c)]. As a result, commercially available HTS devices are currently in the form of magnetic-sensing rather than current-sensing devices.

8.4.5 Gradiometers

Magnetometers are extremely sensitive to the outside environment. That may be acceptable if we are measuring ambient fields. If what is to be measured is close to the detection coil and weak, outside interference may prevent measurements at SQUID sensitivities. If the measurement is of a magnetic source close to the detection coil, a gradiometer coil may be preferred. The field of a magnetic dipole is inversely proportional to the cube of the distance between the dipole and the sensor. It follows that the field from a distant source is relatively uniform in direction and magnitude at the sensor. If we connect in series two identical and exactly parallel loops wound in opposite senses, separated by a distance b (the baseline), we obtain a coil, like that shown in Figure 8.12(b), that will reject uniform fields.

Because the response of a single coil to a magnetic dipole goes as $1/z^3$, an object that is much closer to one coil than the other couples better to the closer coil. Sources that are relatively distant couple equally into both coils. For objects that are closer than $0.3b$, the gradiometer acts as a pure magnetometer, while rejecting more than 99% of the influence of objects more than $300b$ distant (Figure 8.13). In essence, the gradiometer acts as a compensated magnetometer. It is possible to use two gradiometers connected in series opposition, as shown in Figure 8.12(d), to further minimize the response of the system to distant sources. That can be extended to higher

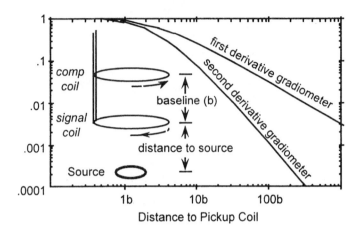

Figure 8.13 Response of gradient coils relative to magnetometer response ($1/z^3$ suppressed).

orders by connecting in series opposition two second-order gradiometers, and so on. Doing so, however, reduces the sensitivity of the instrument to the signal of interest and may not significantly improve the signal-to-noise ratio.

Rejection of distant noise sources depends on having a precise match (or balance, as it is sometimes referred to) between the number of area-turns in the coils. A symmetric gradiometer, like that shown in Figure 8.12(b), requires that $N_{signal} A_{signal} = N_{comp} A_{comp}$, where N is the number of turns and A is the area of the signal and compensation coils. An asymmetric design, like that shown in Figure 8.12(e), has the advantage that the inductance (L_{signal}) of the signal coil(s) is much greater than the compensation coils (L_{comp}); greater sensitivity is achieved than with a symmetric design. Another advantage is that the signal coil diameter is reduced, leading to potentially higher spatial resolution. The optimum conditions for the number of turns in an asymmetric signal coil is given by [22]:

$$(L_{signal} + L_{comp} + L_{input} + L_{leads}) - N_{signal} \frac{\partial}{\partial N_{signal}} (L_{signal} + L_{comp} + L_{input} + L_{leads}) = 0 \quad (8.6)$$

If a gradiometer is perfectly made (balanced), it will reject uniform fields. However, if one coil has a larger effective diameter than the other, the response will be that, not of a perfect gradiometer, but of a gradiometer in series with a magnetometer. Mathematically, the balance, β, can be defined as $V_t \propto G + \beta \mathbf{H}$, where V_t is the system response, G is the coil's response to a gradient field (e.g., dB_Z/dz), and $\mathbf{H}$ is the applied uniform field. If the coils are coplanar, no x- or y-field components should be detected. However, the reality of fabrication, (e.g., tilt due to coil forms constructed from multiple pieces) is such that there may be β's with x and y components.

Typically, coil forms used to wind gradiometers can be machined (grooved) to achieve balances that range from $\beta = 0.01 \sim 0.001$. Planar devices, through photolithography, can achieve lower levels—a factor of 10 or better. Superconducting trim tabs placed within the detection coils can improve β to the parts-per-million level. High degrees of balance can allow a SQUID gradiometer to operate in relatively large (millitesla) ambient fields while maintaining sensitivities in the tens of femtotesla (fT).

For multichannel systems (such as those used in biomagnetism), it is not possible to use externally adjustable trim tabs—the tabs tend to interfere with each other. The use of electronic balancing [23] can provide balance ratios at the parts-per-million level.

8.4.6 Electronic Noise Cancellation

In this situation, portions of (additional) magnetometer reference channel response(s) are summed electronically with the gradiometer's input to balance out its effective magnetometer response. The simplest scheme is to use a second B_Z magnetometer coil and subtract its output, $V(B_Z)$ from the output of the gradiometer, V_t. The actual output is attenuated (by a factor b) so as to exactly cancel the imbalance (β) of the gradiometer. By adjusting $bV(B_Z)$ to equal $\beta \mathbf{H}$, the net result goes to G. Because there can be imbalance in the x and y components, a three-axis set of coils (Figure 8.14) allows compensation of the B_X, B_Y, and B_Z components.

In addition to field noise, gradient fields generated by distant sources can be large enough to mask the signals being measured. Additional improvement can be achieved by the addition of a second gradiometer compensation channel. Thus, the system output can be described as

$$V_{out} \propto \mathbf{G} + \beta \mathbf{H} + b_1 V(B_X) + b_2 V(B_Y) + b_3 V(B_Z) + g_4 V(G_Z) \tag{8.7}$$

where b_1 is the weighting for the B_X component, b_2 is the weighting for the B_Y component, b_3 is the weighting for the B_Z component, and g_4 is the weighting for the gradient reference G_Z (dB_Z/dz) component.

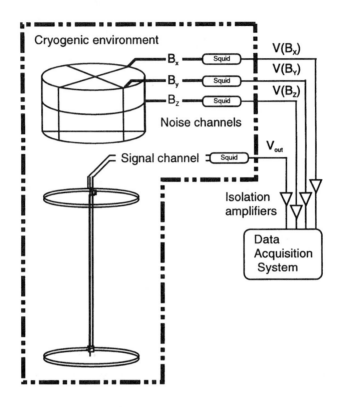

Figure 8.14 First-order gradiometer with three noise cancellation channels.

For a single-axial gradiometer, the use of three magnetometers (B_X, B_Y, B_Z) gave an attenuation of externally generated noise of 12 dB [24]. The addition of an external gradient reference channel improved noise rejection to better than 40 dB. Moving the reference gradiometer 1m away from the signal coil reduced noise rejection by a factor of 4 (12 dB). This gradient channel should be located sufficiently far from the sources being measured so as not to detect significant signal but close enough so that it sees the same gradient noise.

Because the detection coil is not perfectly balanced, ideally we should subtract all field and gradient components, expanding (8.7). To do that would require eight "noise" channels, for example, B_X, B_Y, B_Z, dB_X/dx, dB_Y/dy, dB_X/dy, dB_X/dz, and dB_Y/dz. From those components, all nine elements of the gradient tensor can be created and used to compensate for any imbalance of the detection coil(s). The use of eight-element tensor arrays as reference channels can further improve external noise rejection, with rejection values exceeding 60 dB [23]. A major advantage of electronic balancing is significant improvement in immunity to low-frequency environmental noise.

The simplest way to perform noise cancellation is to simultaneously take data from the detection coil(s) and the noise channels. Then, in postprocessing, digitize the data and determine the weighting factors for each noise channel to minimize any common mode "noise." We can also include time derivatives of the field and gradient components into the cancellation algorithm to minimize effects of eddy-current noise. If there is sufficient processing power, it may be possible to do real-time processing of the noise contribution.

8.5 Refrigeration

The superconducting nature of SQUIDs requires them to operate well below their superconducting transition temperature (9.3K for niobium and 93K for YBCO). Ideally, the cryogenic environment should provide stable cooling (mK or μK depending on $d\Phi/dT$ of the sensor; see Section 8.3.5), have no time-varying magnetic signature, be reasonably compact and reliable, and, if mechanical in nature, introduce neither mechanical vibration nor a magnetic signature into the detection system. The thermal environment for the SQUID sensor and detection coil typically has been liquid helium or liquid nitrogen contained in a vacuum-insulated vessel known as a dewar (Figure 8.15). The cryogen hold time depends on the boil-off rate (heat load) and the inner vessel volume.

8.5.1 Dewars

The major heat load on dewars is due to thermal conduction down the neck tube and magnetometer probe, along with blackbody radiation. The space between the inner and outer walls is evacuated to prevent thermal conduction between room temperature and the cryogen chamber. Within the vacuum space, a thermal shield (anchored to the neck tube) acts to reduce heat transfer by thermal (blackbody) radiation. The thermal shield can be vapor cooled, either by using the enthalpy of the evaporating helium or nitrogen gas or by having the shield thermally connected to a liquid nitrogen reservoir. Dewars with removable sections (e.g., tails) use liquid nitrogen-cooled shields.

If the experiment involves measurements interior to the dewar, then a metallic dewar is preferable. Metallic dewars offer significant shielding

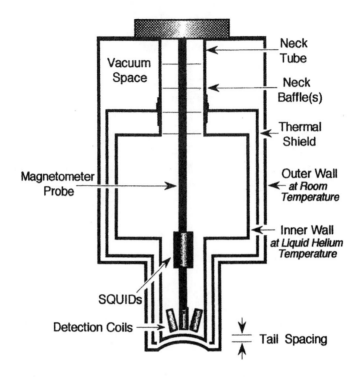

Figure 8.15 Typical design of a fiberglass dewar used for biomagnetic measurements.

from environmental noise at frequencies above 10 ~ 100 Hz. If the system is to measure magnetic fields exterior to the dewar, the dewar must be magnetically transparent and metallic construction is not appropriate. Dewars for external field measurements (Section 8.7) normally are constructed of nonmetallic, low-susceptibility materials to minimize their magnetic interactions with the SQUID sensors and detection coils. Materials used typically are glass-fiber epoxy composites such as G-10. To get the detection coil(s) as close as possible to the object being measured, a "tailed" design is often used. That decreases the forces on the bottom of the dewar and allows the use of thinner end pieces (closer tail spacing). Dewars for biomagnetic measurements often have curved tails to get closer to the head, chest, or abdomen.

The major advantage of high-temperature superconductivity is the simplified cryogenics and reduced spacing between cryogenic regions and room temperature. The thermal load (due to conduction and blackbody radiation) is less, and the heat capacity of what needs to be cooled is larger (implying smaller temperature variations for a given heat load). Because the latent heat per unit volume of liquid nitrogen is ~60 times larger than liquid helium, hold times become months rather than days for an equivalently sized dewar.

8.5.2 Closed-Cycle Refrigeration

As an alternative to the use of liquid cryogens, closed-cycle refrigeration [25] is desirable for several reasons: reduction of operating costs, use in remote locations, operation in nonvertical orientations, avoiding interruptions in cryogen deliveries, safety, and the convenience of not having to transfer every few days. Parameters governing suitability include physical size, absence of periodic replacement of cryogenic fluid, and—most important—vibration and magnetic signature. While cryocoolers [26, 27] can have large cooling capacities, unless hundreds of channels are involved, only milliwatts of cooling capacity are needed to maintain SQUID sensors at their operating temperatures.

The first practical cryocooled SQUID system was the BTi CryoSQUID (Figure 8.16) [28]. Based on a two-stage Gifford-McMahon (GM) refrigerator, the use of a joule-Thompson (JT) stage allowed 4K operation with

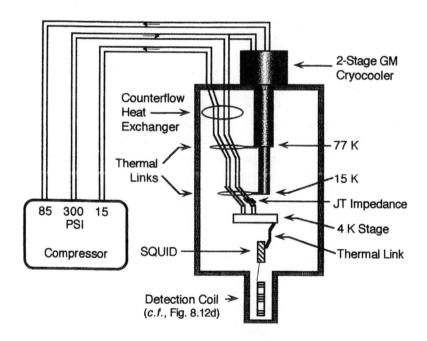

Figure 8.16 CryoSQUID components.

reduced vibration. An electronic comb filter was required to filter the ~1 Hz compressor vibration from the output of the dc SQUID electronics to achieve system performance of 20 fT/$\sqrt{Hz}$. However, the acoustic (audible) noise from the compressor prevented its use in auditory evoked brain measurements.

Although multichannel cryocooled systems [28] have been built, the inherent vibration and magnetic signature of present-day closed-cycle refrigerators prevent their widespread use. However, the development of pulse tube refrigerators [29] offers promise for magnetometer operation with significantly reduced vibration.

8.6 Environmental Noise (Noise Reduction)

The greatest obstacle to SQUID measurements is external noise sources. If the object being measured is within the cryostat (as is typical in most laboratory experiments), metallic shielding can minimize external noise (e.g., act as a low-pass eddy-current shield). The use of gradiometer detection coils (Section 8.4.4) can significantly attenuate the effect of distant noise sources. Superconducting shields essentially eliminate all external field variations. That assumes that any electrical inputs to the experimental region have been appropriately filtered. Powerline or microprocessor clock frequencies can severely degrade performance. Unfortunately, if external objects are to be measured, superconducting shields are not appropriate.

When measuring external fields, the SQUID magnetometer must operate in an environment—the magnetic field of the Earth—that can be 10 orders of magnitude greater than its sensitivity (Figure 8.17). The magnetic field at the surface of the Earth is generated by a number of sources. There exists a background field of ~50 μT with a daily variation of ±0.1 μT. In addition, there is a contribution (below 1 Hz) from the interaction of the solar wind with the magnetosphere. The remaining contributions to external magnetic fields are primarily man-made. They can be caused by structural steel and other localized magnetic materials such as furniture and instruments that distort the Earth's field and result in field gradients, moving vehicles that generate transient fields, electric motors, elevators, radios, televisions, microwave transmitters, and the ever present powerline electromagnetic field and its harmonics.

8.6.1 Gradiometers for Noise Reduction

If the purpose of the measurement is to detect the magnetic field of a relatively close object, the detection coil(s) can be configured as a gradiometer

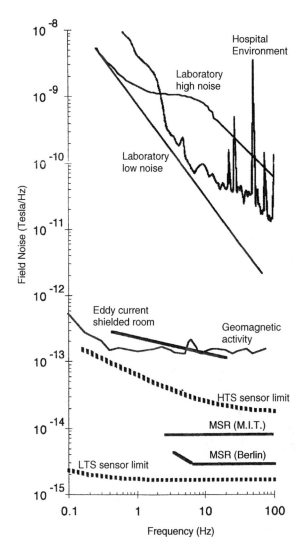

Figure 8.17 Spectra of rms field noise in various environments as a function of frequency (after [30]).

(see Section 8.4.4) whose baseline is larger than the distance from the coil(s) to the object. That can allow rejection of external noise by more than 120 dB with less than a decibel loss of signal. It is standard practice to configure SQUID measurement systems for biomedical and nondestructive evaluation measurements as gradiometers.

8.6.2 Magnetic Shielding

One method to attenuate external noise sources is with an eddy-current shield that generates fields that act to cancel the externally applied fields within the conducting material. The shielding effect is determined by the skin depth, λ—the distance where the field is attenuated by a factor $1/e$. For a sinusoidal varying wave,

$$\lambda = \sqrt{\rho/\pi\mu_0 f} \qquad (8.8)$$

where f is the frequency of the applied field, ρ is the electrical resistivity, and μ_0 is the magnetic permeability of free space. In situations where the wall thickness $t \ll \lambda$, external fields are attenuated by

$$\frac{H_{\text{internal}}}{H_{\text{external}}} = \frac{1}{1 + (2\pi f L/R)} \qquad (8.9)$$

where L is the inductance of the enclosure and R is the resistance along the path of current flow. Unfortunately, induced currents in the shield generate noise. For a cylindrical shape at a temperature T,

$$B_{rms} = \sqrt{\frac{64\pi k_B T t}{l d \rho}} \qquad (8.10)$$

where l is the length and d the diameter of the can. The cutoff frequency is given by $f_{-3\text{dB}} \approx \rho/4\pi t d$. Because of noise considerations, eddy-current shields that are to be placed near the detection coils should be made from relatively poor conductors, such as BeCu. Magnetic shielding is further discussed in Chapter 11.

8.6.2.1 Shielded Rooms

Another approach is to use eddy-current shielding to shield the entire measurement system. An eddy-current room constructed with 2-cm high-purity aluminum walls can achieve shielding >40 dB at 60 Hz, with improved performance at higher frequencies. The equivalent field noise is less than 200 fT/$\sqrt{\text{Hz}}$ at frequencies above 1 Hz.

In the situation where $t \gg \lambda$, the attenuation goes as $(r/\lambda)e^{t/\lambda}$. The need for shielding at lower frequencies has led to the use of magnetically shielded rooms (MSRs). If pure eddy-current shielding were used, it would require wall thicknesses that could exceed 1m or more (below 1 Hz). For a ferromagnetic material, the permeability of the material $[\mu = \mu_0(1 + \chi)]$ replaces μ_0 in (8.8). The shielding is because flux prefers the path with the highest permeability. Since magnetically "soft" materials (e.g., Mumetal®) can have permeabilities that exceed 10^4, the external magnetic flux is routed around the walls, avoiding the interior. The use of multiple shields can act to further shield the interior of a MSR. For the six-layer Berlin MSR (see Figure 8.17), shielding factors exceeded 80 dB at frequencies above 0.01 Hz, with noise levels below 3 $fT/\sqrt{Hz}$. All commercial MSRs combine multiple Mumetal® and aluminum walls.

8.7 Applications

A large number of applications configure the SQUID as a magnetometer (Figure 8.18).

SQUIDs can also be configured to measure a wide variety of electromagnetic properties, as shown in Figure 8.19.

The state of the art in materials processing limits the variety of superconducting input circuits that can be used with HTS SQUIDs. There is no existing method for making superconducting connections to SQUIDs with HTS wire. As a result, commercially available HTS devices are in the form of magnetic-sensing [Figure 8.19(b)] rather than current-sensing devices [Figures 8.19(a) and 8.19(c) through (f)].

8.7.1 Laboratory Applications

Table 8.2 lists the typical capabilities of SQUID-based instruments. The letters in parentheses refer to the corresponding configurations in Figure 8.19. Additional information on laboratory applications of SQUID systems can be found in [19, 31].

8.7.1.1 ac Measurements

The SQUID can also be used as the null detector in an ac bridge circuit (Figure 8.20) to measure both resistive and reactive components of a complex impedance. The unknown impedance Z is excited by a current generated by an oscillator voltage that is attenuated by a precision ratio transformer (λ). The difference between the voltage developed across the unknown impedance Z and that developed in the secondary of a nulling mutual inductor **m** is applied to the input of the SQUID circuit. The primary current in **m** is proportional to the oscillator voltage and is defined by the

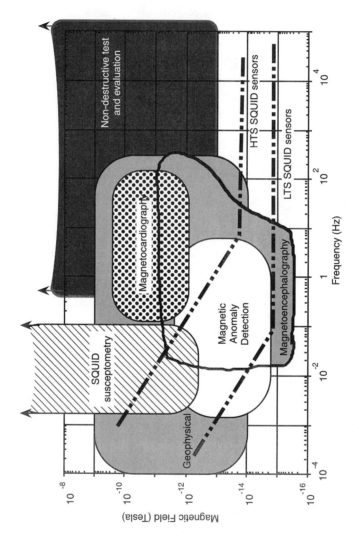

Figure 8.18 Field sensitivities and bandwidths typical of various applications. The lines indicate the sensitivity of commercially available SQUIDs.

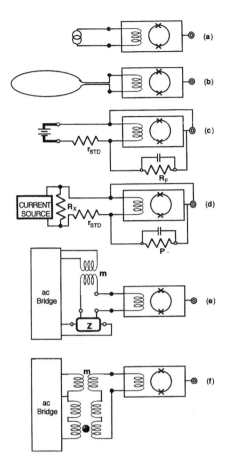

Figure 8.19 (a) ac and dc current; (b) magnetic field; (c) dc voltage; (d) dc resistance; (e) ac resistance/inductance bridge; and (f) ac mutual inductance (susceptibility bridge).

setting of the ratio transformer (α). An additional reactive current is supplied by a second ratio transformer (β), which causes the primary current to be passed through a capacitor rather than a resistor, thus generating a 90-degree phase shift in the voltage applied to **m**. The amplified off-balance signal that appears at the output of the SQUID control electronics can be displayed by means of a lock-in amplifier tuned to the oscillator frequency. Assuming $I_N \approx 1 \text{ pA}/\sqrt{\text{Hz}}$, such a system is capable of measuring self- and mutual inductances between 10^{-12}H and 10^{-3}H with $1:10^6$ part resolution [7, 31].

Table 8.2
Capabilities of SQUID-Based Measurements

Measurement	Sensitivity
Current (a)	10^{-12} A/$\sqrt{Hz}$
Magnetic field (b)	10^{-15} T/$\sqrt{Hz}$
dc voltage (c)	10^{-14} V
dc resistance (d)	10^{-12} Ω
Mutual/self-inductance (e)	10^{-12} H
Magnetic moment (f)	10^{-10} emu

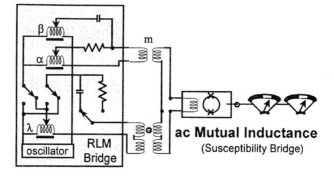

Figure 8.20 Block diagram of ac bridge.

Figure 8.21(a) shows a typical experimental setup for measurement of ac susceptibility. Such a configuration can be used for thermometry by measuring the susceptibility of paramagnetic salts such as $Ce_2Mg_3(NO_3)_{12} \cdot 24\ H_2O$, usually referred to as CMN [32].

8.7.1.2 SQUID Magnetometers/Susceptometers

Instead of using a secondary ac excitation coil [Figures 8.19(f) and 8.21(b)], a dc field can be used to magnetize samples. Typically the field is fixed and the sample moved into the detection coil's region of sensitivity [Figure 8.21(c)]. The change in detected magnetization is directly proportional to the magnetic moment of the sample. Because of the superconducting nature of SQUID input circuits, true dc response is possible.

Commonly referred to as SQUID magnetometers, these systems are properly called SQUID susceptometers. They have a homogeneous superconducting magnet to create a uniform field over the entire sample measuring region and the superconducting pickup loops. The magnet induces a moment

Superconducting Quantum Interference Devices (SQUIDs)

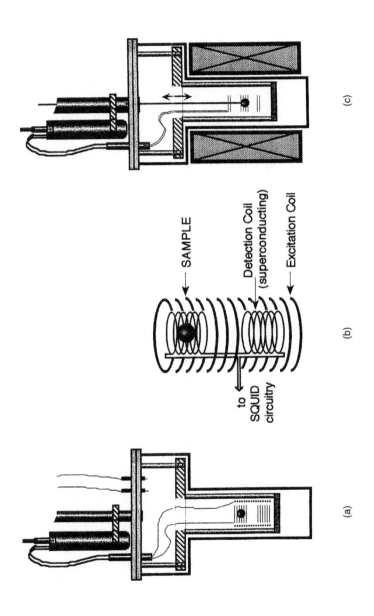

Figure 8.21 Magnetic susceptibility measurement apparatus (liquid helium dewar not shown): (a) ac susceptibility; (b) signal and excitation coil details; and (c) second derivative oscillating magnetometer for dc measurements with external dc field coils.

allowing a measurement of magnetic susceptibility. The superconducting detection loop array is rigidly mounted in the center of the magnet. That array is configured as a gradient coil to reject external noise sources. The detection coil geometry determines what mathematical algorithm is used to calculate the net magnetization. Oppositely paired Helmholtz coils and first and second derivative gradiometers have all been used successfully. Coupling two axial channels of differing gradient order can significantly improve noise rejection.

Sensitivities better than 10^{-8} emu have been achieved, even at applied fields of 9T [Figure 8.21(c)]. Placement of secondary excitation coils can allow ac susceptibility measurements approaching 10^{-8} emu to be made in the presence of a significant dc bias field. Variable temperature capability (1.7K to 800K) is achieved by placing a reentrant cryostat within the detection coils.

8.7.2 Geophysical Applications

SQUID magnetometers are used to measure the Earth's magnetic field (see Figure 8.17) at frequencies ranging between 1 kHz and 10^{-4} Hz. A technique known as magnetotellurics [33] can be used to determine the electrical conductivity distribution of the Earth's crust by measuring the Earth's electric and magnetic fields. Because the Earth is a good electrical conductor compared to the air, the electrical field generated in the ionosphere (due to the solar wind) is reflected at the Earth's surface, with components of both the electric and magnetic fields decaying as they penetrate into the Earth. The decay length or skin depth $\delta = 500\sqrt{\rho\tau}$ (where ρ is the electrical resistivity of the Earth and τ is the period of the electromagnetic wave).

In magnetotellurics, the electric field (as a function of frequency) is related to the magnetic field via an impedance tensor, where $\mathbf{E}(\omega) = Z\mathbf{H}(\omega)$. The impedance tensor ($Z$) contains the four complex elements Z_{xx}, Z_{xy}, Z_{yx}, and Z_{yy} and is related to the resistivity by $\rho_{ij} = 0.2|Z_{ij}(\omega)|^2 \tau$, where Z has units of millivolts per kilometer-nanotesla.

Magnetic anomaly detection utilizes the five unique spatial components of $\nabla \mathbf{B}$ to uniquely locate a magnetic dipole. This method has potential uses in mineralogical surveys and detection of unexploded ordnance.

8.7.3 Nondestructive Test and Evaluation

Magnetic sensing techniques such as eddy-current testing have been used for many years to detect flaws in structures. A major limitation on their

Table 8.3
NDE Measurement Techniques

Imaging
  Intrinsic currents
  Remnant magnetization
    Embedded magnetic sensors
  Flaw-induced perturbations in applied currents
  Johnson noise in metals
  Eddy currents in an applied ac field (flaws)
Hysteretic magnetization due to:
  Cyclic stress (strain)
  Simultaneous dc and ac magnetic fields
Magnetization of paramagnetic, diamagnetic, and ferromagnetic materials in dc magnetic fields

---

sensitivity is the skin depth, shown as (8.8), of metallic materials. Because SQUID sensors have true dc response and superior sensitivity, they can see "deeper" into metallic structures. The dc response also means that SQUID sensors can detect remnant magnetization without the need for externally applied magnetic fields. Their flat frequency response and zero phase distortion allow for a wide range of applications. One potential application of SQUIDs is in detection of stress or corrosion in reinforcing rods used in bridges, aircraft runways, and buildings. Table 8.3 lists some of the measurement techniques for which SQUID sensors can be used.

SQUID magnetometers have been used to make noncontact measurements of electronic circuits [34]—one instrument has better than 10-μm resolution [35]. Such instruments with megahertz bandwidths could be used for circuit board and IC mapping. An excellent overview of SQUID nondestructive evaluation (NDE) research can be found in [36].

8.7.4 Medical Applications

The use of bioelectric signals as a diagnostic tool is well known in medicine, for example, the electrocardiogram (EKG) for the heart and the electroencephalogram (EEG) for the brain [30]. The electrical activity that produces the surface electrical activity that is measured by EEGs and EKGs also produces magnetic fields. The analogous magnetic measurements are known as the magnetocardiogram (MCG) and the magnetoencephalogram (MEG). Other physiological processes also generate electrical activity with analogous magnetic fields (Figure 8.22).

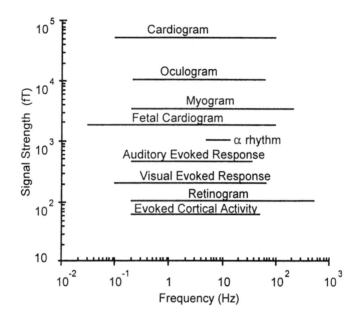

Figure 8.22 Typical amplitudes and frequency ranges for various biomagnetic signals.

Magnetic fields from active electrical sources in the body can be measured passively and external to the body by placing the magnetometer in close proximity to the body's surface. It has been shown that a population of neurons in the brain can be modeled as a current dipole that generates a well-defined magnetic field profile. Mapping of those field profiles can be used to infer the location of the equivalent active dipole site region within millimeters. Using evoked response techniques, the location of signal pathways and information processing centers in the brain can be mapped at different delay times (latencies) following the stimulus.

There are also magnetic measurements for which there are no electrical analogs [37]. These are measurements of static magnetic fields produced by ferromagnetic materials ingested into the body and measurements of the magnetic susceptibility of materials in the body. In particular, information on the quantity and depth of diamagnetic or paramagnetic materials (such as iron stored in the liver) can be obtained by using magnetizing and detection coils of differing sizes in the same instrument and measuring the induced field as a function of distance. This technique is already being used to monitor patients suffering from iron overload diseases such as thalassemia and hæmochromatosis.

The development of the SQUID has allowed the development of noninvasive clinical measurements of biomagnetic fields. The use of gradiometers can allow measurements to be made in unshielded environments at sensitivities below 20 $fT/\sqrt{Hz}$. Typically, however, neuromagnetic measurements are made in room-sized MSRs [38] which allow measurements of the magnetic field of the brain over the entire surface of the head (>150 positions simultaneously). Table 8.4 gives some of the areas in which SQUID magnetometers are currently being used in medical research.

Table 8.4
Applications of Biomagnetism

Neuromagnetism (studies of the brain)
Epilepsy
Presurgical cortical function mapping
Drug development and testing
Stroke
Alzheimer's disease
Neuromuscular disorders
Prenatal brain disorders
Performance evaluation

Magnetocardiography (studies of the heart)
Arrhythmia
Heart muscle damage
Fetal cardiography

Other medical applications
Noninvasive *in-vivo* magnetic liver biopsies
Gastroenterology (studies of the stomach)
Intestinal ischemia
Lung function and clearance studies
Nerve damage assessment

References

[1] Kamerlingh-Onnes, H., Akad. van Wetenschappen, Amsterdam, Vol. 14, No. 113, 1911, p. 818.

[2] Bardeen, J., et al., "Theory of Superconductivity," *Phys. Rev.*, Vol. 108, 1957, p. 1175.

[3] Bednorz, J. G., and K. A. Müller, "Possible High Tc Superconductivity in the Ba-La-Cu-O System," *Z. Phys.*, Vol. B64, 1986, pp. 189–193.

[4] Meissner, W., and R. Oschsenfeld, "Ein Neuer Effekt bei Eiutritt der Supraleiffahigkeit," *Naturwissenschaften*, Vol. 21, 1933, pp. 787–788.

[5] Quin, D. J., and W. B. Ittner, "Resistance in a Superconductor," *J. Appl. Phys.*, Vol. 33, 1962, pp. 748–749.

[6] Josephson, B. D., "Possible New Effect in Superconductive Tunneling," *Phys. Lett.*, Vol. 1, 1962, pp. 251–253.

[7] Giffard, R. P., R. A. Webb, and J. C. Wheatley, "Principles and Methods of Low-Frequency Electric and Magnetic Measurements Using an rf-Biased Point-Contact Superconducting Device," *J. Low Temp. Phys.*, Vol. 6, 1972, pp. 533–610.

[8] Van Duzer, T., and C. W. Turner, *Principles of Superconductive Devices and Circuits*, New York: Elsevier, 1981.

[9] Orlando, T. P., and K. A. Delin, *Foundations of Applied Superconductivity*, Reading, MA: Addison-Wesley, 1991.

[10] Stephens, R. B., and R. L. Fagaly, "High Temperature Superconductors for SQUID Detection Coils," *Cryogenics*, Vol. 31, 1991, pp. 988–992.

[11] Koelle, D., et al., "High-Transition-Temperature Superconducting Quantum Interference Devices," *Rev. Mod. Physics*, Vol. 71, 1999, pp. 631–686.

[12] Clarke, J., "SQUID Fundamentals," in H. Weinstock (ed.), *SQUID Sensors: Fundamentals, Fabrication, and Applications*, Boston, MA: Kluwer Academic Publishers, 1996, pp. 1–62.

[13] Prance, R. J., et al., "Fully Engineered High Performance UHF SQUID Magnetometer," *Cryogenics*, Vol. 21, 1981, pp. 501–506.

[14] SHE model 330X rf SQUID electronics.

[15] Quantum Design model 2000 rf SQUID electronics.

[16] Tesche, C. D., and J. Clarke, "DC SQUID: Noise and Optimization," *J. Low Temp. Phys.*, Vol. 29, 1982, pp. 301–331.

[17] Faley, M. I., et al., "Operation of HTS dc-SQUID Sensors in High Magnetic Fields," *IEEE Trans. Applied Superconductivity*, Vol. 9, 1999, pp. 3386–3391.

[18] Simmonds, M. B., and R. P. Giffard, Apparatus for reducing low frequency noise in dc biased SQUIDs, U.S. Pat. No. 4,389,612, 1983.

[19] Fagaly, R. L., "Superconducting Magnetometers and Instrumentation," *Sci. Prog.*, Oxford, Vol. 71, 1987, pp. 181–201.

[20] Grover, F. W., *Inductance Calculations, Working Formulas and Tables*, New York: Dover, 1962.

[21] Wikswo, J. P., Jr., "Optimization of SQUID Differential Magnetometers," *AIP Conf. Proc.*, Vol. 44, 1978, pp. 145–149.

[22] Ilmoniemi, R., et al., "Multi-SQUID Devices and Their Applications," in D. F. Brewer (ed.), *Progress in Low Temperature Physics*, Vol. XII, Amsterdam: Elsevier, 1989, pp. 1–63.

[23] Vrba, J., "SQUID Gradiometers in Real Environments," in H. Weinstock (ed.), *SQUID Sensors: Fundamentals, Fabrication, and Applications*, Boston, MA: Kluwer Academic Publishers, 1996, pp. 117–178.

[24] Robinson, S. E., "Environmental Noise Cancellation For Biomagnetic Measurements," in S. J. Williamson, et al. (eds.), *Advances in Biomagnetism*, New York: Plenum Press, 1989, pp. 721–724.

[25] Walker, G., *Miniature Refrigerators for Cryogenic Sensors and Cold Electronics*, Oxford, England: Clarendon Press, 1989.

[26] Walker, G., *Cryocoolers*, Vols. 1 & 2, New York: Plenum Press, 1982.

[27] Buchanan, D. S., D. N. Paulson, and S. J. Williamson, "Instrumentation for Clinical Applications of Neuromagnetism," in R. W. Fast (ed.), *Advances in Cryogenic Engineering*, Vol. 33, New York: Plenum Press, 1988, pp. 97–106.

[28] Sata, K., "A Helmet-Shaped MEG Measurement System Cooled by a GM/JT Cryocooler," in T. Yoshimoto, et al. (eds.), *Recent Advances in Biomagnetism*, Sendai, Japan: Tohoku University Press, 1999, pp. 63–66.

[29] Heiden, C., "Pulse Tube Refrigerators: A Cooling Option," in H. Weinstock (ed.), *SQUID Sensors: Fundamentals, Fabrication, and Applications*, Boston, MA: Kluwer Academic Publishers, 1996, pp. 289–306.

[30] Romani, G.-L., S. J. Williamson, and L. Kaufman, "Biomagnetic Instrumentation," *Rev. Sci. Instrum.*, Vol. 53, 1982, pp. 1815–1845.

[31] Sarwinski, R. E., "Superconducting Instrumentation," *Cryogenics*, Vol. 17, 1977, pp. 671–679.

[32] Lounasmaa, O., *Experimental Principles and Methods Below 1K*, London: Academic Press, 1974.

[33] Vozoff, K., "The Magnetotelluric Method in the Exploration of Sedimentary Basins," *Geophysics*, Vol. 37, 1972, pp. 98–114.

[34] Fagaly, R. L., "SQUID Detection of Electronic Circuits," *IEEE Trans. Magn.*, Vol. MAG-25, 1989, pp. 1216–1218.

[35] Kirtley, J., "Imaging Magnetic Fields," *IEEE Spectrum*, Vol. 33, 1996, pp. 40–48.

[36] Jenks, W. G., S. S. H. Sadeghi, and J. P. Wikswo, "SQUIDs for Nondestructive Evaluation," *J. Phys. D: Appl. Phys.*, Vol. 30, 1997, pp. 293–323.

[37] Robinson, S. E., and R. L. Fagaly, "Biomagnetic Instrumentation: Current Capabilities and Future Trends," *Proc. ASME*, Vol. AES-9, 1989, pp. 7–12.

[38] Fagaly, R. L., "Neuromagnetic Instrumentation," in S. Sato (ed.), *Advances in Neurology*, Vol. 54: *Magnetoencephalography*, pp. 11–32, New York: Raven Press, 1990.

9

Other Principles
Pavel Ripka and Luděk Kraus

This chapter covers the unusual types of magnetic field sensors that do not fall into the categories covered by Chapters 2 through 8. *Unusual* here means sensors that are not widely available on the market in 2000. It does not mean that such sensors are rare: Natural magnetic sensors inside animals' bodies are made in enormous quantities, but there are no "datasheets" to explain their design. Researchers have used scientific methods to try to prove that animal navigation is based on such sensors, but finding out how those sensors work or even where are they located is difficult, to say the least. "Biological sensors" are examined in Section 9.3.

Unusual also does not mean that such sensors are unknown. Sometimes they are the subject of numerous scientific papers, then are forgotten, then reappear, in most cases soon to be forgotten again. Sometimes developments in technology, instrumentation, and data processing cause a revival of an almost forgotten principle. One example is the sensor based on the Lorenz force on a current conductor [1]. The micromachined silicon resonating structure with a high Q is activated by an ac current, and the movement is detected by a capacitive method. The potential advantage of such a sensor is a very high dynamic range, up to 1T, with a theoretical nanotesla resolution (if operated with low damping, i.e., in a vacuum).

The sensor market is dynamic and varied; if a sensor type described in the literature is not commercially available, it usually means that the device has serious drawbacks. Thin-film inductance-variation sensors [2], which are based on the change of permeability with a magnetic field, serve as an example. It turned out that those sensors had poor temperature stability of both the sensitivity and the offset; they were replaced by GMI sensors (Section 9.1). Nevertheless, there are application areas for so-called unusual sensors. If we want to measure the field profile along the optical cable or average the field value over a long line, magnetostrictive sensors (Section 9.2) may be a good solution.

Here we have to briefly mention semiconductor sensors other than the most popular Hall sensors, which were covered in Chapter 5. The far most important of them are semiconductor magnetoresistors [3, 4]. They have nonlinear characteristics, and their main application is for position switching and revolution counters (Chapters 10 and 12). Other types of semiconductor magnetic sensors, such as magnetodiodes, magnetotransistors, and carrier-domain devices, are not so practical. Information about their principles can be found in [5, 6].

9.1 Magnetoimpedance and Magnetoinductance

Magnetoinductive effects in ferromagnetic conductors can be used for various sensors. Although the principle has been known for a long time, the usefulness of these effects for practical applications was recognized only recently, and intensive investigations began. Magnetoinductive effects are related to magnetization of a magnetic conductor (wire, strip, thin film, etc.) by a magnetic field, which is produced by an electric current passing through the conductor itself. If the current is varying with time, the magnetic flux in the conductor also varies and induces the electromotive force, which is superimposed to the ohmic voltage between its ends. For instance, in a wire with a circular cross section, the circumferential magnetic field H induced by a constant current with the density j is $H = jr/2$, where r is the distance from the wire axis. For a wire of diameter 1 mm and the current density of 10^6 A/m^2, which is low enough so as not to greatly increase the temperature by the joule heating, the maximum magnitude of a magnetic field on the wire surface is 250 A/m. To get sufficiently high magnetoinductive voltage, which can be easily detected on the ohmic background signal, the circumferential reversal of conductor magnetization must take place in magnetic fields of this order or lower. Therefore, good soft magnetic metals with high circumferential permeability are required for such applications.

The systematic study of magnetoinductive effects in soft magnetic conductors started after the technology for production of amorphous wires had been successfully developed [7]. For example, large magnetoinductive effect has been found in the zero-magnetostrictive amorphous CoFeSiB wire with the circumferential bamboolike domain structure in the outer shell [8]. When an ac current of 1 kHz was applied to the wire, sharp peaks (about 0.2V) were induced on the background ohmic signal by the circumferential magnetization reversal in the outer shell. The amplitude of peaks decreased with an increasing external dc magnetic field. With utilization of this effect, a simple magnetic head (Figure 9.1) was constructed and used for a noncontact rotary encoder and a cordless data tablet [8].

Another magnetoinductive effect observed in soft ferromagnetic metals is giant magnetoimpedance (GMI), which is characterized by a strong dependence of ac impedance on an applied magnetic field (Figure 9.2). This effect is observed only at sufficiently high frequencies and can be explained by means of classical electrodynamics [9, 10]. It is known that RF current is not homogeneous over the cross section of conductor but tends to be concentrated near the surface (skin effect). The exponential decay of current density from the surface into the interior is described by the skin depth,

$$\delta = \sqrt{2\rho/\omega\mu} \qquad (9.1)$$

which depends on the circular frequency of RF current ω, the resistivity ρ, and the permeability μ. In nonferromagnetic metals, μ is independent of frequency and applied magnetic field; its value is close to the permeability of free space μ_0. In ferromagnetic materials, however, the permeability depends on the frequency, the amplitude of the ac magnetic field, and other parameters such as the magnitude and orientation of the bias dc magnetic field, mechanical strain, and temperature. The large permeability of soft magnetic metals and its strong dependence on the bias magnetic field are, in fact, the origin of GMI effect.

According to definition, the complex impedance $Z(\omega) = R + iX$ of a uniform conductor (Figure 9.3) is given by the ratio of voltage amplitude U to the amplitude of a sinusoidal current $I\sin\omega t$ passing through it. The

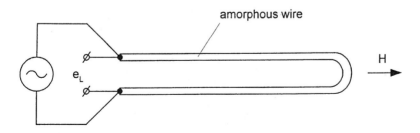

Figure 9.1 Simple magnetoinductive head using an amorphous wire (after [8]).

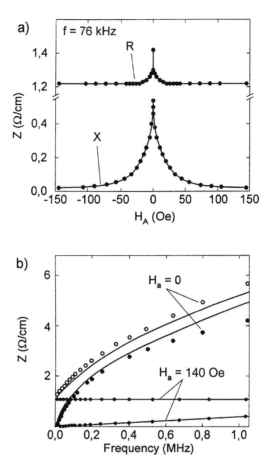

Figure 9.2 GMI of amorphous CoFeSiB wire (after [9]): (a) R resistance and X reactance as functions of applied field; (b) resistance (open circles) and reactance (solid circles) as functions of frequency.

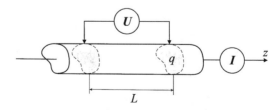

Figure 9.3 Impedance definition.

real part of impedance is called the resistance, and the imaginary part is the reactance. For the conductor of length L and the cross-section area q, the impedance is given by the formula

$$Z = \frac{U}{I} = \frac{LE_z(S)}{q\langle j_z\rangle_q} = R_{dc}\frac{j_z(S)}{\langle j_z\rangle_q} \quad (9.2)$$

where E_z and j_z are the longitudinal components of an electric field and current density, respectively, and R_{dc} is the dc resistance. S is the value at the surface, and $\langle\;\rangle_q$ is the average value over the cross section q. As can be seen, for a uniform current density the impedance is equal to the dc resistance. The definition, (9.2), is valid only for linear elements, that is, when the voltage U is proportional to the current I. It should be noted, however, that a ferromagnetic conductor generally is a nonlinear element. That means the voltage is not exactly proportional to the current; moreover, it also contains higher order harmonics of the basic frequency. Therefore, the term *impedance* should be taken with some precaution.

If the current density $\mathbf{j}(\mathbf{r})$ in the conductor is known, the impedance can be calculated. The current density generally can be obtained by the solution of Maxwell equations with the given material relation between $\mathbf{B}$ and $\mathbf{H}$ for the conductor material. Using the simple material relation $\mathbf{B} = \mu\mathbf{H}$, with constant permeability μ, and a simple conductor cross section, the current density $j_z(\mathbf{r})$ can be calculated (see, for example, [11]). From (9.2), then, we get

$$\frac{Z}{R_{dc}} = kR\frac{J_0(kR)}{2J_1(kR)} \quad (9.3)$$

for the circular wire of radius R or

$$\frac{Z}{R_{dc}} = k\frac{t}{2}\cot\left(k\frac{t}{2}\right) \quad (9.4)$$

for the semi-infinite film of thickness t. J_0 and J_1 are Bessel functions of the first kind, and the propagation constant k is given by the relation $k = (1 - i)/\delta$, where δ is the skin depth given by (9.1). For real ferromagnetic metals, however, the simple relation $\mathbf{B} = \mu\mathbf{H}$ cannot be used, and the theoretical calculation of the current density and the corresponding impedance Z is much more difficult.

The most exact phenomenological approach to the problem is the simultaneous solution of Maxwell equations for a ferromagnetic conductor

$$\text{rot } \mathbf{H} = \mathbf{j} \tag{9.5}$$

$$\text{rot } \mathbf{j} = \frac{1}{\rho} \text{rot } \mathbf{E} = -\frac{\mu_0}{\rho}(\dot{\mathbf{H}} + \dot{\mathbf{M}}) \tag{9.6}$$

$$\text{div } (\mathbf{H} + \mathbf{M}) = \frac{1}{\mu_0} \text{div } \mathbf{B} = 0 \tag{9.7}$$

together with the Landau-Lifshitz equation for the motion of the magnetization vector

$$\dot{\mathbf{M}} = \gamma \mathbf{M} \times \mathbf{H}_{\mathit{eff}} - \frac{\alpha}{M_s} \mathbf{M} \times \dot{\mathbf{M}} \tag{9.8}$$

where γ is the gyromagnetic ratio, M_s the saturation magnetization, $\mathbf{H}_{\mathit{eff}}$ is the effective magnetic field, and α is the Gilbert damping parameter. The effective field can be calculated from the free-energy density of the system, which depends on the particular magnetic domain structure of the sample. That means that the current density depends not only on the material parameters and the conductor geometry but also on the actual magnetic state. Therefore, the exact solution of the problem is practically impossible. Various simplifying assumptions have been used to solve (9.5) through (9.8). At high frequencies (above 1 MHz), the domain wall movements are heavily damped by eddy currents, and only magnetization rotations are responsible for magnetic permeability. Then the minimum calculated skin depth is [12]

$$\delta_m = \sqrt{\frac{\alpha \rho}{\gamma \mu_0 M_s}} \tag{9.9}$$

which is, for soft magnetic amorphous alloys, of the order 0.1 μm and gives the maximum theoretical values of $|Z|/R_{dc}$ of the order 10^3. That theoretical magnitude of GMI can be achieved only in uniaxial materials with the easy direction of the anisotropy exactly perpendicular to the conductor axis and the axial bias field H satisfying the condition

$$H = H_K + N_z M_s + \frac{1}{M_s}\left(\frac{\omega}{\gamma}\right)^2 \tag{9.10}$$

where N_z is the longitudinal demagnetizing factor and H_K is the effective anisotropy field. Any deviation of easy axis from the perpendicular direction or any fluctuation of H_K leads to a substantial reduction of GMI effect.

9.1.1 Materials

In actual soft magnetic metals, the maximum GMI effect experimentally observed up to now is much lower than the theoretically predicted values. Research in the field is focused on special heat treatments of already known soft magnetic metals and on development of new materials with properties appropriate for practical GMI applications. The GMI-curve $\eta(H)$ is defined as

$$\eta(H) = 100\% \times \left(\frac{|Z(H)|}{|Z_0|} - 1 \right) \quad (9.11)$$

where $Z(H)$ is the impedance for bias filed H, measured at a given frequency and constant driving current. Z_0 is the impedance for $H \to \infty$, which should be equal to the impedance of a nonmagnetic conductor with the same cross-section q and the same resistivity ρ. Practically, for Z_0 the value of impedance measured with maximum field $H_{\max}$, available for the given experimental equipment, is used. Some authors use $Z_0 = Z(0)$, but that value depends on the remanent magnetic state, which may not be well defined. The parameters that well characterize the GMI efficiency are the maximum GMI, $\eta_{\max}$, and the maximum field sensitivity, $(d\eta/dH)_{\max}$. Typical values obtained for some soft magnetic conductors are listed in Table 9.1.

Although GMI was first reported for amorphous metals, some crystalline materials also exhibit large GMI. Sometimes the crystalline metals are even better than the amorphous ones. According to the theoretical (9.9), the largest GMI should be obtained in materials with low resistivity ρ, high saturation magnetization M_s, and low damping parameter α. The crystalline metals have the advantage of lower resistivity, but in amorphous metals, better soft magnetic behavior can be obtained because of the lack of magnetocrystalline anisotropy. Because the magnetoelastic contribution to magnetic anisotropy substantially deteriorates the soft magnetic behavior, the nonmagnetostrictive materials also show the best GMI performance.

Amorphous cobalt-rich ribbons, wires, and glass-covered microwires are good candidates for GMI applications. The low magnetostriction and the easy control of magnetic anisotropy by appropriate heat treatment are the advantages of these materials; the disadvantage is high resistivity. Soft

Table 9.1
Materials for GMI Sensors

Material	Comment	η_{max} (%)	$(d\eta/dH)_{max}$ (% m/A)	Frequency (MHz)	Reference
Amorphous ribbon, $Co_{68.25}Fe_{4.5}Si_{12.25}B_{15}$	Joule heated	400	—	1	[13]
Amorphous wire, $Co_{68.15}Fe_{4.35}Si_{12.5}B_{15}$	Joule heated	220	22	0.09	[14]
Amorphous microwire, $Co_{68.15}Fe_{4.35}Si_{12.5}B_{15}$	Glass covered, joule heated	56	0.73	0.9	[15]
FINEMET wire	Annealed 600°C	125	—	4	[16]
Sandwich film, CoSiB/ SiO_2/Cu/SiO_2/CoSiB	RF-sputtered in magnetic field	700	3.8	20	[17]
Textured Fe-3%Si sheet	—	360	—	0.1	[18]
$Ni_{80}Fe_{20}$ plated on BeCu wire	—	530	4.8	5	[19]
CoP multilayers electroplated on Cu wire	Twisted	230	—	0.09	[20]
Mumetal stripe	Vacuum annealed	310	0.26	0.6	[21]

magnetic nanocrystalline metals exhibit GMI behavior similar to amorphous metals. Their somewhat higher M_s and lower ρ can lead to a small improvement. The low resistivity and bulk dimensions of crystalline soft magnetic alloys lead to better performance, especially at low driving frequencies (<1 MHz). The presence of large magnetocrystalline anisotropy (e.g., in iron-silicon alloys), however, requires a high texture of crystalline grains and proper adjustment of the driving current and dc bias field directions [18]. Excellent GMI behavior was found in combined conductors consisting of a highly conductive nonmagnetic metal core (such as Cu or CuBe) with a thin layer of soft magnetic metal on the surface [22, 14, 19]. An insulating interlayer between the core and the magnetic shell, in sandwich thin-film structures, results in further improvement of GMI behavior [23, 17]. The thin-film structures, which can be used in integrated circuits and the glass-covered microwires, from which simple sensing elements for electrotechnical devices can be easily constructed, seem to be particularly promising for wide exploitation of GMI.

Not only η_{max} and $(d\eta/dH)_{max}$ but also the particular shape of $\eta(H)$ curve are important for sensor applications. The shape of a GMI curve can be controlled by induced magnetic anisotropy and/or bias dc current. For wires and ribbons with transversal magnetic anisotropy, the double-peak GMI curve with the maxima close to $\pm H_K$ is observed. If the easy direction is parallel to the conductor axis, the single peak at $H = 0$ is present (as in Figure 9.2). In this case, however, the η_{max} sharply decreases with increasing anisotropy field [12]. Helical anisotropy, induced in amorphous wires by torque stress or torque annealing, combined with a bias dc current, results in an asymmetric GMI curve [24]. Such a curve can be exploited by a linear field sensor.

9.1.2 Sensors

The high sensitivity of magnetoimpedance to external dc or low frequency ac field (here *low frequency* means the frequency that is at least one order lower than the driving frequency) can be used for magnetic field sensors and other sensors based on the change of a local magnetic field (such as displacement, electric current). The high driving frequency, which must be used to get sufficient sensitivity of sensors, involves many problems like parasitic displacement currents in the circuits connecting the magnetoimpedance (MI) element with the signal source and the measuring unit, impedance mismatching, the presence of reflected signals, and so forth. To avoid those problems, oscillation circuits are used, such as the Colpitts oscillator and the resonance multivibrator, with the MI element as the circuit inductance.

Figure 9.4 illustrates the Colpitts oscillator, utilizing a resonance of the inductance of MI element and the capacitances C_1 and C_2 [25]. For oscillation frequencies of the order of 100 MHz, the GMI signal can be increased several times. Because the field dependence of the oscillator output signal roughly follows the GMI curve, it is a nonlinear function of applied field. To get a linear field sensor, a pair of MI elements were used in a multivibrating oscillator circuit, as shown in Figure 9.5(b) [26]. The two MI elements, connected in the two symmetric branches of the multivibrator, were biased with opposite dc fields H_b so that in the range of applied fields $-H_b < H_{ex} < H_b$ the output voltage was nearly linear, as shown in Figure 9.5(a). The bias field H_b, however, requires small magnetizing solenoids wound around the MI elements. That complication can be avoided if the asymmetrical GMI effect in twisted wires with dc bias current is used [24]. The bias current in the pair of twisted MI elements flows in the opposite directions with respect to the applied dc field. A linear characteristic similar to Figure 9.5(a) is then obtained in a certain range of applied field.

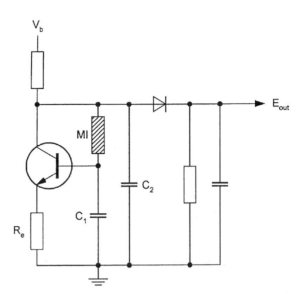

Figure 9.4 The Colpitts oscillator with MI element (after [25]).

In the usual field sensor, the MI elements, in the two oscillator branches, are arranged parallel, and the bias fields H_b are opposite. If the bias fields for the parallel elements are in the same direction, or if the elements are arranged in series with the opposite bias fields, a gradient field sensor can be obtained [27]. Because the MI elements may be as small as 1 mm, very localized weak magnetic fields can be detected. These types of sensors can be used, for example, for the detection of stray fields caused by cracks in steel sheets and for magnetic rotary encoders of high resolution.

Miniature magnetic field sensors based on GMI effect have been used for various applications. They are especially appropriate for medical applications as small permanent magnet movement sensors for the control of human physiological functions. They also can be used for automation and control in industry. Although GMI sensors are quite new and their development is not yet finished, their low prices and high flexibility probably will lead to a wide exploitation in the near future.

9.2 Magnetoelastic Field Sensors

A number of mechanical variables such as force and torque are measured using magnetostriction; those variables are briefly described in Chapter 12. This chapter looks strictly at the sensors of magnetic field.

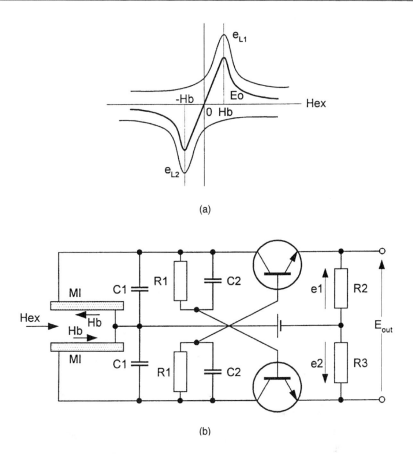

Figure 9.5 Linear field sensor with multivibrator: (a) the field dependence of the output voltage and (b) the circuit diagram. (After [26].)

To reach high sensitivity, magnetostrictive field sensors are often excited at longitudinal mechanical resonance. The response of magnetostrictive sensors is, in general, not dependent on the field sign, so the characteristic is odd and the sensitivity for low fields is very small. To obtain a linear response, either the dc bias field or ac modulation should be applied. In the former case, the sensor stability critically depends on the stability of that bias field. The ac modulation technique requires extra coil and a PSD, which increases the sensor complexity. Another disadvantage of magnetostrictive sensors is sensitivity to temperature changes and mechanical vibrations.

9.2.1 Fiber-Optic Magnetostriction Field Sensors

The new wave of fiber-optical technology has also brought new magnetic field sensors [28]. Some of those based on the Faraday effect were described in Chapter 6. In this chapter, we mention the application of optical fibers coupled to magnetostrictive material. The sensing element is either tape or wire glued to the fiber [29] or a layer deposited on the fiber surface. The strain of the magnetostrictive transducer caused by an applied magnetic field causes a length change of the optical fiber, which is sensed by the interferometer. The block diagram of the sensor based on an all-fiber Mach-Zehnder interferometer is shown in Figure 9.6. Although such a magnetometer is complicated and has limited stability (10-nT drift in 10 hours at constant temperature is reported in [30]), it may find application for multipoint measurement along telecommunication cables. An array of fiber-optic magnetostriction magnetometers for underwater detection of vessels is described in [31]. The sensors are powered with copper cable, but all the signals are

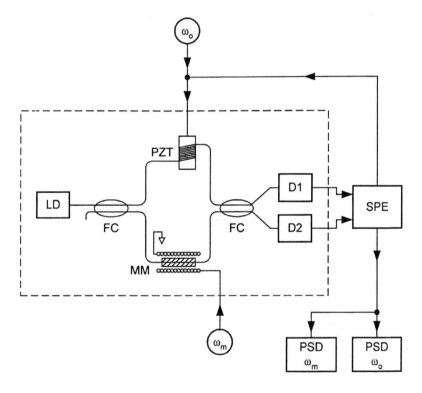

Figure 9.6 Fiber-optic magnetostriction magnetic field sensor based on Mach-Zehnder interferometer consists of a laser diode (LD), fused couplers (FC), optical detectors (D1, D2), magnetostrictive material coupled to the fiber signal arm (MM), and a piezoceramic element for active compensation of the reference arm (PZT) (after [30]).

transmitted optically. The sensors with 5-cm-long magnetostrictive amorphous strips exhibit 30 pT/$\sqrt{Hz}$@1Hz noise, which is higher than that of the fluxgate sensor of the same length (Chapter 3).

9.2.2 Magnetostrictive-Piezoelectric Sensors

Magnetostrictive sensors may have piezo excitation or detection [32]. The sensor geometry is shown in Figure 9.7.

The "piezo-driven" type is excited by ac voltage applied to the electrodes of the piezo element, which is coupled, usually through a viscous fluid, to the high magnetostriction ferromagnetic core. The sample vibration causes a change of the core properties, which are detected by the solenoid coil [33]. The hysteresis can be removed by magnetic shaking [34]. Such sensors may have a 1-nT resolution and a linear range of 1 to 100 μT, but poor temperature stability. Disk-shaped sensors of this type can measure the field in two perpendicular directions [35].

The second type of piezomagnetostrictive sensor uses magnetic excitation of the core by an ac-supplied solenoid coil. The piezoelectric element is again interfaced to the core, but here it serves as a detector. Another possibility is to excite the core by the electric current flowing through it, so no coil is necessary [32]. If the current flowing through the core has frequency f, the size changes, and thus the output signal has frequency of $2f$. That is an advantage over other types of piezomagnetostrictive sensors, because the excitation signal can easily be filtered out of the output. The common disadvantage of all the sensors excited by the current through the core is that high amplitude of the excitation field is close to the surface; the field in the middle is zero, which may cause hysteresis and perming. Serious problems are also associated with the current contacts.

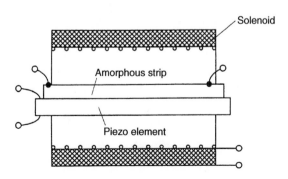

Figure 9.7 Magnetostrictive-piezoelectric magnetic field sensor.

9.2.3 Shear-Wave Magnetometers

The last type of magnetoelastic magnetic field sensors is based on the magnetic field dependence of the elastic modulus E (ΔE effect), which causes the change of acoustic wave velocity [36]. The basic sensor is shown in Figure 9.8. The piezotransmitter is driven by an RF source (1.8 MHz) and creates an acoustic wave that propagates at 2.6 km/s along the ribbon toward the piezoreceiver. The device works in continuous mode; the transmitted and received signals are processed in the phase comparator, which has a voltage output proportional to the phase difference and thus to the field-dependent velocity. The basic characteristic is odd [Figure 9.9(a)], and the phase noise caused by mechanical and temperature disturbances is high.

Thus the modulation technique is used: The measured dc field is superposed by an LF ac field produced by the solenoid. The LF ac signal is phase modulated, as shown in Figure 9.9(a). With changing slope of the first odd curve, the phase of the response is reversed. If the receiver output is demodulated by another PSD, the resulting response is even, as shown in Figure 9.9(b). The sensor is completed by field feedback (Figure 9.10), which improves the performance similarly as in fluxgate magnetometers [37]. The resulting sensor has a noise of 100 pT/$\sqrt{Hz}$@1Hz, but large offset tempco of 8.4 nT/K.

9.3 Biological Sensors

Magnetic orientation was first proven in birds, later in some kinds of fish, sea turtles, and honeybees. A lot of information can be found in [38].

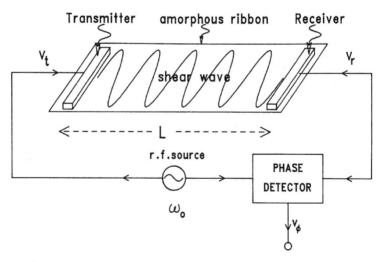

Figure 9.8 The principle of the shear-wave magnetometer.

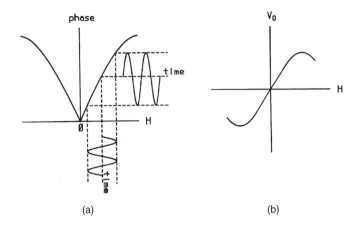

Figure 9.9 Principles of the ac phase modulation: (a) phase characteristics and (b) even characters obtained by modulation.

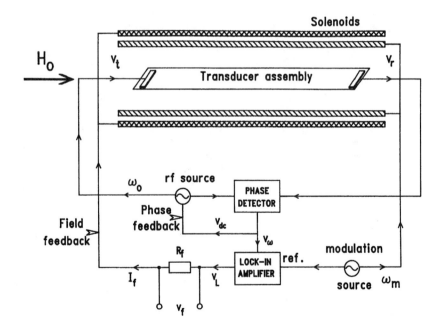

Figure 9.10 Complete shear-wave feedback magnetometer.

9.3.1 Magnetotactic Bacteria

Magnetotactic bacteria are found in saltwater and freshwater sediments. In most cases, the bacteria use the Earth's magnetic field to find the vertical direction; their density is too low to use gravitation. They are anaerobic, so if they are stirred up into the surface, oxygen-rich water, they move downward into their favorable environment. Magnetotactic bacteria in the southern hemisphere are reversely polarized from those from the northern hemisphere. They can be reversed by a strong magnetic field. Although the bacteria actively move, the magnetic alignment along the field lines is a passive process. The magnetic response is caused by chains of magnetite (Fe_3O_4) or greigite (Fe_3S_4). Some bacteria actually produce magnetic minerals for themselves from nonmagnetic FeS_2 [39].

9.3.2 Magnetic Orientation in Animals

Migrating birds and carrier pigeons use the Earth's field for navigation, as well as the Sun and stars. The magnetic sensor in birds is very different from man-made compasses. A bird can perceive the field direction but not the polarity. Reversing the field direction causes no response. This behavior is called *inclination compass*: According to the inclination, birds can distinguish between the poleward direction (i.e., when field lines point to the ground) and equatorward direction (i.e., when field lines point to the sky). The birds are magnetically disoriented if the field has a horizontal direction. Birds can use their compass only if the magnetic field is constant within about ±5% to 10% of the steady level, but they can slowly adapt to another constant level.

Sea turtles also have an inclination compass, while salmon have a polarity compass, which works well in a horizontal field. The magnetic compass orientation can be innate (as in first migration of young birds) or acquired (as in local orientation and homing). Magnetic orientation normally is only one component in the animal integrated navigation system; other components include Sun and star compass, inclination sensors, internal clock, memory for visual landmarks, and other factors like odors.

There are indications that some animals (e.g., carrier pigeons) may even use the information about local magnetic gradients and anomalies like man-made sophisticated autonomous missiles.

Our knowledge of how animals sense the magnetic field is weak. Three realistic principles have been suggested: induction sensors, chemical sensors, and force sensors (based on magnetized particles).

It is very likely that fish use induction sensors. Movement in the Earth's field induces voltages inside the body, which are strong enough to be detected by the fish's electroreceptors.

Another hypothesis involves chemical sensors, mainly photopigments. Although their mechanism is unclear, there are behavioral and physiological indications that, in some cases, the magnetic field information comes from the eye and depends on visible light. Magnetic particles are present mainly as single domain, but they may form chains. They may interact with membranes and neurons; however, the complete evidence of such mechanisms has not yet been given.

References

[1] Kadar, Z., et al., "Magnetic-Field Measurements Using an Integrated Resonant Magnetic-Field Sensor," *Sensors and Actuators A*, Vol. 70, 1988, pp. 225–232.

[2] Hoffman, G. R., "Some Factors Affecting the Performance of a Thin Film Inductance Variation Magnetometer," *IEEE Trans. Magn.*, Vol. 17, No. 6, 1981, pp. 3367–3369.

[3] Heremans, J., "Magnetic Field Sensors for Magnetic Position Sensing in Automotive Applications," *Proc. Conf. on Properties and Applications of Magnetic Materials*, Chicago, IL, May 1997, Session 1.

[4] Heremans, J., "Solid State Magnetic Sensors and Applications," *J. Phys. D*, Vol. 26, 1993, p. 1149.

[5] Boll, R., and K. J. Overshott (eds.), "Magnetic Sensors," *Sensors*, Vol. 2, Veiden, Germany: VCH, 1989.

[6] Roumenin, C. S., *Solid-State Magnetic Sensors*, Lausanne, Switzerland, Elsevier, 1994.

[7] Ogasawara, I., "Amorphous Magnetic Wire and Applications," *INTERMAG '90 Satellite Symp.*, London, Apr. 23, 1990, not published.

[8] Mohri, K., et. al, "Magneto-Inductive Effect (MI Effect) in Amorphous Wires," *IEEE Trans. Magn.*, Vol. 28, 1992, pp. 3150–3152.

[9] Beach, R. S., and A. E. Berkowitz, "Giant Magnetic Field Dependent Impedance of Amorphous FeCoSiB Wire," *Appl. Phys. Lett.*, Vol. 64, 1994, pp. 3652–3654.

[10] Panina, L. V., and K. Mohri, "Magneto-Impedance Effect in Amorphous Wires," *Appl. Phys. Lett.*, Vol. 65, 1994, pp. 1189–1191.

[11] Landau, L. D., and E. M. Lifshitz, *Electrodynamics of Continuous Media*, Oxford, England: Pergamon, 1975.

[12] Kraus, L., "Theory of Giant Magneto-Impedance in the Planar Conductor With Uniaxial Magnetic Anisotropy," *J. Magn. Magn. Mater.*, Vol. 195, 1999, pp. 764–778.

[13] Tiberto, P., et al., "Giant Magnetoimpedance Effect in Melt-Spun Co-Based Amorphous Ribbons and Wires With Induced Magnetic Anisotropy," *J. Magn. Magn. Mater.*, Vol. 196–197, 1999, pp. 388–390.

[14] Costa-Kramer, J. L., and K. V. Rao, "Influence of Magnetostriction on Magneto-Impedance in Amorphous Soft Ferromagnetic Wires," *IEEE Trans. Magn.*, Vol. 31, 1995, pp. 1261–1265.

[15] Kraus, L., et al., "The Influence of Joule-Heating on Magnetostriction and GMI Effect in a Glass-Covered CoFeSiB Microwire," *J. Appl. Phys.*, Vol. 85, 1999, pp. 5435–5437.

[16] Knobel, M., et al., "Giant Magneto-Impedance Effect in Nanostructured Magnetic Wires," *J. Appl. Phys.*, Vol. 79, 1999, pp. 1646–1654.

[17] Morikawa, T., et al., "Enhancement of Giant Magneto-Impedance in Layered Film by Insulator Separation," *IEEE Trans. Magn.*, Vol. 32, 1996, pp. 4965–4967.

[18] Carara, M., and R. L. Sommer, "Giant Magneto-Impedance in Highly Textured (110)[001] FeSi3%," *J. Appl. Phys.*, Vol. 81, 1997, pp. 4107–4109.

[19] Beach, R. S., et al., "Magneto-Impedance Effect in NiFe Plated Wire," *Appl. Phys. Lett.*, Vol. 68, 1996, pp. 2753–2755.

[20] Favieres, C., et al., "Giant Magnetoimpedance in Twisted Amorphous CoP Multilayers Electrodeposited Onto Cu Wires," *J. Magn. Magn. Mater.*, Vol. 196–197, 1999, pp. 224–226.

[21] Nie, H. B., et al., "Giant Magnetoimpedance in Crystalline Mumetal," *Sol. State Comm.*, Vol. 112, 1999, pp. 285–289.

[22] Hika, K., L. V. Panina, and K. Mohri, "Magneto-Impedance in Sandwich Film for Magnetic Sensor Heads," *IEEE Trans. Magn.*, Vol. 32, 1996, pp. 4594–4596.

[23] Senda, M., et al., "Thin-Film Magnetic Sensor Using High Frequency Magneto-Impedance (HFMI) Effect," *IEEE Trans. Magn.*, Vol. 30, 1994, pp. 4611–4613.

[24] Kitoh, T., K. Mohri, and T. Uchiyama, "Asymmetrical Magnetoimpedance Effect in Twisted Amorphous Wires for Sensitive Magnetic Sensors," *IEEE Trans. Magn.*, Vol. 31, 1995, pp. 3137–3139.

[25] Uchiyama, T., et al., "Magneto-Impedance in Sputtered Amorphous Films for Micro Magnetic Sensor," *IEEE Trans. Magn.*, Vol. 31, 1995, pp. 3182–3184.

[26] Mohri, K., "Applications of Amorphous Magnetic Wires to Computer Peripherals," *Mater. Sci. and Eng. A*, Vol. 185, 1994, pp. 141–146.

[27] Bushida, K., et al., "Amorphous Wire MI Micro Magnetic Sensor for Gradient Field Detection," *IEEE Trans. Magn.*, Vol. 32, No. 5, pp. 4944–4946.

[28] Bucholtz, F., "Fiber Optic Magnetic Sensors," Chap. 12 in E. Udd (ed.), *Fiber Optic Sensors*, New York: Wiley, 1991.

[29] Koo, K. P., et al., "A Compact Fiber-Optic Magnetometer Employing an Amorphous Metal Wire Transducer," *IEEE Photonics Tech. Letters*, Vol. 1, 1989, pp. 464–466.

[30] Koo, K. P., et al., "Stability of a Fiber-Optic Magnetometer," *IEEE Trans. Magn.*, Vol. 22, No. 3, 1986, pp. 141–144.

[31] Bucholtz, F., et al., "Demonstration of a Fiber Optic Array of Three-Axis Magnetometers for Undersea Application," *IEEE Trans. Magn.*, Vol. 31, No. 6, 1995, pp. 3194–3196.

[32] Prieto, J. L., et al., "Magnetostrictive-Piezoelectric Magnetic Sensor With Current Excitation," *Proc. SMM 1999, J. Magn. Magn. Mater.*, Vol. 215–216, 2000, pp. 756–758.

[33] Marmelstein, M. D., "A Magnetoelastic Glass Low-Frequency Magnetometer," *IEEE Trans. Magn.*, Vol. 28, No. 1, 1992, pp. 36–56.

[34] Prieto, J. L., et al., "Reducing Hysteresis in Magnetostrictive-Piezoelectric Magnetic Sensors," *IEEE Trans. Magn.*, Vol. 34, No. 6, 1998, pp. 3913–3915.

[35] Prieto, J. L., et al., "New-Type of 2-Axis Magnetometer," *Electronics Letters*, Vol. 31, No. 13, 1995, pp. 1072–1073.

[36] Squire, P. T., and M. R. J. Gibbs, "Shear-Wave Magnetometry," *IEEE Trans. Magn.*, Vol. 24, No. 2, 1988, pp. 1755–1757.

[37] Kilby, C. F., P. T. Squire, and S. N. M. Willcock, "Analysis and Performance of a Shear-Wave Magnetometer," *Sensors and Actuators A*, Vol. 37–38, 1993, pp. 453–457

[38] Wiltschko, R., and W. Wiltschko, *Magnetic Orientation in Animals*, Berlin, Germany: Springer, 1995.

[39] Mann, S., et al., "Biomineralization of Ferrimagnetic Greigite and Iron Pyrite in a Magnetotactic Bacterium," *Nature*, Vol. 343, 1990, pp. 258–261. More information on www.calpoly.edu/~rfrankel.

10

Applications of Magnetic Sensors
Pavel Ripka and Mario H. Acuña

10.1 Biomagnetic Measurements

Magnetic trackers (Chapter 12) are used to determine the position of medical tools inside the body (endoscope, colonoscope, biopsy needle, etc.) and to observe biomechanical motions (eyelid movement, articulatory movement, etc.). Trackers are also used in biomechanical feedback systems for the handicapped. The radiation-free magnetic method to measure gastric emptying was developed by Forsman [1]. The research subjects ingested magnetic particles of maghemite (γ-Fe_2O_3) in the form of pancakes. Before each measurement, the intragastric powder was magnetized by a 40-mT field generated by an air coil pair. The first-order gradient in the direction perpendicular to skin surface was measured by a 140-mm-base-length fluxgate gradiometer. The field gradient was scanned by moving the tested person on a pneumatically driven bench. Measuring the particle remanence has advantages over the susceptibility method in that the required amount of the magnetic tracer is smaller (400 mg for remanence method instead of 15g for susceptibility), and position error is smaller (signal $\sim 1/r^3$ instead of $1/r^6$). The drawback is the tendency of magnetized particles to form clusters and also a high required field (40 mT was not sufficient). The magnetic method proved to be sufficiently sensitive to replace radiolabeled tests in case of repeated measurements or pregnancy.

Magnetopneumography is a magnetic method that can detect ferromagnetic dust deposited in human lungs by using its magnetic moment after dc magnetization [2, 3]. It can be used for examination of welders, grinders, and other metal workers [4]. Ferromagnetic dust is also found in the lungs of miners and those in other, similar professions, but the scaling and interpretation of the field values is much more difficult because the magnetic properties of the dust are so variable [5]. Study of magnetic properties of the tissue and extracted dust samples from the lungs of exposed workers had shown that even the dust and aerosols originating from nonmagnetic stainless steel can be detected as they become ferromagnetic during the welding or grinding process [6]. The dc field necessary to sufficiently magnetize the dust in lungs is 100–200 mT. The magnetization device should be either a superconducting coil or an electromagnet with a yoke; lower field values from air coils give irreproducible results.

The SQUID magnetometers traditionally were used for biomagnetic measurements. Recently it was shown that fluxgate gradiometers (Chapter 3) can be made sufficiently sensitive for this application [7]. Fluxgate is much cheaper, can be made portable, and requires no cryotechnique.

Magnetopneumographic system consists of three parts:

- A magnetization device, which generates a magnetic field of required magnitude in the whole volume of human lungs;
- A gradiometer, which maps the remanent magnetic field of the lungs;
- Software that solves the inverse problem, that is, estimates the size, location, and density of the dust deposit.

Figure 10.1 shows the magnetic field distribution of a local dust deposit modeled by 250 mg of magnetite distributed in 256 cm^3 in the central part of the lung model. The deposit position is shown and the coordinate system defined in Figure 10.1(a). The sample was magnetized by a 100-mT dc field, and the remanent field in the z direction (perpendicular to skin) was measured as a function of distance, as shown in Figure 10.1(b). The magnetic field distribution in the xy-plane is shown in Figure 10.1(c).

Figure 10.2 shows the situation when the dust is homogeneously deposited in the whole lung volume. It is clear how fast the spatial resolution drops with distance from the measured object.

Figure 10.3 shows the field gradient distribution of a professional welder [8]. The concentration of the dust is decreasing toward the waist, which is caused by the filtration mechanism of lungs.

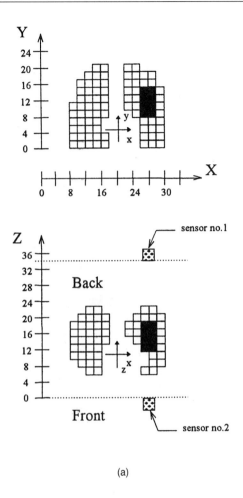

(a)

Figure 10.1 Magnetic fields of a local deposit of magnetite in the model of human lungs (250 mg of magnetite in 256 cm^3): (a) size and location; (b) magnetic field in z direction as a function of the sensor distance; shown are the measured and calculated values (dipole $1/r^3$ model); and (c) magnetic field distribution in the xy-plane (measured values) in the distance $z = 0$ (skin surface).

10.2 Navigation

For centuries, the magnetic compass was one of the main navigation devices [9]. A compass should be gimbaled (leveled) and its reading corrected by a declination angle, because the magnetic north is not on the Earth's rotational axis. Declination changes with the position on Earth and slowly changes with time. The angle between the Earth's field vector and the horizontal plane

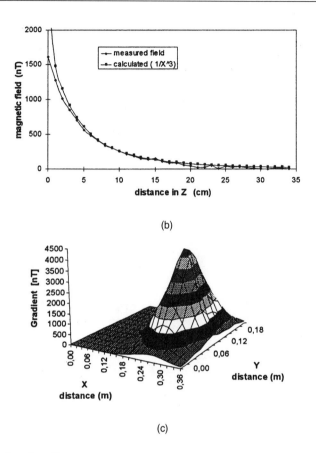

Figure 10.1 (continued).

is called dip or inclination (Figure 10.4). This angle is close to 90 degrees in the polar regions, which causes well-known navigation problems. Local field anomalies also cause fatal errors.

The limiting factor of the precision of magnetic navigation is often the magnetic field of the vehicle, which can be caused by magnetic materials and dc current loops. Although dc current loops can be minimized by design (using twisted conductors instead of a chassis as one conductor), they still are a serious problem in solar cells. A lot of magnetic navigation systems allow compensation for field distortion caused by ferromagnetic parts, but they need periodic recalibration. The Earth's field is too weak to saturate the magnetic parts in the vehicle. Therefore, their magnetization characteristics are nearly linear with a constant (hard) component and a field proportional (soft) component.

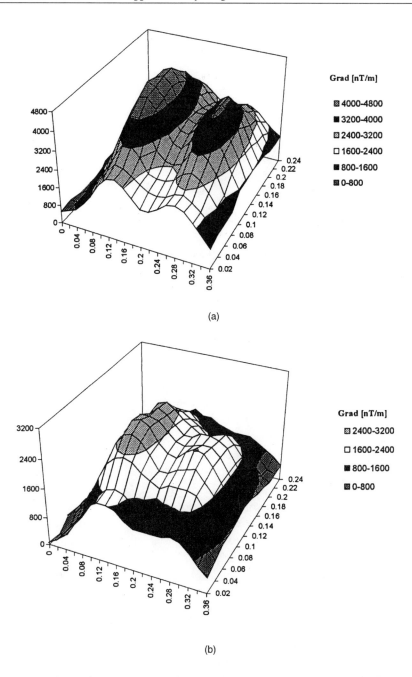

Figure 10.2 Gradient distribution in the xy-plane 6 cm from the lung model containing 4g of magnetite in the distance: (a) z = 6 cm; and (b) z = 8 cm.

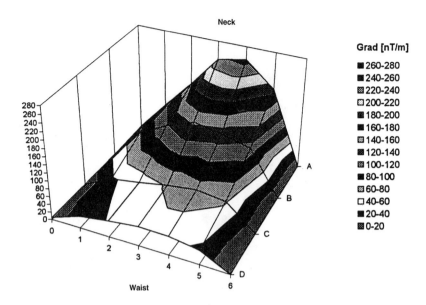

Figure 10.3 Magnetic field gradient measured *in vivo* on a welder. The grid was 5 cm.

The hard component (remanent magnetization) is caused by permanent magnets and remanence of magnetized iron parts; its effect is equivalent to the offsets of field sensors. The soft component (induced magnetization) is caused by the permeability of ferromagnetic parts; its effect is equivalent to a change of sensitivity of the field sensors [10]. The vehicle magnetometer readings as a function of direction are shown in Figure 10.5. If the host is magnetically clean, the polar plot is a circle, and individual sensor outputs are orthogonal sinewaves, as shown in Figure 10.5(a). The soft component causes the polar circle to become ellipsoid; the hard component causes the center of the ellipsoid to be displaced from the 0,0 point, as shown in Figure 10.5(b). After the simple compensation (four variables: two sensitivities and two offsets), some distortion of the polar plot still exists, as shown in Figure 10.5(c). That may be caused by field gradients, contributions to a vehicle horizontal field from vertical Earth's field components due to tensor character of effective (i.e., apparent) permeability, and other effects. Some systems allow more complex compensation that uses more than the mentioned four correction coefficients.

Traditional pivoting needles have serious constraints: The sensor cannot be separated from the display, it cannot make automatic corrections of the reading, the instrument has a moving part, and it is sensitive to vibrations (so for use in vehicles, it should be damped by liquid to slow the response

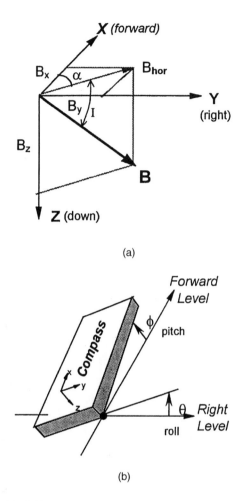

Figure 10.4 (a) Earth's field vector **B** components: vertical component B_z and horizontal component B_{hor}, which is a vector sum of B_x and B_y. α is azimuth or heading; I is inclination or dip. (b) Pitch and roll are the inclinations of the magnetometer from a horizontal plane. (After [10].)

to fast maneuvers). Pewatron manufactures a patented compass sensor that contains a small rotating magnetic needle, the position of which is measured by Hall sensors. The sensor has an electrical output, but the precision is limited.

Fluxgate compasses are popular devices used in aircraft, ships, and cars. They are still a part of modern navigation systems, together with global positioning systems (GPSs) and inertial sensors; nowadays the compass is

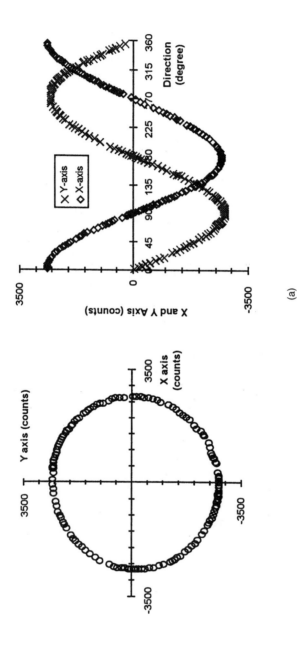

Figure 10.5 Vehicle magnetometer readings as a function of direction: (a) without disturbances; (b) disturbed by the vehicle magnetic field; and (c) corrected for soft (scale) and hard (offset) components. (From [10].)

Applications of Magnetic Sensors

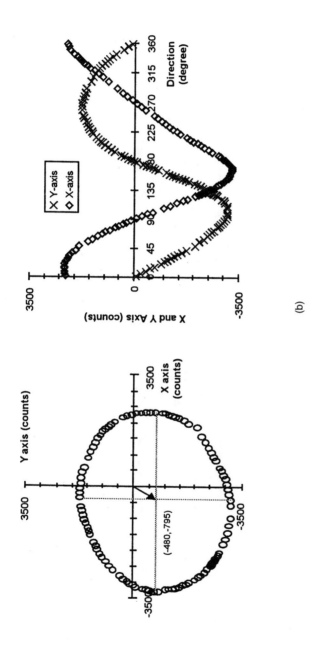

Figure 10.5 (continued).

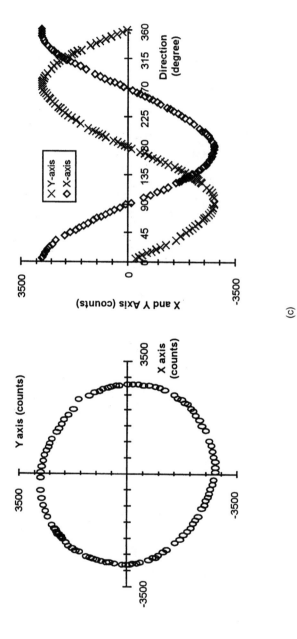

Figure 10.5 (continued).

used mostly as a backup during the loss of the GPS signal. The standard type of fluxgate compass consists of a leveled (gimbaled) double-axis, single ring-core sensor with a pair of orthogonal pickup coils (see Chapter 3). The 0.1-degree precision is easily achievable in a wide temperature range, but a lot of cheap and low-power devices have an accuracy of 0.5 or 1 degree, which is still sufficient for consumer products such as compass watches [11].

The attitude sensor, which is used for autonomous underwater vehicles and was developed by Fowler [12], has three individually gimbaled fluxgate sensors. The pitch and roll information is provided by air coils fixed to the frame that sense the ac field generated by fluxgates.

The strapdown or electronically gimbaled compass has a three-axial sensing head fixed to the host. The sensing head inclination is measured by two inclinometers usually measuring the roll (angle between trajectory and horizontal plane) and pitch (inclination in direction orthogonal to trajectory). [See Figure 10.4(b).] The horizontal field component and the heading information are then calculated from the magnetometer and inclinometer readings.

Fluxgate compass sensors with excitation and processing electronics can be integrated on a single chip or multichip module [13]. Other integrated fluxgate sensors were described in Chapter 3.

Precision Navigation manufactures single-core fluxgate sensors, which are part of a relaxation oscillator [14]; such auto-oscillation magnetometers are described in Chapter 3. Precision Navigation calls the sensors "magnetoinductive sensors," although they have nothing to do with magnetoimpedance or magnetoinductance effects.

Magnetoresistors are new competitors of fluxgates for compass applications. The high-temperature dependence of their sensitivity (~600 ppm/°C) is suppressed by ratiometric output of two perpendicular sensors. Using the flipping mechanism (Chapter 4), offset temperature variations may be below 0.25 nT/°C, which corresponds to an angular error below 0.1 degree in the military temperature range. Honeywell manufactures a digital compass module HMR 3000 [15], which has 0.1-degree resolution; the specified 0.5-degree accuracy is only an rms value, not a worst-case error; typical repeatability is 0.3 degree. Although the manufacturer specifies the typical offset drift as ±114 ppm/°C, which corresponds to 23 nT/°C, the measured average value of the offset temperature coefficient was 0.2 nT/°C (with the set/reset ratio flipping pulses continuously on) [16].

Navigation of mobile robots as a specific discipline is covered in [17]. The magnetic methods include a magnetic compass, artificial landmarks made by permanent magnets, magnetic guides made of ac-powered induction loops, RF beacons, and also the tracking methods described in Chapter 12.

10.3 Military and Security

10.3.1 UXO

An unexploded ordnance (UXO) contains dangerous explosives, propellants, or chemical agents. It may be unexploded either through malfunction or intentionally by design. The size ranges from several millimeters (pistol munition) to several meters (missiles and bombs). UXOs are a serious hazard and should be located and cleaned from former military areas and battlefields. In many European and Asian cities, a search for unexploded bombs from World War II still should be made before the start of any construction work.

The most effective method of locating buried ordnance is by mapping the magnetic field. The instruments, which are carried or vehicle-towed [18, 19], are either scalar magnetometers (proton, Overhauser, or cesium vapor; see Chapter 7) or vector fluxgates (Chapter 3). The remanent magnetization of ordnance is usually small; for example, airborne bombs are demagnetized by ground impact. The magnetic signature does not depend much on the ferrous mass—objects with a thick iron shell and nonmagnetic central part behave similarly as if they were solid. The signature is proportional to outer volume, shell thickness, relative permeability, length-diameter ratio, and orientation in the geomagnetic field [20]. Prolate objects in vertical or north-south horizontal position are easily detectable; if they are positioned exactly east-west, their total field signature may be very low. Figure 10.6 shows calculated magnetic signature of a 155-mm projectile in middle latitudes (65-degree declination of the Earth's field). If the 150-cm-deep object is located horizontally in north-south direction, the signature has a typical local peak maximum and valley, as shown in Figure 10.6(a), while the total field signature of a vertically located projectile has only a peak, as shown in Figure 10.6(b).

The fieldwork techniques for locating UXO are similar to those used for geophysical prospection (see Section 10.7). The realistic detectable field signature is 1 nT, although the instrument noise is sometimes declared to be much better. A large, 2.4m-long bomb can be easily located from more than a 6m distance, while a 0.5-inch-caliber projectile is hardly detectable from 20 cm. A database of real ordnance signatures at various geographical locations is available from the U.S. Army Corps of Engineers [21].

Electromagnetic methods utilize eddy currents to measure the difference between the conductivity of the soil and buried objects. The source coil is supplied by either a variable-frequency sinewave (frequency-domain systems) or a pulsed source (time-domain systems). The field from an eddy current

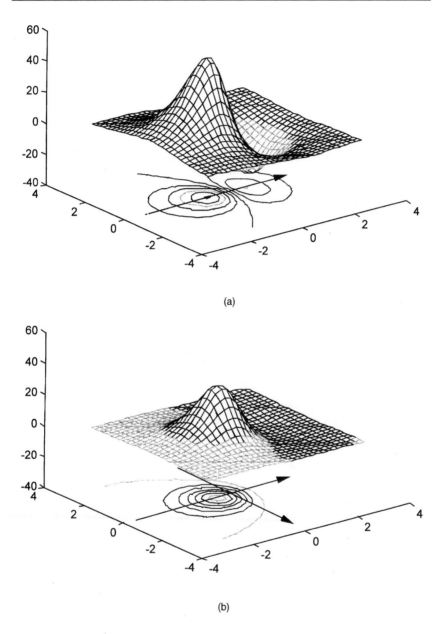

Figure 10.6 Magnetic signature (total field change as measured by scalar magnetometer) of a prolate spheroid model of a 155-mm projectile. The projectile center of the mass is 150-cm deep. Earth's magnetic field inclination is 65 degrees. The projectile in (a) has a horizontal north-south direction; in (b) the direction is vertical (from [20]).

is usually detected by coil (which may be identical to the transmitter coil in time-domain systems). Using an array of GMR sensors for that purpose may increase the system sensitivity and resolution [22]. (Giant magnetoresistors are described in Section 4.2.) Another electromagnetic method is ground-penetrating radar, which directs short electromagnetic pulses into the ground and monitors the reflections. These methods can detect nonferrous or even nonmetal objects, such as plastic landmines. Similar magnetic and electromagnetic methods are used for weapon detection.

10.3.2 Target Detection and Tracking

Magnetic sensors are used to detect and locate vehicles and to track the autonomous missiles or intelligent ammunition during the flight. Fluxgate sensors used for projected grenades should withstand acceleration of several thousands of gravitational force (g) when they are fired.

Submarine detection is a special topic. Underwater sensor arrays of field sensors can detect very small field gradients, because the sensors (usually fluxgates) are kept at a constant temperature and can be made long. The countermeasures include demagnetization of whole submarines and extensive use of nonmagnetic materials such as titanium.

10.3.3 Antitheft Systems

Goods in shops and books in libraries contain small antitheft labels that can be detected at the exit. The labels should be invisible, give strong specific response, and, if possible, enable easy deactivation. Various electromagnetic antenna resonators, magnetoelastic resonators (Section 10.6), and Wiegand wires (Chapter 12) are used. Deactivation can be performed by, for example, burning the flat coil resonator with a strong pulse of high-frequency electromagnetic energy or by demagnetizing the magnetically hard part of the Wiegand wire. Some exhibited pieces of art have an attached permanent magnet. Several magnetic sensors (fluxgates or magnetoresistors) detect the field change that accompanies any movement.

Similar methods are used for identification and authorization, but here the label may be visible. Although magnetic credit cards and door keys are gradually being replaced by smarter chip systems, which can contain more information and communicate wirelessly, magnetic strips will remain on disposable tickets. Magnetic ink patterns are found on some bank notes.

Wiegand wires are popular for keys (e.g., Sensorcard by HID Corp.), because they also work remotely.

Reed contacts together with permanent magnets are still the most common sensors used for door switches. They need no power and can be made resistant against disabling (Chapter 12).

10.4 Automotive Applications

The main automotive application of magnetic sensors is position sensing with a permanent magnet and a moving magnetic circuit. The basic configurations are proximity sensors, tooth sensors, analog contactless potentiometers, and brushless motors with permanent magnet rotors. The sensor types used are Hall sensors, InSb magnetoresistors, and AMR and GMR magnetoresistors. The main selection criterion is temperature stability in the required temperature range of −20°C to +150°C. InSb sensors (produced by Emcore) were found to be best in regard to temperature stability if used with SmCo magnets [23]. Position sensors are described in Chapter 12.

Car navigation systems use magnetic compasses: fluxgates [24] and recently magnetoresistive ones [25] (see Section 10.2). Other magnetic navigation methods are used in unmanned vehicles and experimental cars [17].

Stationary magnetometers are used in traffic monitoring and control to detect passing vehicles and eventually recognize their type. The basic device is an induction loop placed under the road surface. Determination of the vehicle type (car-van-truck-trailer) has an efficiency between 75% and 95%; it fails at low speeds like those in traffic jams [26]. The dc magnetic sensors give more information: Using a three-axial fluxgate or magnetoresistive magnetometer buried under the road surface allows recognition of the vehicle type. Figure 10.7 shows the components of the magnetic field of a small car. A single sensor can be used to monitor the car's presence (e.g., in garages); two separated sensors can measure the direction and driving speed. In a noisy environment, correlation techniques should be used [16, 25, 27].

10.5 Nondestructive Testing

Magnetic methods of nondestructive evaluation can be used either to monitor material state and properties (such as residual stresses) or to find defects. An overview can be found in [28]. The material properties are tested by using the Barkhausen effect, magnetoacoustic emission, monitoring of the hysteresis loop, and magnetoelastic methods. Material inhomogeneities, cracks, and

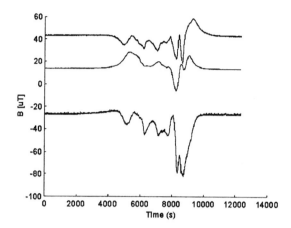

Figure 10.7 Magnetic field of a car: upper trace, vertical component; middle trace, component in the movement direction; lower trace, perpendicular horizontal component (from [15]).

other defects are monitored by dc methods: magnetic particle inspection and magnetic flux leakage method, or by ac eddy currents. Magnetic flux leakage is usually measured by fluxgate sensors. One of the fathers of this method and the author of classic papers was F. Förster, who founded a company that is one of the leading producers of magnetic testing equipment.

Eddy-current methods can be used for nonmagnetic metals. Sasada [29] proposed an eddy-current probe for detection of small defects in nonmagnetic conducting media. The probe consists of excitation and pickup figure-eight coils wound on ferrite. The excitation frequency is 10 kHz or 60 kHz. The coils are cross-coupled, that is, if no defect is present, the voltage induced in the pickup coil is zero. An array of such sensors with a common excitation coil can be scanned to map the sample surface.

Another method used to evaluate materials is magnetic imaging. Two devices are of importance: the magnetic force microscope (MFM) and the scanning SQUID microscope (SSM). An MFM measures the force between the magnetized tip and the specimen surface; it has 50-nm spatial resolution and 10-μT sensitivity. SSM has 1-nT sensitivity, but spatial resolution is only 10 μm [28].

10.6 Magnetic Marking and Labeling

Magnetoelastic labels use resonance of longitudinal vibrations in high-magnetostriction strips. Iron-rich amorphous materials (metallic glasses) are ideal for this application, because they have a high magnetoelastic coupling coefficient, so they are able to transform most of the elastic energy into magnetic energy and vice versa [30]. The vibrations can be excited and also detected by remote coils. Some materials also exhibit a large ΔE effect (change of Young's modulus with the bias dc field), so that the resonating frequency depends not only on the strip dimension but also on the applied dc field. This property can be used for position sensing of a vibrating strip in a known dc field gradient. Multibit tags for object identification based on a cantilever array are described in [31].

Magnetic marking of steel ropes, pipes, and rails is similar to magnetic storage technology. The sensors for reading should be more sensitive, because the medium-sensor distance is large and the marked object is made of construction steel, which is not optimized for magnetic performance. An example is the magnetic stripping of steel winding ropes in British mines. The reading device operates at speeds from zero to 18 m/s, the distance between the marks is 20 cm, and the system operates in shaft depths up to 1,250m [32].

Magnetic bar codes can be used in dirty environments and be made invisible. A microhead, made of two perpendicular U-shaped amorphous wire cores, is able to detect a 1-mm pitch strips made of magnetic ink. Each core is wrapped by a coil: One coil is excited by 100-kHz current, and the other coil detects the sample field. If no strip is present, the output is zero because of the symmetry [33]. Miniature fluxgate sensors (Chapter 3) or magnetoresistors (Chapter 4) also are used for this purpose.

10.7 Geomagnetic Measurements: Mineral Prospecting, Object Location, and Variation Stations

The problems associated with the Earth's field magnetometry and the instrumentation used for geomagnetic research are reviewed by Stuart [34]. Breiner [35] has written a practical handbook for using portable magnetometers for geological and archeological prospecting, magnetic mapping, and the measurement of magnetic properties of rocks. Instrumentation, measurement, and calibration techniques for magnetic observatories are described in a book issued by the International Association of Geomagnetism and Aeronomy (IAGA) [36].

The Earth has a crust, a mantle, and a metallic core. While most of the core is liquid, the inner part is solid. Complex processes are associated with the increase of the inner core ("freezing") together with the Earth's rotation drive, the Earth's so-called magnetic dynamo, which is believed to cause the Earth's magnetic field [37]. The Earth's field (Figure 10.8) has a dipole character; in northern Canada, about 1,000 km from the geographical north pole, there is the north magnetic pole: Paradoxically, it is a south pole of an equivalent bar magnet, because it attracts the north pole of the magnet needle. The Earth's field is changing in time: At present, the amplitude is decreasing by 0.1% each year, the pole is drifting westward by 0.1 degree/year. The tilt of the dipole axis, which is 10.4 degrees in 2000, is decreasing by 0.02 degrees/year [37].

The Earth's field at the poles has a vertical direction (90-degree inclination) with a magnitude of about 60 μT. In the equatorial region, the direction is horizontal (0-degree inclination) with a magnitude of about 30 μT. Local anomalies are associated with remanent or induced magnetization of rocks. A 400-km-long anomaly near Kursk, Russia, has a top vertical field of 180 μT, and the change in declination is as large as 180 degrees. In Kiruna, Sweden, the magnetite ore caused a vertical field of 360 μT. Smaller-size anomalies of geological origin are numerous and cause serious navigation problems. Similar or even larger fields are associated with man-made iron structures.

The daily (i.e., diurnal) variations are of the order of 10 to 100 nT, as can be seen in Figure 10.9(a). They are caused by a solar tide of ionized gas in the ionosphere, which creates electric currents. Micropulsations, as shown in Figure 10.9(b), have periods of 10 ms to 1 hr, with amplitudes up to 10 nT. Magnetic storms, graphed in Figure 10.9(c), occur several times per month with durations of up to several days and amplitudes of several hundreds of nanoteslas. Storms are caused by the interaction between

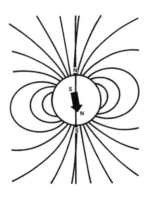

Figure 10.8 The Earth's magnetic field. (From [35].)

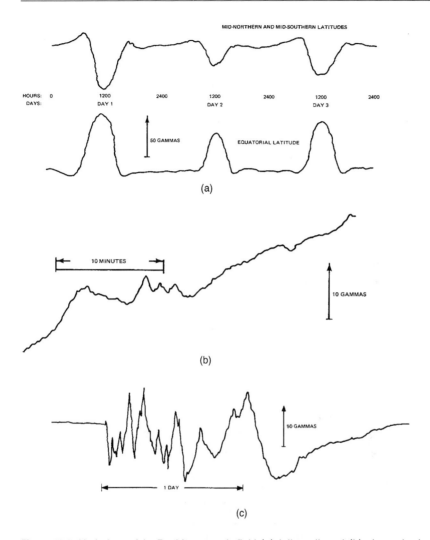

Figure 10.9 Variations of the Earth's magnetic field: (a) daily or diurnal; (b) micropulsations; and (c) magnetic storm (from [35]). 1 gamma = 1 nT (traditional unit).

the coronary plasma bubble in the solar wind and magnetosphere and secondary (mega-ampere) currents associated with a flow of charged particles in the magnetosphere. A typical storm begins with a sudden field jump followed by a long-term depression of the field caused by the secondary ring current.

Surface magnetic mapping is performed mainly by a walking operator (nonmagnetic vehicles are rarely used), who should be free of magnetic materials. The sensor is usually carried on a 2m stick; sometimes a 4m long stick is used so the sensor is removed from locally disturbing fields of surface materials [35].

Airborne magnetic surveys for prospection of ores are performed in low heights (300m or less); geophysical mapping is made at higher altitudes (3,000m to 10,000m) or from low-orbit satellites (~500 km). To reduce the influence of stray fields from the aircraft, the sensor is often towed.

The instruments most often used in geophysical measurements are scalar resonance magnetometers; classic proton magnetometers were often replaced by Overhauser magnetometers (Chapter 7). Optically pumped instruments are more expensive, but they generally offer a high bandwidth together with an excellent precision so they are used for airborne surveys, in which the field changes may be very fast. If the vector measurement is required, the most usual instrument is the fluxgate magnetometer (Chapter 3).

The Earth's field variations during the surveys should be monitored and corrections made. Another possibility is to use gradiometers. They compensate not only for natural field variations but also for noise from distant sources. Gradiometers are advantageous for localization of small and close objects. Most often, the vertical gradient of the total field is estimated as a difference between the readings of two scalar resonant magnetometers (separated by a distance of 90 cm or more). The advantage of a vertical gradient is that the vertical direction is easily defined. Also, at medium and higher latitudes, where the vertical Earth's field component is stronger, vertical gradient is usually rich in information about buried objects. Some gradiometer systems (mainly for search applications) use fluxgate sensors, either two single-axis sensors mounted vertically or multisensor systems.

Geophysical methods are also used to locate buried man-made ferromagnetic structures like pipelines, tanks, drums, and UXO (Section 10.3). Most pipelines have a strong remanent field, which changes at the pipe sections which have a different magnetic history. The pipelines are often magnetized after a dc magnetic defectoscopy. Sometimes the pipes are demagnetized before welding, because the strong remanent field deflects the welding arc. Long horizontal pipe produces a field that decreases as $1/r^2$. Short objects have a dipole field character; the field decreases as $1/r^3$ (compare the field of a straight dc current conductor, which decreases with $1/r$). As a rule of thumb, 1 kg of iron makes a 1-nT field in the distance of 3m, while a single piece of 1,000-kg of iron or long 6-inch pipeline makes 1 nT at the distance of 40m [35]. Figure 10.10 shows the total field profiles of induced dipoles at various inclinations.

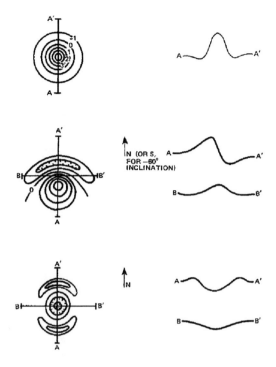

Figure 10.10 The total field profiles of induced dipoles at various inclinations: in a polar region, where inclination is 90 degrees (permanent dipole has a similar field signature); at 60-degree inclination; and at 0-degree inclination, close to equator, where the Earth's field has horizontal direction (from [35]).

The permanent dipole has a constant field signature independent of the Earth's field magnitude and direction, but the total field measured by a scalar magnetometer is the vectorial sum of the Earth's field and dipole field so that it changes with their angle. Figure 10.11 shows the problem: if the small dipole field **M** is perpendicular to the high Earth's field **F**, the resulting change in total field may be very small at equatorial (horizontal)

Figure 10.11 Total field profiles of permanent dipole **M** not parallel to Earth's field **F** (from [35]).

field. That is one reason why vectorial fluxgate magnetometers are used for some search applications, although they have worse stability than resonant magnetometers. Three-axis vectorial magnetometers give more information, so the interpretation is much easier. Figure 10.12 illustrates how the total field anomaly width increases with object depth. A simple but useful rule is that the total field anomaly width is approximately 1 to 3 times the depth.

The common instruments used for the location of reinforcing steel and small-diameter steel gas lines are based on eddy currents. They allow estimation of the rod or pipe location if the depth is less than 12 cm and the separation is more than 5 cm. It is difficult to estimate the rod diameter if the depth is greater than 5 cm. A fluxgate magnetometer system developed for this purpose by McFee has depth range between 24 and 60 cm, depending on the rod type. The rod diameter cannot be reliably estimated due to large variable remanence [38].

Magnetic observatories for long-term measurement of the Earth's field variations form a world network called Intermag. The observatories are on magnetically clean locations, far from gradients and interferences. The measured data can be found on the World Wide Web [39]. The standard instrumentation is a resonant magnetometer and stable thermostated triaxial fluxgate. Periodical absolute calibration is performed with the help of nonmagnetic theodolite with a mounted single-axis fluxgate sensor [36].

Magnetic susceptibility and remanent magnetization of rock samples can be measured by a magnetometer, if the sample is slowly rotated in the Earth's field. Susceptibility is also measured in ac bridges; the sample inserted into the air coil slightly changes its impedance. Remanent magnetization also can be measured by rotating sample magnetometers: The sample rotates inside the magnetic shielding, and the ac field is measured by an induction coil or a fluxgate magnetometer (Molspin system [40]). High values of the sample remanent magnetization are caused by heating effects. During cooling through the Curie temperature, the sample is magnetized by the Earth's field. The heating may be natural, as in igneous rocks, or artificial, as in baked pottery, bricks, and other archeological objects subjected to fire. The

Figure 10.12 Depth/amplitude behavior of dipole anomalies (from [35]).

objects "remember" the magnetization direction, which may be different from today's Earth's field, because the position of magnetic poles changes with time.

10.8 Space Research

The accurate measurement of the ambient magnetic field vector and its orientation in space is recognized as a basic requirement for space physics research. The range of field strengths to be measured in exploratory missions may cover up to nine orders of magnitude ($5 \cdot 10^{-3}$ nT to $2 \cdot 10^{6}$ nT), while an angular determination accuracy of the order of 1 degree is generally sufficient for most studies. In special cases, accuracies of the order of arcseconds are required for detailed mapping surveys of planetary magnetic fields. The time resolution of the measurements ranges from 1 second to several hundred samples per second, depending on scientific objectives [41]. The vast majority of magnetic field measuring instruments used to date have been of the vector type, which measure three orthogonal components of the local field referenced to an inertial coordinate system. Scalar-type instruments, which measure only the magnitude of the ambient field without providing directional information, have been used for specialized missions that require their intrinsic absolute accuracy (e.g., geomagnetic field surveys). The most commonly used vector magnetometers are of the fluxgate type.

10.8.1 Deep-Space and Planetary Magnetometry

In the early years of the space program, measurements of Earth's magnetic field were collected with rockets and balloons. The instruments measured the strength of the equatorial electrojet, the auroral current system, and other high-altitude magnetic phenomena. Using space flight measurement techniques adapted from instruments developed around World War II, early space probes established the cometlike morphology of the distant Earth's magnetic field and discovered many of the features and boundaries of the magnetosphere (the bow shock, magnetopause, and geomagnetic tail). Dolginov et al. [42, 43], Sonnett [44], Heppner and Cahill [45], and Ness [46] were among the first investigators to equip the early probes with magnetic field measuring instruments and carry out measurements in the Earth's magnetosphere and interplanetary medium.

Contemporary space missions such as Pioneer Venus, Mariners, Voyager, Helios, Ulysses, Giotto, Galileo, Phobos, Vega, NEAR, and Mars

Global Surveyor have carried out magnetic field measurements around most of the planets in our solar system, in the interplanetary medium, and near comets and asteroids. Planetary magnetic fields like those of Earth, Jupiter, and Saturn are believed to be generated by currents driven by thermal convection in the interface between their mantles and liquid metallic cores. Uranus and Neptune are assumed not to have formed metallic cores, and their magnetic fields are thought to be generated closer to the surface, where electrical currents can flow in the high-conductivity liquid crust [47–50]. In terrestrial planets, Venus does not possess an intrinsic magnetic field, while Mercury is magnetized by the remains of an ancient dynamo, which is decaying with time [51]. Mars Global Surveyor recently established that Mars does not currently possess an intrinsic magnetic field but that it had one in its early history. That field left a significant portion of Mars's crust strongly magnetized [52–54] and reflecting large-scale tectonic processes that are yet to be fully understood.

Only two types of measurements from an orbiting or flyby spacecraft provide clues about the interior of a planetary body, either its structure or thermal history: the effects of gravity on the spacecraft orbit or trajectory and the internal magnetic field strength and geometry. Many missions utilize magnetic field measurements for engineering applications. For instance, Earth-orbiting spacecraft apply magnetic field information to attitude determination and control, spacecraft momentum management, and scientific instrument pointing. The Earth's magnetic field provides one of the basic "natural" forces that modern systems and spacecraft utilize to establish their orientation with respect to a reference frame when inertial systems are too complex or costly to implement.

While numerous discoveries have been made in researching the sources and behavior of planetary magnetic fields, there is still much to be learned. Magnetic field measurements are essential to complement onboard energetic charged particle measurements to aid understanding of the behavior of plasmas in the solar system and energetic trapped particles around magnetized planets. Electrically charged particles move easily along the local magnetic field line but have great difficulty moving transverse to it. Hence, magnetic field lines are important tracers of particle motion geometry, trapping, and drift around magnetized bodies.

10.8.2 Space Magnetic Instrumentation

Both scalar and vector instruments have been used onboard spacecraft to measure magnetic fields in space, but vector magnetometers are far more common because of their ability to provide directional information. This section presents a brief discussion of their advantages and disadvantages and the general problem of performing sensitive magnetic field measurements onboard an orbiting spacecraft or planetary probe. A comprehensive review of early space research magnetometers can be found in [1].

10.8.2.1 Scalar Magnetometers

The polarize/count cycle of the typical proton precession magnetometer requires 1 second or more in traditional designs, and the liquid sample volume is relatively large and massive, particularly when the polarizing coil mass is considered. Liquids that can operate over a wide temperature range are necessary, and the polarizing power required to generate the 10-mT (or more) polarizing field is significant; useful signals can be obtained only for ambient fields larger than approximately 20,000 nT. Those limitations have restricted the use of classic proton precession instruments onboard spacecraft, although substantial use has been made in sounding rocket applications, in which short flight durations in the Earth's field are typical. Recent developments like the Overhauser-effect proton precession magnetometer, which uses an indirect technique to "polarize" the sample and generate a continuous Larmor precession signal, promise further advances in this area. (See Chapter 7 for more about scalar magnetometers.)

Optically pumped magnetometers are another class of magnetic field measuring instruments that have found considerable application onboard spacecraft, both as scalar as well as vector instruments [55–59].

10.8.2.2 Vector Magnetometers

Vector magnetometers are, by a large margin, the most widely used type of instrument for magnetic field measurements onboard spacecraft, balloons, and sounding rockets. In addition to providing information about the field strength, they also indicate the direction and the sense of the ambient field. Triaxial orthogonal arrangements of single-axis sensors are used to measure the three components of the ambient field in a coordinate system aligned with the sensor magnetic axes. In contrast to proton precession and alkali vapor scalar magnetometers, whose accuracy is determined by quantum mechanical constants, vector magnetometers must be calibrated against known magnetic fields, both in strength and direction. Their output for zero field, scale factor, and stability with temperature and time depend on electrical component values, which may drift as the instruments age or are exposed to the effects of the space environment [1, 60–62]. In spite of those shortcomings, vector instruments are capable of measuring magnetic fields over a large dynamic range ($5 \cdot 10^{-3}$ nT to more than $2 \cdot 10^6$ nT), are lightweight, and consume little power. In addition, they are capable of operation over a wide temperature range and have proved to be extremely reliable and extremely resistant to the destructive effects of intense radiation from solar flares and energetic trapped particles in planetary magnetospheres. Ultraprecise instruments have been used to map the Earth's magnetic field from orbit with unprecedented accuracy, both in magnitude as well as direction: 5 nT and 3 arc-seconds, respectively [63, 64].

10.8.3 Measurement of Magnetic Fields Onboard Spacecraft

The instruments described so far are carried into space onboard spacecraft that include complex systems of mechanical, electrical, and electronic components. Those components, unless carefully controlled, have the potential to generate strong magnetic fields of their own. Batteries, solar arrays, motors, wiring, and other materials must be especially designed and selected to minimize the generation of stray magnetic fields that will affect measurements. The design and implementation of a magnetically "clean" spacecraft meeting the stringent requirements of an interplanetary mission is a demanding task that has tested the fiber of many seasoned project managers and engineers. Because it is practically impossible to reduce the stray spacecraft magnetic field to the small levels required for sensitive measurements, the use of long booms to place the magnetic sensors away from the main body of the spacecraft is commonplace (Figure 10.13). This technique exploits the fact that magnetic fields decrease rapidly away from their source, proportionally to $1/r^3$ as a minimum, where r is the distance to the source.

The booms must be rigid and preserve the required alignment between the magnetic sensors and the attitude-determination sensors mounted in the main spacecraft body, which limits their practical length. For a given spacecraft design, the tradeoff between boom length and level of magnetic cleanliness required in the main body is a major decision that must take into account many conflicting requirements, including whether the spacecraft is spin stabilized or three-axis stabilized. Details of magnetic interference, spacecraft testing, and magnetic cleanliness programs can be found in [1]. In sensitive missions, a magnetics control program is usually carried out to ensure that the sources of stray magnetism onboard the spacecraft are eliminated or minimized to the required levels [65]. Such programs originally emphasized the control of all spacecraft-generated fields, either static or dynamic. With the advent of wide dynamic range instruments and high-

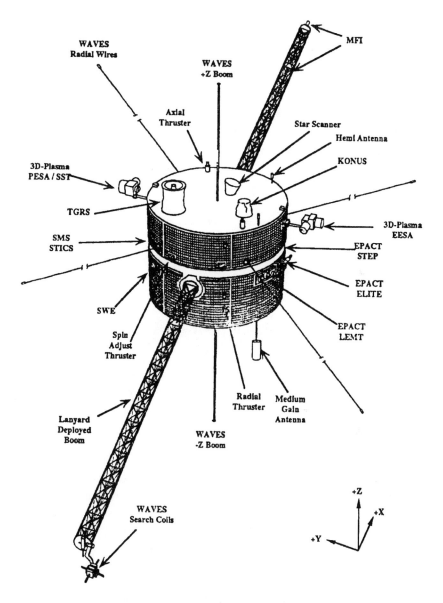

Figure 10.13 Satellite with booms for magnetic sensors.

resolution ADCs, they now emphasize control of dynamic fields over static fields, which can be calibrated out before launch.

The dual magnetometer technique was introduced in 1971 by Ness et al. to ease the problem of making sensitive magnetic field measurements in the presence of a significant spacecraft field [66]. The method is based on the experimental observation that beyond a certain distance most spacecraft-generated magnetic fields decrease as expected for a simple dipole source located at the center of the spacecraft ($1/r^3$). Thus, it can be shown that if two magnetometer sensors are used, mounted along a radial boom and located at distances r_1 and r_2, it is possible to uniquely separate the spacecraft-generated magnetic field from the external field being measured. If we denote as $\mathbf{B}_1$ and $\mathbf{B}_2$ the vector fields measured at radial locations 1 and 2, with $r_2 > r_1$, the ambient field and the spacecraft field at location r_1 are given by

$$\mathbf{B}_{amb} = (\mathbf{B}_2 - \forall \mathbf{B}_1)/(1 - \forall) \qquad (10.1)$$

$$\mathbf{B}_{s/c1} = (\mathbf{B}_1 - \mathbf{B}_2)/(1 - \forall) \qquad (10.2)$$

where

$$\mathbf{B}_1 = \mathbf{B}_{amb} + \mathbf{B}_{s/c1} \qquad (10.3)$$

$$\mathbf{B}_2 = \mathbf{B}_{amb} + \mathbf{B}_{s/c2} \qquad (10.4)$$

$$\mathbf{B}_{s/c2} = \forall \mathbf{B}_{s/c1} \qquad (10.5)$$

$$\forall = (r_1/r_2)^3 \qquad (10.6)$$

Note that (10.5) and (10.6) imply that the spacecraft field can be correctly represented by a dipole centered on the main body.

The dual magnetometer method is illustrated schematically in Figure 10.14. The spacecraft magnetic field decreases with distance due to the finite size of the sources, which are assumed to be located at the center of the spacecraft, leading to the existence of a gradient between the two magnetometer sensors. On the other hand, the ambient field being measured is identical at both sensors because its gradient is insignificant over the dimensions of the spacecraft and boom. Thus, each sensor measures a different mixture of spacecraft and ambient field. For the special case of a dipolar spacecraft field, the fields due to the spacecraft at the two sensor locations are related by a simple proportionality constant. Thus, the ambient and spacecraft field can be separated analytically as shown above.

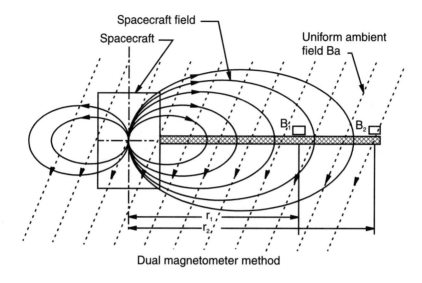

Figure 10.14 Dual magnetometer.

A particular advantage of the dual magnetometer method is that it allows the unambiguous real-time identification and monitoring of changes in the spacecraft field. In addition, the use of two sensors provides a measure of redundancy that has proved extremely useful in long-duration missions, like those to the outer planets.

10.8.3.1 Coordinate Systems, Power Spectrum, Zero Levels, and Alignments

So we can interpret the measurements in a physical sense and carry out the intended research, the measurements must be expressed in a coordinate system that is different from that of the magnetic field sensors [67]. Therefore, coordinate transformations are required that must take into account not only the orientation of the spacecraft in inertial space but also the internal rotations associated with booms and other instruments and sensors onboard and possibly their time dependence. In general, it is difficult to simulate accurately on the ground the zero-g flight environment for booms and appendages; hence, sensor alignment must be verified in flight. Many techniques have been developed for that purpose, but they all have in common the need for spacecraft maneuvers or rotations about principal axes to establish sensor alignment with respect to the same. Spinning spacecraft are preferred for space physics applications because the rotational motion helps the scanning of ambient particle distribution functions through a large solid angle.

In the case of magnetometers, however, the spacecraft spin modulates the ambient magnetic field perpendicular to the spin axis, and additional bandwidth is required to transmit that information to the ground unless onboard despin data processing is used. This is a classic example of data processing and compression algorithms, which are frequently used to reduce the bandwidth required to transmit the information to ground [68–70].

On spinning spacecraft, the data need to be despun to an inertial coordinate system using spin-phase information generally provided by the spacecraft attitude control system. This is a nontrivial task if the spin tone and its harmonics contributions to the overall power spectrum are to be minimized. The study of the frequency spectrum of dynamic perturbations of the magnetic field is a powerful tool used to identify the types and characteristics of waves and other time-variable phenomena detected by spacecraft instruments. The outputs from dc magnetic field sensors typically are digitized with high-resolution ADCs to resolve small amplitude fluctuations superimposed on larger background fields. Space instruments may include onboard FFT processors and other digital implementations of data compression techniques to reduce the bandwidth required to transmit the information to ground [71, 68].

In spite of the many advantages of fluxgate magnetometers, the stability of its output for zero field input and the sensor noise have been and will continue to be the major challenges for sensitive measurements. In the case of zero levels, several techniques have been developed for their determination in flight for both spinning and nonspinning spacecraft, techniques that minimize the problems associated with drifts. The interplanetary magnetic field is characterized by frequent changes in direction rather than magnitude, which can be used to advantage to statistically estimate effective magnetometer zero levels. Spinning spacecraft and spacecraft maneuvers help resolve zero levels in the rotation plane by modulating the ambient signal.

In many space missions, the motion of the medium and the spatial boundaries in response to the time variability of the solar inputs significantly affect the interpretation of single spacecraft magnetic field measurements. The instruments may detect a rapid change in the ambient conditions, but the observation cannot be interpreted unambiguously because the crossing of a spatial boundary or a simple temporal variation yields similar signatures in the data. Therefore, simultaneous multispacecraft measurements are required to remove those ambiguities from the observations. Simultaneous observations by two spacecraft provide ambiguity resolution only along a straight line (single dimension), while four spacecraft are required to provide full three-dimensional resolution [72].

References

[1] Forsman, M., "Gastric Emptying of Solids Measured by Means of Magnetised Iron Oxide Powder," *Medical & Biological Engineering & Computing*, Vol. 36, 1998, pp. 2–6.

[2] Cohen, D., "Measurements of the Magnetic Fields Produced by the Human Heart, Brain and Lungs," *IEEE Trans. Magn.*, Vol. 11, 1975, pp. 694–700.

[3] Stroink, G., D. Dahn, and F. Brauer, "A Lung Model to Study the Magnetic Field of Lungs of Miners," *IEEE Trans. Magn.*, Vol. 18, 1982, pp. 1791–1793.

[4] Moilanen, M., et al., "Measurement of the Magnetic Properties of Metal Dusts and Fumes," *IEEE Trans. Magn.*, Vol. 18, 1982, pp. 788–791.

[5] Ripka, P., et al., "Magnetic Properties of the Respired Particles," *3rd Jap.-Cz.-Slov. Joint Seminar on Applied Electromagnetics*, Prague, Czech Republic, July 1995.

[6] Ripka, P., et al., "Fluxgate Magnetopneumography," *Proc. Imeko World Congress*, Osaka, Japan, 1999, Vol. 8, pp. 161–165.

[7] Ripka, P., and P. Navratil, "Fluxgate Sensor for Magnetopneumometry," *Sensors and Actuators A*, Vol. 60, 1997, pp. 76–79.

[8] Ripka, P., et al., "Magnetic Lung Diagnostics Using Fluxgate," *Proc. Imeko World Congress*, Vienna, Vol. VII, 2000, pp. 101–104.

[9] Barber, G. W., and A. S. Arrott, "History and Magnetics of Compass Adjusting," *IEEE Trans. Magn.*, Vol. 24, Nov. 1988, pp. 2883–2885.

[10] Caruso, M. J., "Applications of Magnetoresistive Sensors in Navigation Systems," Honeywell, SSEC, 12001 State Highway 55, Plymouth, MN 55441, http://www.ssec.honeywell.com.

[11] Ripka, P., "Improved Fluxgate for Compasses and Position Sensors," *J. Magn. Magn. Mater.*, Vol. 83, 1990, pp. 543–544.

[12] Fowler, J. T., and A. D. Little, "New Technology Magnetic Attitude Sensors in Autonomous Underwater Vehicle Applications," *Proc. IEEE Symp. Autonomous Underwater Vehicle Technology*, Washington, D.C., 1990, pp. 263–269.

[13] Tangelder, R., "Smart Sensor System Application: An Integrated Compass," *Proc. ED&TC Conf.*, New York: IEEE Computer Society Press, 1997, pp. 195–199.

[14] Cantrell, T., "Do You Know the Way to San Jose?" *Computer Applications J.*, No. 53, Dec. 1994, pp. 72–77. More information at www.precisionnavigation.com.

[15] Honeywell HMR 3000 Digital Compass Module, datasheets and application notes at http://www.ssec.honeywell.com/products/magsensor_data.html.

[16] Ripka, P., "Magnetic Sensors for Traffic Control," *Proc. ISMCR 99*, Tokyo, Japan, 1999, Vol. 10, pp. 241–246.

[17] Borenstein, J., H. R. Everett, and L. Feng, *Navigating Mobile Robots*, Wellesley, MA: A. K. Peters, 1995.

[18] UXO information on the Web, April 1999, Jefferson Proving Ground, Highway 421, Madison, IN 47250, www.jpg.army.mil.

[19] Young, R., and L. Helms, "Applied Geophysics and the Detection of Buried Munitions," U.S. Army Engineering and Support Center, Huntsville, 74 Washington Ave, N., Suite 7, Battle Creek, MI 49017-3804, www.hnd.usace.army.mil/oew/tech/rogppr1.html.

[20] Altshuler, T. W., "Shape and Orientation Effects on Magnetic Signature Prediction for UXO," *Proc. UXO Forum*, 1996.

[21] Geophysical Signature Database, U.S. Army Engineering and Support Center, Huntsville, Battle Creek, MI, www.scainc-ma.com/cgi-bin/db1.htm.

[22] Wold, R. J., et al., "Development of a Handheld Mine Detection System Using a Magnetoresistive Sensor Array," *Proc. SPIE Detection and Remediation Technologies for Mines and Minelike Targets IV*, Orlando, FL, Apr. 5–9, 1999, pp. 113–123.

[23] Heremans, J., "Magnetic Field Sensors for Magnetic Position Sensing in Automotive Applications," *Proc. Conf. Properties and Applications of Magnetic Materials*, Chicago, IL, May 1997, Session 1.

[24] Peters, T. J., "Automobile Navigation Using a Magnetic Flux-Gate Compass," *IEEE Trans. Vehicle Technology*, VT-35, 1986, pp. 41–47.

[25] Caruso, M. J., and L. S. Withanawasam, "Vehicle Detection and Compass Applications Using AMR Magnetic Sensors," Honeywell, SSEC, 12001 State Highway 55, Plymouth, MN 55441, http://www.ssec.honeywell.com.

[26] Gajda, J., and M. Stencel, "Determination of Road Vehicle Types Using an Inductive Loop Detector," *Proc. Imeko World Congress*, CD-ROM, Tampere, Finland, 1997, Topic 21, Paper 501, pp. 1–8.

[27] Lao, R., and D. Czajkowski, "Magnetoresistors for Automobile Detection and Traffic Control," *Sensors*, Apr. 1996, pp. 70–73.

[28] Jiles, D., "Introduction to Magnetism and Magnetic Materials," Chap. 16 in *Magnetic Methods for Materials Evaluation*, London: Chapman & Hall, 1998.

[29] Sasada, I., and N. Watanabe, "Eddy Current Probe for Nondestructive Testing Using Cross-Coupled Figure-Eight Coils," *IEEE Trans. Magn.*, Vol. 31, 1995, pp. 3149–3151.

[30] Barandiaran, J. M., and J. Gutierrez, "Magnetoelastic Sensors Based on Soft Amorphous Magnetic Alloys," *Sensors and Actuators A*, Vol. 59, 1997, pp. 38–43.

[31] Schrott, A. G., and R. J. Gutfeld, "Magnetic Arrays and Their Resonant Frequencies for the Production of Binary Codes," *IEEE Trans. Magn.*, Vol. 34, 1998, pp. 3765–3771.

[32] Lewis, D. C., and H. Ormondroyd, "Magnetic Striping of Steel Ropes," *Mining Technology*, June 1978, pp. 220–227.

[33] Watanabe, N., I. Sasada, and N. Asuke, "A New High Density Magnetic Bar Code System," *J. Appl. Phys.*, Vol. 85, 1999, pp. 5462–5464.

[34] Stuart, W. F., "Earth's Field Magnetometry," *Reports on Progress in Physics*, Vol. 35, 1972, pp. 803–881.

[35] Breiner, S., "Applications Manual for Portable Magnetometers," GeoMetrics, Sunnyvale, CA, 1973.

[36] Jankowski, J., and C. Sucksdorff, *Guide for Magnetic Measurements and Observatory Practice*, Warsaw: IAGA, 1996.

[37] Russel, C. T., and J. G. Luhmann, "Earth: Magnetic Field and Magnetosphere," in J. H. Shirley and R. W. Fainbridge (eds.), *Encyclopedia of Planetary Sciences*, New York: Chapman and Hall, 1977.

[38] McFee, J. E., et al., "A Magnetometer System to Estimate Location and Size of Long, Horizontal Ferrous Rods," *IEEE Trans. Instr. Meas.*, Vol. 45, 1996, pp. 153–158.

[39] Intermagnet: The International Real-Time Magnetic Observatory Network, c/o Geological Survey of Canada, National Geomagnetism Program, 7 Observatory Crescent, Ottawa, Ontario, Canada, K1A 0Y3, http://www.intermagnet.org.

[40] Stephenson A., and L. Mollyneux, "A Versatile Instrument for the Production, Removal and Measurement of Magnetic Remanence at Different Temperatures," *Meas. Sci. Technol.*, Vol. 2, 1991, pp. 280–286.

[41] Ness, N. F., "Magnetometers for Space Research," *Space Science Reviews*, Vol. 11, 1970, pp. 459–554.

[42] Dolginov, S. S., L. N. Zhuzgov, and N. V. Pushkov, "Preliminary Report on Geomagnetic Measurements Carried Out From the Third Soviet Artificial Earth Satellite," *Artificial Earth Satellites*, Vol. 2, 1959, pp. 63–67.

[43] Dolginov, S. S.., L. N. Zhuzgov, and V. A. Selyutin, "Magnetometers in the Third Soviet Earth Satellite," *Iskusstvennyye Sputniki Zemli*, Vol. 4, 1960, pp. 135–160 (*Artificial Earth Satellites*, Vol. 1–2, New York: Plenum Press, 1960, pp. 358–396).

[44] Sonnet, C. P., "The Distant Geomagnetic Field, 2, Modulation of a Spinning Coil EMF Magnetic Signals," *J. Geophys. Res.*, Vol. 68, 1963, pp. 1229–1232.

[45] Cahill, L. J., "The Geomagnetic Field," in D. P. LeGalley and A. Roser (eds.), *Space Physics*, New York: Wiley, 1964, pp. 301–349.

[46] Ness, N. F., "The Earth's Magnetic Tail," *J. Geophys. Res.*, Vol. 70, 1965, pp. 2989–3005.

[47] Ness, N. F., et al., "Magnetic Fields at Uranus," *Science*, Vol. 233, 1986, pp. 85–89.

[48] Ness, N. F., et al., "Magnetic Fields at Neptune," *Science*, Vol. 246, 1989, 1473–1478.

[49] Connerney, J. E. P., M. H. Acuña, and N. F. Ness, "The Magnetic Field of Uranus," *J. Geophys. Res.*, Vol. 92, 1987, pp. 234–248.

[50] Connerney, J. E. P., M. H. Acuña, and N. F. Ness, "The Magnetic Field of Neptune," *J. Geophys. Res.*, Vol. 96, 1991, pp. 023–19042.

[51] Ness, N. F., "The Magnetic Fields of Mercury, Mars, and Moon," *Ann. Rev. Earth Planet. Sci.*, Vol. 7, 1979, pp. 249–288.

[52] Acuña, M. H., et al., "Magnetic Field and Plasma Observations at Mars: Initial Results of the Mars Global Surveyor Mission," *Science*, Vol. 279, 1998, pp. 1676–1680.

[53] Acuña, M. H., et al., "Global Distribution of Crustal Magnetism Discovered by the Mars Global Surveyor MAG/ER Experiment," *Science*, Vol. 284, 1999, pp. 790–793.

[54] Connerney, J. E. P., et al., "Magnetic Lineations in the Ancient Crust of Mars," *Science*, Vol. 284, 1999, pp. 794–798.

[55] Slocum, R. E., and F. N. Reilly, "Low Field Helium Magnetometer for Space Applications," *IEEE Trans. Nucl. Sci.*, Vol. NS-10, 1963, pp. 165–171.

[56] Farthing, W. H., and W. C. Folz, "Rubidium Vapor Magnetometer for Near Earth Orbiting Spacecraft," *Rev. Sci. Instrum.*, Vol. 38, 1967, pp. 1023–1030.

[57] Slocum, R. E., P. C. Cabiness, and S. L. Blevins, "Self-Oscillating Magnetometer Utilizing Optically Pumped He," *Rev. Sci. Instrum.*, Vol. 42, 1971, pp. 763–766.

[58] Slocum, R. E., "Zero-Field Level-Crossing Resonances in Optically Pumped $2^3S_1He^4$," *Phys. Rev. Let.*, Vol. 29, 1972, pp. 1642–1645.

[59] Smith, J. E., and C. P. Sonnett, "Extraterrestrial Magnetic Fields: Achievements and Opportunities," *IEEE Trans. Geo. Electr.*, Vol. GE-14, 1976, pp. 154–171.

[60] Acuña, M. H., "Fluxgate Magnetometers for Outer Planets Exploration," *IEEE Trans. Magn.*, Vol. 10, 1974, pp. 519–523.

[61] Acuña, M. H., et al., "Mars Observer Magnetic Fields Investigation," *J. Geophys. Res.* Vol. 97, 1992, pp. 7799–7814.

[62] Acuña, M. H., et al., "The NEAR Magnetic Field Investigation—Science Objectives at Asteroid Eros 433 and Experimental Approach," *J. Geophys. Res.*, Vol. 102, 1997, pp. 23751–23759.

[63] Acuña, M. H., et al., "The Magsat Vector Magnetometer: A Precision Fluxgate Magnetometer for the Measurement of the Geomagnetic Field," NASA TM79656, 1978.

[64] Duret, D. N., et al., "Overhauser Magnetometer for the Danish Oersted Satellite," *IEEE Trans. Magn.*, Vol. 31, 1995, pp. 3197–3199. Other information also on web.dmi.dk/projects/oersted.

[65] Harten, R., and K. Clark, "The Design Features of the GGS WIND and POLAR Spacecraft," *Space Science Rev.* 71, 1995, 4 vols., C. T. Russell (ed.), pp. 23–40.

[66] Ness, N. F., et al.," Use of Two Magnetometers for Magnetic Field Measurements on a Spacecraft," *J. Geophys. Res.*, Vol. 76, 1971, pp. 3564–3573.

[67] Russell, C. T., "Geophysical Coordinate Transformations," *Cosmic Electrodynamics*, Vol. 2, 1971, pp. 184–196.

[68] Lepping, R. P., et al., "The WIND Magnetic Field Investigation," *Space Science Rev.* 71, 1995, 4 vols., ed. by C. T. Russell, pp. 207–229.

[69] Russell, C. T., et al., "The GGS/Polar Magnetic Fields Investigation," *Space Science Rev.* 71, 1995, 4 vols., ed. by C. T. Russell, pp. 563–582.

[70] Behannon, K. W., et al. "Magnetic Field Experiment for Voyagers 1 and 2," *Space Sci. Rev.*, Vol. 21, 1977, pp. 235–245.

[71] Acuña, M. H., et al., "The Global Geospace Science Program and Its Investigations," *Space Science Rev.* 71, 1995, 4 vols. ed. by C. T. Russell, pp. 5–21.

[72] www-istp.gsfc.nasa.gov/istp/cluster/home_page.html. Also Balogh, A., et al., "The Magnetic Field Investigation on Cluster," ESA SP-1103, European Space Agency, 1998, pp. 15–20.

# 11

Testing and Calibration Instruments
Ichiro Sasada, Eugene Paperno, and Pavel Ripka

Magnetic sensor specifications were discussed in Chapter 1. This chapter describes equipment for sensor testing and calibration and also discusses some techniques in more detail.

There are two methods for the calibration of the sensor sensitivity and angles between the sensor axes in three-axial magnetometers. The first one uses the Earth's field, the second one needs calibration coils. Using the Earth's field requires precise positioning of the sensor under test, and it can be used only for low-field sensors such as fluxgate and AMR sensors. The absolute precision achievable is about 10 nT (down to 1 nT at magnetic observatory). The procedure requires a precise reference proton or cesium magnetometer (Chapter 7), a nonmagnetic theodolite, a magnetically clean and quiet location, and a lot of patience. The calibration procedures are described in geophysical literature such as [1]. The second method needs precise coils and current sources. The calibration coils are described in Section 11.1.

For the measurement of sensor noise and offset, a place with very low magnetic field is required. Shielding of the ambient field (geomagnetic field of about 50,000 nT, its variations up to 500 nT, and magnetic noise originating in human activities and ranging from 10 nT at magnetically silent locations to 1 mT in industrial environments) can be performed by various means. A ferromagnetic shielding is usually made in the form of a multilayer cylinder from Permalloy or amorphous materials. This type of shielding is available from several manufacturers in a large selection of sizes up to shielded rooms, which are used for biomagnetic measurements using SQUIDs (Chapter 8).

Magnetic shieldings are often custom-made for specific requirements. Typical shielding for calibration of offset and noise of magnetic sensors has a cylindrical shape with an internal diameter of 15 to 20 cm. The shielding typically has a length of 60 cm to 1m, the access is from the top side, and the bottom side is closed. The best location of the sensor under test is about 10 to 15 cm from the bottom lid. The calibration shielding typically has six layers of 1-mm-thick Permalloy (Mumetal®), and it should be once annealed and periodically demagnetized by the air coil supplied by 50 Hz current. The demagnetization coil is inserted inside and slowly pulled out. The demagnetization field should be sufficiently strong to saturate all the shielding layers. Such cylindrical shieldings have been made by Schoensted and other manufacturers and have become standard equipment in geophysical laboratories. The field inside typically is units of nanoteslas, and it is caused more by remanence than by uncompensated rest of the external field. The literature on magnetic shielding is fragmented. Section 11.2 gives both the basic theory and practical rules for design, maintenance, and application of ferromagnetic shieldings.

A superconducting shield based on the Meissner effect (Chapter 8) was described by Brown [2]. It had an attenuation factor of the order of 10^{-7} in transverse and 10^{-9} in the radial direction. At present, superconducting shieldings are used only as a part of cryogenic instruments; much cheaper ferromagnetic shieldings are sufficient for testing and operation of room-temperature devices. Hechtfischer showed that if the superconducting cylinder is cut by slit, the field inside is not "frozen;" his device can be used for homogenization of a calibration magnetic field [3].

Another way to make a magnetic vacuum is to actively compensate the field by means of a three-dimensional system of coils. Section 11.1 discusses that approach. The ac magnetic fields can be suppressed by eddy-current shieldings made of thick conductive material; 45-mm-thick aluminum plates were used in the Tampere shielded room [4].

The absolute value of the sensor offset can be measured by rotating the sensor through 180 degrees. While in the presence of the Earth's field, this method requires precise positioning using theodolite and complex compensation of the axis nonalignment [1]; it becomes very simple when used in the low residual field inside the shielding. A similar technique of flipping the sensor was used at the AMPTE subsatellite for inflight offset calibration. Some satellites and rockets rotate with respect to their axes to stabilize the trajectory. The rotation modulates the component of the field perpendicular to the axis, so the rotation can be used for the same calibration purpose. A method of inflight magnetometer calibration on a spinning spacecraft is described in [5].

11.1 Calibration Coils

The formulas for the magnetic field of a long solenoid and a current loop were given in Chapter 1, in (1.7). For a finite cylindrical solenoid, the field can be calculated by integration [6]. In its center, it is equal to

$$H = \frac{NI}{\sqrt{4R^2 + l^2}} \quad (11.1)$$

where l is the coil length and R its radius, N is the number of turns, and I is the coil current.

The field at the ends of the solenoid is approximately one-half that value; precisely, it is:

$$H = \frac{NI}{2\sqrt{R^2 + l^2}} \quad (11.2)$$

The field at any point on the axis can be calculated using (11.2) if we formally divide the solenoid at that point into two shorter solenoids and add their contributions.

Solenoids are used for compensation windings of, for example, fluxgate sensors. If the compensation solenoid is not long enough, the field nonhomogeneity together with nonlinearity of the sensor may cause even the feedback-compensated sensor to have nonlinear response.

Solenoids are not very practical calibration devices because access to their center, where the sensor under test should be located, is complicated. The most common calibration device is the Helmholtz coil pair, shown in Figure 11.1(a), which consists of two identical, thin, circular coils separated by a distance of their radius R. The 100-ppm homogeneity is achieved inside a cylinder that has 10% dimension of the coils.

The field in the middle of the circular Helmholtz coils is

$$H = NI(1.6/\sqrt{5})/R \cong 0.7155 NI/R \quad (11.3)$$

which corresponds to $B = 899.18 \ NI/R (nT/A)$.

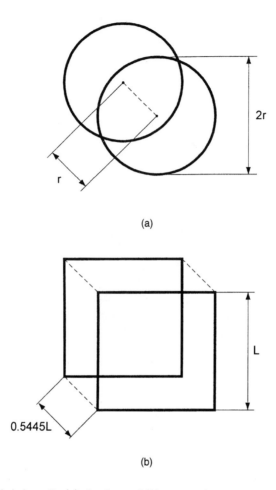

Figure 11.1 Helmholtz coils: (a) circular; and (b) rectangular.

The field gradient in the midpoint of Helmholtz coils is zero. If we connect the two coils antiserially, the field in the middle is zero and we have there a region of a constant field gradient.

Rectangular coils, shown in Figure 11.1(b), are easier to realize in large sizes, and their field is only slightly less uniform [7] than that of circular coils; in fact, they give larger working volume for modest values of uniformity. The distance between the square coils for the best homogeneity should be $0.5445\ L$, where L is the length of the side of the square.

The field in the midpoint of such a rectangular Helmholtz coil pair is

$$B = 1628.7 NI/L\,(\text{nT/A}) \tag{11.4}$$

Alldred and Scollar calculated a system of four rectangular coils that gives more than 10 times better homogeneity than two coils [8]. The coil system is shown in Figure 11.2. The coil dimensions are calculated for rational current ratios, which can be obtained by a realistic number of turns. Such a system was built and calibrated in the Nurmijarvi magnetic observatory for $N_1/N_2 = 21:11$, $L_1 = 0.9552 L_2$, $d_1 = 1.0507 L_2$, and $d_2 = 0.28821 L_2$ [9]. The whole system consists of three orthogonal sets of four square coils with side length ranging from 1.6m to 2.2m. It produces a uniform field with 10-ppm accuracy in a spherical volume of 30-cm diameter (100 ppm for 40-cm diameter). The coil directions were found by using nonmagnetic theodolite with 0.5 arc-min precision, and the individual coil constants were calibrated by proton magnetometer with 10-ppm accuracy. Using coils supplied by computer-controlled current sources and distant reference magnetometer monitoring the variations of the Earth's field, a precise system for calibration of magnetometers was built. Such a system should be located at a magnetically clean and silent location in a nonmagnetic thermostated room. To ensure geometric stability, the coil system should be mounted on a concrete pillar. The ideal, but very expensive, material for large coils and their support system is quartz; coil support made of glass was used for

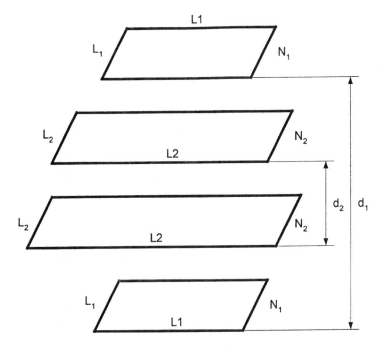

Figure 11.2 Rectangular coil system of Alldred and Scollar (after [8]).

compensation system in Pruhonice [10]. Wooden constructions are sensitive to humidity, and aluminum has large temperature dilatation, but it is most often used together with temperature stabilization [1].

Numerous other field coil systems are described in the literature. Some of them allow all the coils to have the same area, but they require exotic current ratios into the coil sections, which are difficult to keep stable. It should be noted that the only realistic design is that all the coils in the given direction are connected in series and a precise and stable relationship between the individual loop currents is maintained by the number of their turns.

Sometimes the magnetic field outside the coil system must be calculated, for example, to estimate the influence on sensors in their vicinity. The current loop has a magnetic moment of $m = NIS$, and from a large distance it can be considered as a magnetic dipole. So that the field in the distance of r is

$$B_r = 2\mu_0 m \cos\phi / 4\pi r^3 \ldots \text{radial component} \qquad (11.5)$$

$$B_\phi = \mu_0 m \sin\phi / 4\pi r^3 \ldots \text{tangential component}$$

where ϕ is the angle between the coil axis and the direction to the measured point. In a Helmholtz coil pair, the magnetic field is twice as large. If we want to create a magnetic moment (e.g., for a calibration of induction magnetometers), a compact one-layer solenoid with a length-diameter ratio of $l/R = \sqrt{2}/2$ is theoretically the most convenient coil shape.

Ideal spherical coils generate a homogeneous field in the whole inside volume. Real spherical shell coil consisting of 50 turns on each hemisphere has a 100-ppm homogenous field in 67% of the volume (compared to 9% for Helmholtz coils) and a 10-ppm homogeneous field in 50% of the volume (5% for Helmholtz coils) [11]. Spherical coils are difficult to manufacture, and the access to the inside volume is complicated. They are used in proton magnetometers (Chapter 7) and also as a triaxial CSC system for feedback compensation in a fluxgate magnetometer (Chapter 3).

Probably the largest magnetic field calibration facility is at NASA Goddard Space Flight Center. The coil system is of the Braunbeck type, consisting of 12 circular coils (four for each axis). The 10-ppm homogeneity volume has a 1.8-m diameter. The largest coil is 12.7m in diameter, the access window is 3 · 3 m. The coils are wound of the aluminum wire on the aluminum structure and they are temperature compensated. The system is able to compensate the Earth's field using a magnetic observatory, which is located in the distance of about 100 m in the direction with a low natural gradient. Independently to field cancellation, the system is able to generate an artificial field up to 60 μT with an arbitrary direction [12].

Several rules should be observed when making coil calibrations. At first it is necessary to adjust and determine the sensor position precisely. In most cases, the directional response of a vector sensor is cosine type, so 100-ppm precision requires 0.8-degree alignment; 10-ppm precision requires 0.25-degree alignment. The adjustment should be made magnetically, because the sensor geometric axis may differ from its magnetic sensitivity axis. The best solution is to adjust the sensor position using perpendicular coils, for which the angular response is much sharper. The final correction is then made numerically, taking into account the estimated angles between the calibration coils.

Large calibration coils have large time constants. It is therefore important to wait after each field change until the current is constant. It is necessary to monitor the field interferences during the measurement and repeat each calibration step several times. The environmental noise and other variations caused, for example, by the noise of the calibration current or mechanical vibrations can be suppressed by using a long integration time of the voltmeters, simultaneous measurements of all the variables, and averaging. According to our experience, 2-ppm resolution in sensitivity measurement in a normal laboratory requires 2-sec integration time for each voltage measurement and averaging from 100 readings. Simultaneously triggered 6-1/2 digit voltmeters such as HP 34401 are sufficient for that purpose [13]. Such measurements are time-consuming, and it is difficult to keep all the parameters constant. An alternative is to use low-frequency ac calibration field and synchronous detection at the output. Modern digital lock-in amplifiers are fast enough even if the reference signal is very slow. The resolution and accuracy of ac calibration is limited; it can be increased by compensation methods.

Coils for ac calibration should satisfy additional requirements. First of all, the coil self-resonant frequency should be well above the required frequency range. The coil-supporting structure should be nonconductive, to prevent eddy currents. Also, all metal objects in the coil vicinity should be avoided. It is even important to check the ground conductivity; the reasonable minimum is 5,000 Ωm.

In general, the calibration coil should be 5 to 20 times larger than the sensor being tested. The only exceptions are the air-coil induction magnetometers, which are perfectly linear so that errors caused by the calibration field nonhomogeneity can be corrected for. The correction factors were calculated by Nissen and Paulsson; the correction is only 1% if the radius of the induction coil being tested is one-half the radius of Helmholtz coils [14].

11.1.1 Field Compensation Systems

Field compensation devices use coil systems to cancel variations of the external fields, mainly the Earth's natural field variations. Such systems work only when the field sources are distant. The requirements for the field gradient depend on the required quality of the field stabilization. One example may be a simple compensation system for computer monitors, which suppresses the field variations below 1 μT; it can work in a normal office environment with strong field sources located in ~20m distance. The field sensor may be an AMR magnetoresistor mounted directly on the monitor.

On the other hand, the system for thermal demagnetization of rock samples requires a magnetic vacuum below 0.5 nT; such a system should be located in a nonmagnetic building kilometers from the interference sources. The field sensors are nonmagnetic air-powered rotation magnetometers located inside the compensation system [10] or fluxgate sensors, which should be outside the coils [15].

Systems that are able not only to compensate the field changes or cancel the field completely but also to create independent artificial calibration field usually have the field sensors outside the coils. The sensor distance should be sufficiently large so that the field from the coil system is negligible; the external field gradients between the coil system and the sensors should be very small.

Active environmental noise-compensation systems for magnetically shielded rooms use induction coils, SQUIDs, or fluxgates as noise-reference sensors [4]. The shielding factor improvement of more than 40 dB in the frequency range of 0.1 to 10 Hz is achievable [16].

11.2 Magnetic Shielding

Magnetic shields are often an important part of high sensitivity magnetic measurements, testing, and calibration. This section reviews the theory and methods for efficient shielding static or slowly varying magnetic fields.

11.2.1 Magnetic Shielding Theory

Magnetic shielding provides a low-reluctance path guiding the magnetic flux around the region to be shielded (Figure 11.3). For the sake of simplicity,

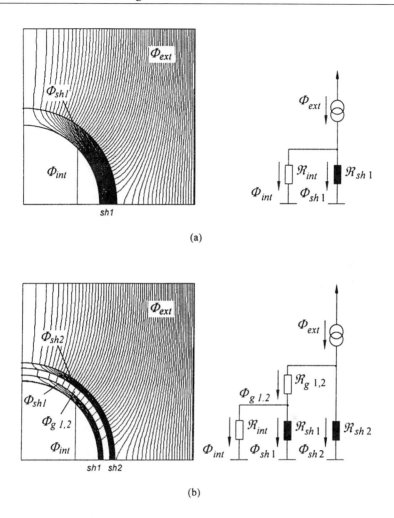

Figure 11.3 Transverse magnetic shielding: (a) single-shell and (b) double-shell cylindrical shields and their equivalent magnetic circuits. (Note that residual flux, Φ_{int}, is uniform.) The flux distributions are obtained numerically for $\mu = 100$.

it is generally assumed that a magnetic shield is placed in a uniform external field, and the permeability does not depend on the magnetic induction. Frequency, f, of the ambient field is assumed to be low enough to satisfy the quasi-static conditions: the skin depth $\delta \approx 1/\sqrt{\pi f \mu_0 \mu \sigma}$, is greater than the thickness t of a shield with the permeability μ and conductivity σ.

The shielding effectiveness is described by the two main parameters: the shielding factor, defined as the ratio of the external field to the residual field (the field within the shielded region), and the uniformity of the residual field. It is obvious that the shielding effectiveness depends on the geometry, thickness, and permeability of the magnetic shell. Less obvious is that an introduction of air gaps between the shielding shells (see Figure 11.3) can greatly increase the shielding factor [17]. Mathematical solutions for the transverse magnetic field penetration into so-called ideal, single- and multiple-shell shields illustrate the preceding statements quantitatively.

11.2.2 Transverse Magnetic Shielding

Shielding assemblies such as a set of concentric spheres or infinitely long cylinders are considered ideal because they provide a uniformly reduced residual field (see Figure 11.3) [18].

For a single shell of thickness t, diameter D, and a constant relative permeability μ, the transverse shielding factor is given as follows [18]:

$$S_T \approx 1 + G\frac{\mu t}{D} \tag{11.6}$$

where $G = 4/3$ for a sphere and $G = 1$ for a cylinder. Exact mathematical solutions for the static magnetic field penetration into multiple-shell structures are cumbersome [19–21]. In the case of high permeability, $\mu \gg 1$, and a relatively small thickness, $t/D \ll 1$, exact solutions can be reduced to simple forms. The transverse shielding factor for a double-shell shield becomes

$$S_{t\text{double}} \approx 1 + S_1 + S_2 + S_1 S_2 \left(1 - \frac{W_1}{W_2}\right) \tag{11.7}$$

where $S_i \approx G\frac{\mu_i t_i}{D_i}$ is the transverse shielding factor for an individual shell, W_1 and W_2 are the volume of a sphere and cross-sectional area of a cylinder defined by the outer surface of layer 1 and inner surface of layer 2, respectively. The total shielding factor for the n-fold-shell shield is calculated in a similar way [21].

The transverse shielding factor for a finite-length cylindrical shield with open ends is reduced by the fringing fields penetrating through the openings [22]. Roughly, fringing fields are attenuated by a factor 10^3 per diameter distance from the open end [23].

Equation (11.7) represents the main principle of magnetic shielding with multiple shells [17]: Decoupling the shells ($W_1/W_2 \ll 1$) allows a multiplicative rather than an additive increase in shielding. As a result, the shielding factor for a set of thin concentric shells can be much greater than that for a single thick shell built with the same amount of material.

With thin multiple shells, the best minimum weight arrangement is when diameters of the shells grow in geometric progression, $D_j = \alpha D_i$, and the shielding material is distributed evenly among the shells [18, 19, 24]. Maximum shielding is obtained when the successive diameter ratio, α, is roughly 1.3 to 1.4 for spheres and 1.5 to 1.6 for cylinders [19]. A simple estimation of shielding with multiple-shell shields is suggested in [24]. A simple diagrammatic method of writing the shielding formulas is suggested in [25].

The transverse shielding factors and the residual field distribution for a set of elliptical cylinders are derived in [26]. If the external field is applied along the major axis of an elliptic cylinder with a 0.62–0.66 aspect ratio, then the shielding factor is increased, compared to a corresponding circular cylinder.

Cylinders with rectangular cross sections are treated in [27]. The transverse shielding factor for square cylinders can be calculated according to (11.6), where the diameter D equals the side of the square, $G \approx 0.70$ for a square cylinder in a perpendicular external field, and $G \approx 0.91$ for a square cylinder in a diagonal external field.

11.2.3 Axial Magnetic Shielding

11.2.3.1 Axial Shielding With Single-Shell Shields

In many cases, cylindrical shields cannot be positioned so that the axis of the shield is perpendicular to the ambient magnetic field (see Figures 11.4–11.6). Reference [22] suggests estimating the residual field within an axial (longitudinal) shield with the intrinsic field in a solid ellipsoid of revolution with the same aspect ratio and with an equivalent permeability averaged over the interior of the cylinder. Neglecting the effect of the openings, the axial shielding factor for an open cylinder is described in [22] as follows.

$$S_{a\,\text{open}} \approx 1 + N \cdot \mu_{\text{equiv}} = 1 + 4N \cdot \mu t/D \qquad (11.8)$$

where N is the demagnetizing factor of a general ellipsoid [28], and $\mu_{\text{equiv}} = 4\mu t/D$ [Figure 11.4(a)]. Both circular and rectangular cylinders

The effect of the openings is neglected

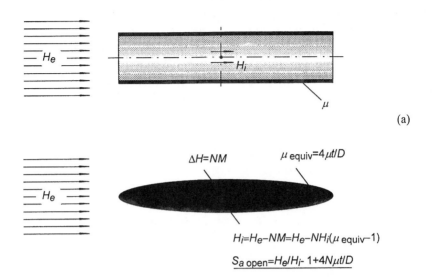

(a)

The effect of the openings is considered

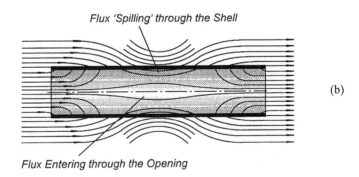

(b)

Figure 11.4 Axial shielding with open-ended cylinders. (a) Neglecting the field penetrating through the openings, the residual field can be approximated by the intrinsic field of an ellipsoid. (b) Simplified flux distribution around and within an open-ended cylinder.

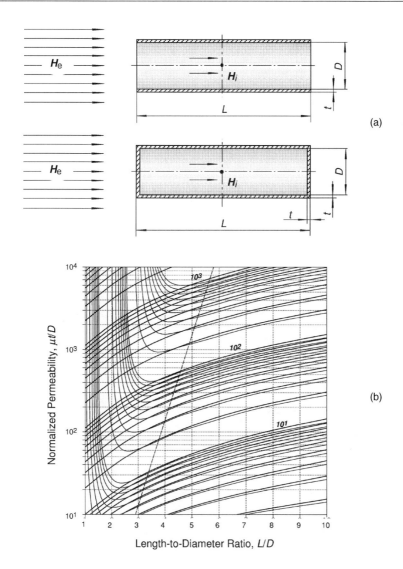

Figure 11.5 Axial shielding with single cylinders: (a) open-ended and closed cylinders; and (b) a numerically obtained chart (contour plot) describing the axial shielding factors (bold numbers) of open-ended cylinders (thin lines) and closed cylinders (thick lines).

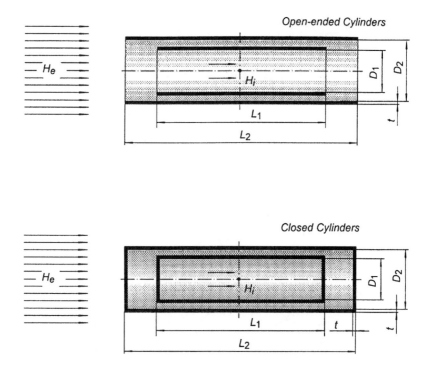

Figure 11.6 Double-shell cylinders: open-ended and closed cylindrical shields.

can be treated in a similar manner with the values of N for corresponding ellipsoids [22]; see also (2.22) through (2.24).

In the case of circular cylinders, it may be more accurate—and numerical calculations prove it [29]—to use demagnetizing factors calculated for rod [30] rather than for ellipsoid.

If the effect of the openings is still neglected, closed cylinders have a reduced axial shielding factor compared to (11.8) because their caps attract excess magnetic flux [22]:

$$S_{a\text{closed}} \approx \frac{1 + 4N \cdot \mu t/D}{1 + 0.5/(L/D)} \qquad (11.9)$$

Numerical calculations [31] suggest a different formula, giving 15% to 20% lower shielding factors:

$$S'_{a\text{closed}} \approx \frac{1 + 4N_{\text{rod}} \cdot \mu t/D}{1 + (L/D)/100} \qquad (11.10)$$

Let us now consider the effect of the openings. The openings dramatically reduce the axial shielding with relatively short open-ended cylinders. Fringing fields penetrating through the open ends [see Figure 11.4(b)] decay approximately exponentially and are attenuated by a factor of about 10^2 per diameter distance from the open end [22, 23].

The axial shielding factors for both open and closed cylinders ($D/t = 100$) versus the length-to-diameter ratio, L/D, and normalized permeability, $\mu t/D$, are shown in Figure 11.5 as a chart (contour plot) that is obtained numerically by a standard ANSYS® software package employing a finite element method. The chart illustrates behavior of the shielding factor as being affected by the shields' geometry and permeability. The two following geometry-related effects cause a decrease in the axial shielding factor: the effect of the openings for open-ended shields and decreasing the axial shielding factor by increasing the length for both open-ended and closed shields.

Although open cylinders generally are considered to provide a less uniform residual field, numerical calculations show that this situation is abruptly changed when the shields' aspect ratios approach the values depicted by the dashed line in the chart in Figure 11.5(b). In the region near the dashed line, the residual field becomes nearly uniform over a relatively wide shielded area and even outperforms the uniformity within corresponding closed cylinders.

11.2.3.2 Axial Shielding With Double-Shell Shields

Unfortunately, existing quasi-analytical estimations of axial shielding with multishell shields [21, 22, 25] predict results that are not in close agreement [21, 31]. On the other hand, employing special charts [32] describing shielding performance for the most widely applicable, canonical shielding structures can be a relatively accurate and time-saving alternative for many scientists and shield designers.

The charts shown in Figure 11.7 cover the case of double-shell open-ended cylinders (see Figure 11.6) with the aspect ratio of the outer cylinder, $L_2/D_2 = 5$; the ratio of diameters, $D_1/D_2 = 0.6$, and 0.9; the ratio of lengths, L_1/L_2, from 0 to 1; the normalized permeability, $\mu t/D_2$, from 10^1 to 10^4; and a constant thickness, $t = D_2/100$.

Figure 11.8 shows the case of closed cylindrical shields. There is no effect of the openings in this case, and the behavior of the axial shielding factor is much simpler. Shorter shields are considered in this case—compared to the case of open shields—because there normally is no need to make closed cylinders longer than four diameters.

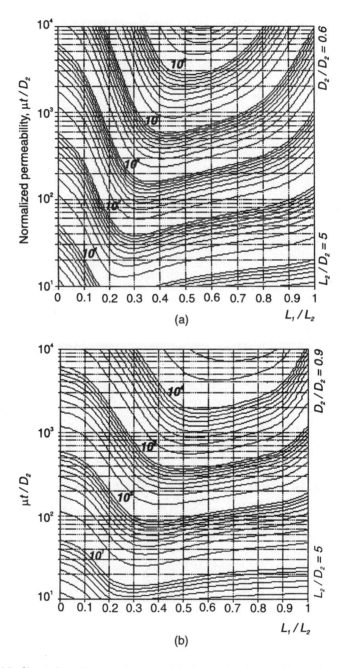

Figure 11.7 Charts for estimating the axial shielding factors for double-shell open-ended (see Figure 11.6) cylindrical shields.

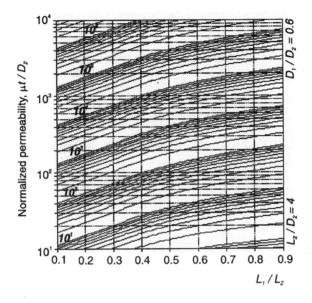

Figure 11.8 Charts for estimating the axial shielding factors for double-shell closed (see Figure 11.6) cylindrical shields.

Compared to the case of single cylinders (see Figure 11.5), two new geometry-related factors affect the axial shielding factor for double-shell open-ended shields: decoupling the shells by spaces between them and screening the ends of a shorter inner shell by a longer outer shell. Both factors increase the shielding factor.

It is important to note that shortening the inner shell is not always effective. For instance, it is worth choosing the same length for short shields with $L_2/D_2 \leq 3$, especially for high-permeability materials. In such a case, it is more important to reduce the effect of the openings by increasing the L_1/D_1 ratio of the inner shell rather than to screen the ends of the shell [32].

As in the case of single-shell shields, a better uniformity of the residual field can be obtained within a properly constructed double-shell open-ended shield [33].

11.2.4 Flux Distribution

Although a linear case is usually assumed in calculations, it is important to remember that the permeability depends strongly on the magnetic induction. Assuming that $\mu \gg 1$, the maximum flux density within the shielding material can be estimated according to [21], and corresponding permeability can be obtained from the normal magnetization curve given by manufacturers. Particular attention should be given to make the shielding shells—especially the outermost one—thick enough to avoid the dramatic decrease in the permeability when the material is saturated.

Considering $\mu \gg 1$, the maximum flux density, $B_{m\,\text{transv}}$, within the material for a single-shell transverse cylindrical shield can be estimated as D/t times greater than the ambient field, B_0 [21]. Figure 11.9 shows how the ratio $B_{m\,\text{axial}}/B_{m\,\text{transv}}$ depends on the L/D ratio ($B_{m\,\text{axial}}/B_{m\,\text{transv}}$ is the ratio of the maximum flux density within the material for an axial single-shell cylindrical shield, to the same parameter for the corresponding transverse shield).

11.2.5 Annealing

Generally, once fabrication has been completed, the shielding shell should be annealed [34]. Annealing relieves mechanical stresses induced during fabrication. As a result, the shielding material reaches its optimum permeability. A standard annealing cycle for crystalline Permalloy materials includes heating above 1100°C in dry hydrogen for several hours and then cooling slowly. After the annealing has been completed, the shield or its parts should

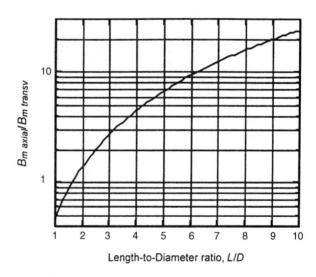

Figure 11.9 Dependence of the ratio of maximum flux density, $B_{m\,\text{axial}}$, within the material for axial single-shell cylindrical shields to the same parameter for corresponding transverse shields, $B_{m\,\text{transv}}$, on the length-to-diameter ratio, L/D.

be handled with care. Even a little mechanical stress can seriously reduce the permeability.

Modern, commercially available shielding materials, such as cobalt-based thin amorphous ribbons, are much less sensitive to mechanical stresses. Materials of this type make an excellent alternative to avoid the expensive annealing cycle in the manufacturing process. Care should be taken to avoid the saturation.

11.2.6 Demagnetizing

The lowest achievable dc field within a magnetic shield depends strictly on the magnetic history of the material after the annealing has been completed [23, 34]. The remnant magnetization of the shielding material can be reduced and the shielding factor against dc magnetic fields can be greatly increased by demagnetizing (degaussing) the shield *in situ*. Although demagnetization generally can be done magnetically or thermally, the magnetic method is usually easier. The magnetic state of the material is cycled by a field decaying slowly from the amplitudes that saturate the material to zero. Typically, the minimum dc residual field obtainable is 1 nT to 10 nT and is limited by stray fields from imperfections of the material's granular structure [21].

11.2.7 Enhancement of Magnetic Shielding by Magnetic Shaking

Magnetic shaking is an inexpensive and effective method to significantly increase the permeability and shielding performance against LF magnetic fields. Because of the application of a relatively strong, HF magnetic field, magnetic shaking keeps the domain walls within the material in continuous vibrant motion and allows the material to be more responsive to a slowly varying, weak ambient field. In other words, it is like sliding friction and static friction, the energy needed to move the domain walls is supplied by a fairly strong shaking field, therefore the permeability seen for the slowly varying low-level magnetic field becomes high [35, 36].

The most important key to effective magnetic shaking is the selection of the proper shielding material. Soft amorphous ferromagnets, such as Metglas® 2705M, with small magnetostriction and a highly rectangular hysteresis loop, seem to be the best choice. With Metglas® 2705M, shaking increases the effective incremental permeability 200- to 300-fold, resulting in μ from $4 \cdot 10^5$ to $6 \cdot 10^5$ [35]. Magnetic shaking is not effective if applied to Permalloy, in which incremental permeability is enhanced only several times [37].

Shaking efficiency strongly depends on the orientation of the magnetic anisotropy of the shielding material, namely, the anisotropy axis should be aligned along the corresponding shielding direction to achieve significant shielding enhancement [38]. The direction of the shaking field is less important. Shaking enhancement is observed with fields that are both parallel and perpendicular to the anisotropy direction. In the latter case, however, the shaking amplitude should be about 10 times greater, ~30 A/m versus ~3 A/m, to obtain maximum shaking enhancement [38].

References

[1] Jankowski, J., and C. Sucksdorff, *Guide for Magnetic Measurements and Observatory Practice*, Warsaw, Poland: IAGA, 1996.

[2] Brown, R. E., "Device for Producing Very Low Magnetic Fields," *Rev Sci. Instr.*, Vol. 39, 1968, pp. 547–550.

[3] Hechtfischer, D., "Homogenisation of Magnetic Fields by Diamagnetic Shields," *J. Phys. E: Sci. Instr.*, Vol. 20, 1987, pp. 143–146.

[4] Malmivuo, J., et al., "Improvement of the Properties of an Eddy Current Magnetic Shield With Active Compensation," *J. Phys. E: Sci. Instr.*, Vol. 20, 1987, pp. 151–164.

[5] Farrell, W., et al., "A Method of Calibrating Magnetometers on a Spinning Spacecraft," *IEEE Trans. Magn.*, Vol. 31, 1995, pp. 966–972.

[6] Kraus, J. D., *Electromagnetics*, 2nd ed., New York: McGraw-Hill, 1984.

[7] Firester, A. A., "Design of Square Helmholtz Coil System," *Rev. Sci. Instrum.*, Vol. 37, 1966, pp. 1264–1265.

[8] Alldred, J. C., and I. Scollar, "Square Cross Section Coils for the Production of Uniform Magnetic Fields," *J. Sci. Instrum.*, Vol. 44, 1967, pp. 755–760.

[9] Pajunpaa, K., V. Korepanov, and E. Klimovich, "Calibration System for Vector DC Magnetometers," *Proc. Imeko XIV World Congress*, Tampere, Finland, Vol. 4, 1998, pp. 97–102.

[10] Prihoda, K., et al., "MAVACS—A New System for Creating a Non-Magnetic Environment for Paleomagnetic Studies," *Geologia Iberica*, Vol. 12, 1988–1989, pp. 223–227.

[11] Everett, J. E., and J. E. Osemeikhian, "Spherical Coils for Uniform Magnetic Fields," *J. Sci. Instrum.*, Vol. 43, 1966, pp. 470–474.

[12] Spacecraft Magnetic Test Facility, NASA report X-754-83-9, Goddard Space Flight Center, Greenbelt, MD, Apr. 1984.

[13] Ripka, P., and A. Platil, "Analysis of Data From Three-Axis Magnetometer Calibrator," *Proc. SMC Conf.*, St. Petersburg, Russia, 1998, Vol. 2, pp. 60–66.

[14] Nissen, J., and L. E. Paulsson, "Influence of Field Inhomogeneity in Magnetic Calibration Coils," *IEEE Trans. Instrum. Meas.*, Vol. 45, 1996, pp. 304–306.

[15]   McElhinny, M. W., et al., "A Large Volume Magnetic Field Free Space for Thermal Demagnetization and Other Experiments in Paleomagnetism," *Pure and Appl. Geophys.*, Vol. 90, 1971, pp. 126–130.

[16]   ter Brake, H. J. M, R. Huonker, and H. Rogalla, "New Results in Active Noise Compensation for Magnetically Shielded Rooms," *Meas. Sci. Technol.*, Vol. 4, 1993, pp. 1370–1375.

[17]   Rücker, A. W., "On the Magnetic Shielding of Concentric Spherical Shells," *Phil. Mag.*, Vol. 37, 1894, pp. 95–130.

[18]   Thomas., A. K., "Magnetic Shielded Enclosure Design in the DC and VLF Region," *IEEE Trans. Electromagnetic Compatibility*, Vol. 10, 1968, pp. 142–152.

[19]   Wills, A. P., "On the Magnetic Shielding Effect of Trilamellar Spherical and Cylindrical Shells," *Phys. Rev.*, Vol. 9, 1899, pp. 193–213.

[20]   Wadey, W. G., "Magnetic Shielding With Multiple Cylindrical Shields," *Rev. Sci. Instrum.*, Vol. 27, 1956, pp. 910–916.

[21]   Sumner, T. J., J. M. Pendlebury, and K. F. Smith, "Conventional Magnetic Shielding," *J. Phys. D: Appl. Phys.*, Vol. 20, 1987, pp. 1095–1101.

[22]   Mager, A., "Magnetic Shields," *IEEE Trans. Magn.*, Vol. 6, 1970, pp. 67–75.

[23]   Freake, S. M., and T. L. Thorp, "Shielding of Low Magnetic Fields With Multiple Cylindrical Shells," *Rev. Sci. Instrum.*, Vol. 42, 1971, pp. 1411–1413.

[24]   Dubbers, D., "Simple Formula for Multiple Mu-Metal Shields," *Nucl. Instrum. Methods A*, Vol. 243, 1986, pp. 511–517.

[25]   Gubser, D. U., S. A. Wolf, and J. E. Cox, "Shielding of Longitudinal Magnetic Fields With Thin, Closely Spaced, Concentric Cylinders of High Permeability Material," *Rev. Sci. Instrum.*, Vol. 50, 1979, pp. 751–756.

[26]   Ohara, T., K. Koyama, and K. Imai, "Magnetic Shield Using Elliptic Cylindrical Shells," *Bulletin Electrotechnical Laboratory*, Vol. 45, 1981, pp. 20–34.

[27]   Mager, A., "Magnetostatische Abschirmfactoren von Zylindern mit rechteckigen Querschnittsformen," *Physica*, Vol. 80B, 1975, pp. 451–463.

[28]   Osborn, J. A., "Demagnetizing Factors of the General Ellipsoid," *Phys. Rev.*, Vol. 67, 1945, pp. 351–357.

[29]   Paperno, E., and I. Sasada, "Optimum Axial Efficiency of Open Cylindrical Magnetic Shields," submitted for publication to *IEEE Trans. Magn.*

[30]   Kobayashi, M., and Y. Ishikawa, "Surface Magnetic Charge Distributions and Demagnetizing Factors of Circular Cylinders," *IEEE Trans. Magn.*, Vol. 28, 1992, pp. 1810–1814.

[31]   Paperno, E., H. Koide, and I. Sasada, "A New Estimation of the Axial Shielding Factors for Multi-Shell Cylindrical Shields," *J. Appl. Phys.*, Vol. 87, 2000, pp. 5959–5961.

[32]   Paperno, E., "Charts for Estimating the Axial Shielding Factors of Open-Ended Cylindrical Shields," *IEEE Trans. Magn.*, Vol. 35, 1999, pp. 3940–3943.

[33]   Paperno, E., I. Sasada, and H. Naka., "Self-Compensation of the Residual Field Gradient in Double-Shell Open Ended Cylindrical Axial Magnetic Shields," *IEEE Trans. Magn.*, Vol. 35, 1999, pp. 3943–3945.

[34] Bozorth, R. M., *Ferromagnetism*, New York: Van Nostrand, 1951.
[35] Sasada, I., S. Kubo, and K. Harada, "Effective Shielding for Low-Level Magnetic Fields," *J. Appl. Phys.*, Vol. 64, 1988, pp. 5696–5598.
[36] Sasada, I., et al., "Low-Frequency Characteristic of the Enhanced Incremental Permeability by Magnetic Shaking," *J. Appl. Phys.*, Vol. 67, 1990, pp. 5583–5585.
[37] Kelhä, V. O., R. Peltonen, and B. Rantala, "The Effect of Shaking on Magnetic Shields," *IEEE Trans. Magn.*, Vol. 16, 1980, pp. 575–578.
[38] Paperno, E., and I. Sasada, "The Effect of Magnetic Anisotropy on Magnetic Shaking," *J. Appl. Phys.*, Vol. 85, 1999, pp. 4645–4647.

12

Magnetic Sensors for Nonmagnetic Variables

Pavel Ripka, Ichiro Sasada, and Ivan J. Garshelis

This chapter briefly covers the many magnetic sensors that measure variables other than the magnetic field. They are still called "magnetic" because they are based on magnetic principles. These sensors are not a primary topic of this book, so the aim of this chapter is that of an overview and to refer to valuable information sources rather than to fully explain all the fine details and design rules. Some of these devices use principles similar to those of magnetic field sensors (e.g., position sensors with permanent magnet target). The measured physical quantities may in principle affect the magnetic properties of sensing material or change parameters of the magnetic circuit.

We try to concentrate on principles that are already used in devices or that are likely to become widely used. Further reading on basic principles and interfacing can be found in [1–4].

12.1 Position Sensors

Magnetic position sensors are widely used in industry. They sense either linear or angular position. Some of them have linear output, others have digital output: either bistable, like proximity switches, or encoded, like incremental and absolute position sensors. The third group combines digital (rough) and analog (fine) outputs, for example, Inductosyn.

The target, whose position is measured, may be a permanent magnet (induction sensors), soft magnetic material (LVDT = linear variable differential transformer, variable reluctance sensors), or just electrically conducting material (eddy-current sensors). Some of these devices work on dc, others on ac principles, and another group needs both ac excitation and dc biasing. The requirements of industrial position sensors are very different from fine specifications of scientific magnetometers: Instead of parts-per-million accuracy, we speak about percentage points, but the most important parameters are ruggedness, temperature stability, and electromagnetic compatibility (EMC). They have a much larger market than optoelectronic sensors, which are sensitive to contamination such as grease or dust in the sensing path.

12.1.1 Sensors With Permanent Magnet

Position sensors of this type consist of a magnetic field sensor that measures the field of a permanent magnet. The magnet is either connected to the target or the target is made of ferromagnetic material and the magnet is attached to the sensor; in both cases, moving the target changes the sensed field.

The sensor may be just a passive induction coil; in such cases, only the movement is sensed, not the static position. This type of position sensor is called induction sensor, speed sensor, or magnetic pickup.

All the other types use dc magnetic sensors: most often Hall, AMR, GMR, semiconductor magnetoresistors, and rarely fluxgate or similar devices. The specific details of these sensors can be found in their respective chapters in this book. Here we mention only common problems and some design examples.

12.1.1.1 Nonlinearity of the Field Dependence

A small magnet far from the sensor behaves like a dipole, so that $B \sim 1/x^3$. Although this nonlinearity may be numerically corrected, it still may cause gross errors for large distances. Some sophisticated designs have been suggested to reduce the nonlinearity of the displacement sensors; an example is shown in Figure 12.1.

Angular shaft sensors usually have one magnet attached to the end of the shaft. If only one sensor is used, the response is a sinewave, so it can be considered linear in the ±30-degree range. Some sensors use two perpendicular sensors, which then have sine and cosine response so that the measured angle can be calculated in the whole 360-degree range. A lot of contactless potentiometers and "dc resolvers" are based on that principle. GMR sensors are popular for angular sensing, since they can be made insensitive to magnet distance. Another design developed by Moving Magnet Technologies SA uses a radially magnetized magnet ring made of bonded NdFeB as rotor.

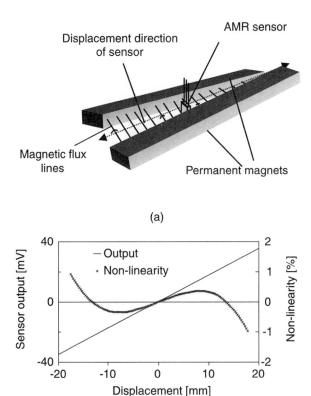

Figure 12.1 (a) Linear displacement position sensor with permanent magnet bars and (b) output voltage. (From [5].)

The linear range (for 0.1% error) is thus increased to 170 degrees even if one sensor is used [6].

If the end of the shaft is not accessible, the situation is more complicated. Figure 12.2 shows one possible solution: Two ring magnets magnetized parallel to the shaft are mounted eccentrically, one attached to the shaft, the other fixed. The field between the rings is sensed by a pair of magnetoresistors. The sensor outputs are shown in Figure 12.2(b); the calculated angular position has linearity error below 0.5 degree, as shown in Figure 12.2(c).

12.1.1.2 Induction Position Sensors

Induction position sensors are passive devices based on induction effect. Two possible configurations excited by permanent magnet are shown in Figure 12.3.

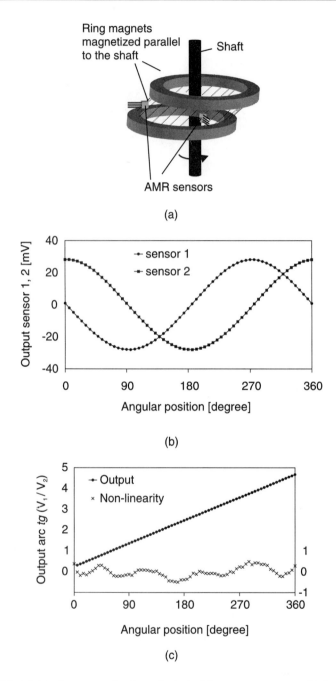

Figure 12.2 (a) Angular sensor for through shaft; (b) sensor outputs; and (c) calculated angular position (in volts) and linearity error (in degrees). (From [5].)

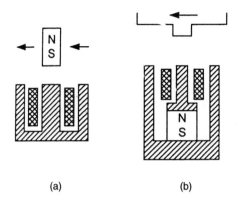

Figure 12.3 Permanent magnet-driven induction sensors: (a) speed sensors or (b) magnetic pickups.

Induction sensors with a moving magnet are also called speed sensors. The target is a permanent magnet and the sensing element is an induction coil, often in the magnetic circuit. These types of sensors are based on induction effect, and they consume no energy. The serious drawback, however, is that the amplitude of the output depends on the target speed. The sensor fails at low speeds, so it has limited use.

Variable reluctance sensors with fixed permanent magnets are also called magnetic pickups. The sensing element is again induction coil, the dc field is also generated by a permanent magnet, which is now fixed, and the coil flux change is produced by changing the position of the soft magnetic target, which is again often a part of the dc magnetic circuit. The induction coil detects the perturbation of the dc magnetic field caused by movement of the target. The applications include geartooth sensing in shaft speed measurements and antilock brake systems (ABSs); because of low sensitivity at low speeds, they are not suitable for car ignition timing systems.

12.1.2 Eddy-Current Sensors

Eddy-current sensors are also called inductive or inductance sensors. Their target should be electrically conducting but not necessarily ferromagnetic. In that aspect, they differ from variable gap sensors, which measure the displacement of the ferromagnetic part of an ac magnetic circuit. Eddy-current–based instruments measure displacement, alignment, dimensions, and vibrations and also identify and sort metal parts in industrial applications. Eddy-current sensors with ac excitation have no lower limit on target speed.

The ac magnetic field is created by the sensor coil fed by an oscillator. The coil is usually tuned by parallel capacitor, and the LC circuit oscillates at the resonant frequency. If the conducting target is present, the eddy currents (mostly on the target surface) create a secondary magnetic field, decreasing the coil flux and thus the effective coil inductive reactance. The sensor sensitivity depends on the target conductivity. An excellent target is aluminum; the recommended thickness is greater than 0.3 mm, and the recommended diameter is 2.5 to 3 times the diameter of the sensor coil [7]. In any case, the target thickness should be larger than the skin depth δ.

$$\delta = \frac{1}{\sqrt{0.5\omega\mu\sigma}} \tag{12.1}$$

where σ is electrical conductivity.

In the case of the ferromagnetic target, the situation is complicated, because the coil inductance is simultaneously increased by the target permeability $\mu > 1$. Figure 12.4 shows the relative eddy-current output for various target materials.

Because of the complex dependence on the target geometry and material, the position measuring system usually should be individually calibrated. For ferromagnetic targets, we may expect increased sensitivity to target axial displacement, to temperature changes, and worse repeatability and long-term stability. On the other hand, ferromagnetic targets increase the measuring range. The usual measuring range is up to 30% of the coil diameter. A lower range of 5% or 10% coil diameter is recommended for high-precision measurements, while in low-accuracy applications (e.g., switching), the range can be more than 50% of the coil diameter. In any case the minimum coil-to-target distance is usually 10% of the range; this distance is often defined by the thickness of the sensor housing.

The sensing coils are often accompanied by a ferromagnetic circuit, which focuses their ac field into one direction so the sensor is insensitive to conducting materials from the side or from the back. The working frequencies range from kilohertz to 1 MHz, the core material is usually ferrite, and typical shape is pot core (Figure 12.5).

The sensor diameter ranges from 8 mm to 150 mm with sensing distances between about 2 mm and 100 mm. The analytical solution of eddy-current sensors usually is not possible. A number of software modeling tools based on finite-element method are available. Improved core shapes are shown in Figure 12.6. Figure 12.7 shows the measured Q factors of the 47 turn coils wound on cores from Figure 12.5 and Figure 12.6 at the

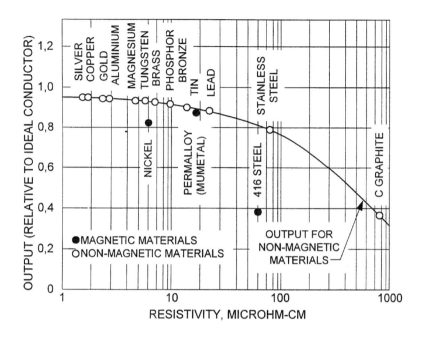

Figure 12.4 Influence of the target material on relative sensitivity of the eddy-current sensor (after [7]).

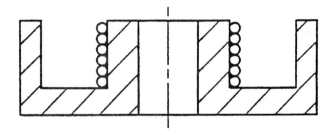

Figure 12.5 Standard pot core used for eddy-current sensors. The diameter of this core is 22 mm, and the height is 6.7 mm (from [8], © 1997 IEEE).

300 kHz frequency versus the target distance [8]. It has been shown that the flat face of the transducer increases the value of B in the axial direction. The larger radiating surface increases the measuring range up to double the value of the pot core. Magnetic short circuits by the core itself or by ferromagnetic housing should be avoided. Eddy-current sensors working at higher frequencies should be wound from litz (stranded) wire.

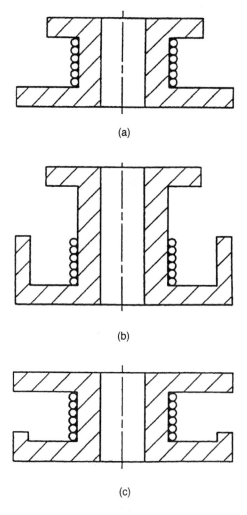

Figure 12.6 Alternative shapes of ferrite cores for eddy-current sensors (from [8], © 1997 IEEE).

The sensors work either in FM mode, when the oscillator frequency changes with target position (relaxation oscillator), or in AM mode, when the variable is the oscillator amplitude [9].

This principle is also often used in bipolar-output proximity switches (see Section 12.5); in that case the electronics may be much simpler. The most popular type of eddy-current proximity switch is the "killed oscillator" (also called the "blocking oscillator"): A metallic object moving close to the coil adds load to the oscillator, which stops the oscillation.

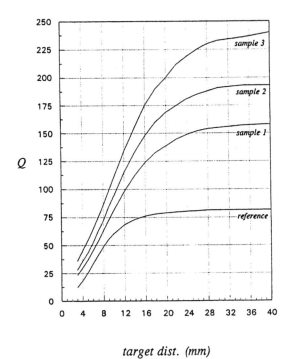

target dist. (mm)

Figure 12.7 Measured Q for reference pot (Figure 12.5) and alternative cores of which samples 1, 2, and 3 are shown in Figure 12.6(a), (b), and (c), respectively, (from [8], © 1997 IEEE).

Miniature eddy-current sensors use flat air coils. The proximity sensor based on a differential relaxation oscillator (Figure 12.8) is described in [10]. The oscillator frequency changes with the position of the target. A 1-mm by 1-mm coil on top of an integrated CMOS readout circuit was developed for short-range applications with limited accuracy. A 3.8-mm side coil temperature-compensated sensor is working in the output frequency range of 3 to 4.5 MHz. The dependence of the output frequency on distance of the aluminum target before and after temperature compensation is shown in Figure 12.9. The measured distance of the aluminum target changes with temperature by less than ±1% (for an aluminum target 1 mm from the sensor) in the whole −20°C to +80°C industrial temperature range.

12.1.3 Linear Transformer Sensors

12.1.3.1 LVDT

The linear variable differential transformer (LVDT) is probably the most popular magnetic position sensor. Because of zero friction, the device is

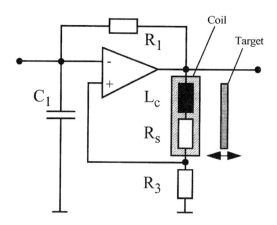

Figure 12.8 Inductive (eddy current) proximity sensor based on differential relaxation oscillator. The output frequency is a function of the conducting target position (from [10]).

highly reliable. It is based on variation of mutual inductance between the primary and two secondary windings caused by the movement of the ferromagnetic core (Figure 12.10).

The LVDT has a high reproducibility, and practical resolution may be better than 0.1% or below 1 μm, linearity up to 0.05%, and temperature coefficient of sensitivity typically 100 ppm/°C. The linear range is 30% to 85% of the device length. Standard measurement ranges are from 200 μm to 50 cm. The excitation frequency is usually between 50 Hz and 20 kHz [2]. The output signal is usually processed by PSD (i.e., synchronous rectifier, lock-in amplifier), sometimes ratiometric processing is used. It is possible to integrate the complete sensor electronics, including the excitation generator into the sensor housing.

The differential variable inductance transducer (DVRT®; also called a "half-bridge LVDT") has only two windings; core position is measured by differential inductance. DVRT sensors with a 1.5-mm outside diameter and 60-nm resolution are available.

12.1.3.2 Variable Gap Sensors

Variable gap (also "variable reluctance") sensors are based on a change of the air gap in a magnetic circuit (between core and armature) of inductor or transformer. They usually are less precise than LVDTs, but because of design reasons, they often are used in conjunction with mechanical transduc-

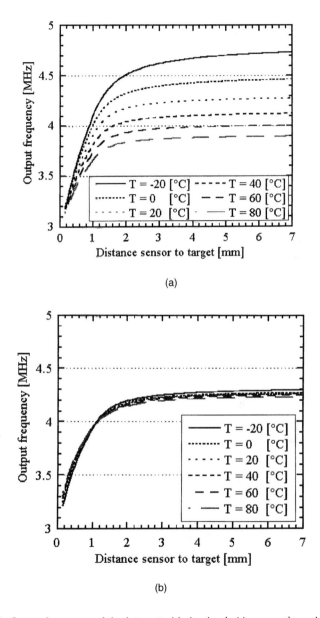

Figure 12.9 Output frequency of the integrated inductive (eddy current) proximity sensor as a function of the distance of the aluminum target at various temperatures: (a) before and (b) after temperature compensation (from [10]).

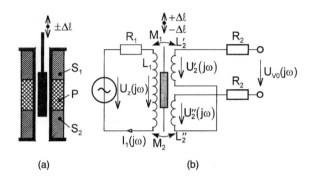

Figure 12.10 LVDT (a) cross section and (b) schematic diagram (from [11]).

ers to measure pressure, strain, force, torque, and other mechanical variables that can be converted into mechanical displacement [4].

12.1.3.3 PLCD Sensor

The permanent magnetic linear contactless displacement sensor (PLCD) is shown in Figure 12.11. The sensor consists of a long magnetic strip core (5) with homogeneous secondary winding (3). The primary winding has two sections (1, 4) connected antiserially, which are supplied with ~4 kHz

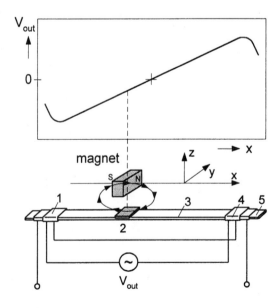

Figure 12.11 PLCD sensor (from [12]).

sinewave. A permanent magnet in the core vicinity causes localized core saturation (2), which effectively divides the core into two halves whose lengths determine the signal induced into the secondary winding. The induced voltage is synchronously rectified to obtain linear output. The typical resolution is 0.2%, and linearity is 1% of the range (which is between 20 mm and 150 cm). The main advantage is that the device is tolerant to changes of air gap between the magnet and the core.

12.1.3.4 Inductosyn

Inductosyn® consists of two parallel flat meander coils: scale and slider. The slider usually has two windings ("sine" and "cosine") shifted by 1/4 mechanical period (pitch). Inductive coupling between scale and slider coils measures the displacement. Inductosyn combines the advantages of incremental sensors (increment is one pitch) and analog sensors (sinewave dependence of the output voltage allows interpolation of the fine position with a resolution of up to pitch/65,000). The scale winding is usually supplied by ac current of typically 10 kHz frequency, and the voltages induced in the sine and cosine slider coils are processed. However, it is also possible to supply the slider sine and cosine coils by quadrature (sine and cosine) voltages and to process the voltage induced in the ruler ("stator") coil. Standard pitch size is 2 mm, and ruler length may range from 25 cm to 36m or more [13]. Inductosyns are also made rotary. Multiple patterns can be combined in one device to increase the incremental resolution by employing techniques known from optical encoders [such as the $N/(N-1)$ method].

12.1.4 Rotation Transformer Sensors

Although rotation transformers are sometimes considered to be archaic, they still find application in extreme conditions, because they are more rugged than optical encoders.

12.1.4.1 Synchros

Synchros are electromechanical devices that replicate the rotor position in a distant location. They have three stator windings displaced by 120 degrees. They combine the properties of sensor and actuator; a typical application is an antenna rotator.

12.1.4.2 Resolvers

Resolvers have windings displaced by 90 degrees. The outputs are sine and cosine voltages, which are often processed by specialized resolver-to-digital converters. Brushless resolvers use another rotational transformer to supply the rotor. Resolvers can be made to withstand temperatures from 20K to 200°C, radiation of 10^9 rads, acceleration of 200g (battleship cannons, punching devices), vacuum, and extreme pressures.

Some manufacturers (e.g., Pewatron) use the term *linear resolver* for linear position sensors, which also have sin/cos outputs, but which are based on two ac-supplied magnetoresistive elements.

12.1.5 Magnetostrictive Position Sensors

Magnetostrictive position sensors measure the time of flight of a strain pulse to sense a position of a moving permanent magnet (Figure 12.12). The sensing element is a wire or a pipe from the magnetostrictive material (sonic waveguide). The devices are based on the Wiedeman effect: If the current passes through the waveguide and perpendicular dc magnetic field is present, the torsional force is exerted on the waveguide.

The device works so that after the current pulse is applied, the torsional force is generated in the location of the permanent magnet. The torsional strain pulse travels at ~3 km/s along the waveguide, and it is detected by the small induction coil at the sensor head [14]. The hysteresis may be as low as 0.4 μm, uncorrected linearity is 0.02% FS, and some devices have an internal linearization and temperature compensation. Maximum sensor length is about 4m. These sensors are manufactured by Gemco-Patriot, MTS, and Balluff.

Other devices based on a delay line principle are suggested in [15]. The mechanical strain in the delay line is caused by the current pulse in the perpendicular movable conductor. The acoustic pulse is again detected in a small axial induction coil close to the end of the delay line due to inverse magnetostriction effect. The delay line position sensors have an accuracy of approximately 1 mm, so they may be suitable to measure distances of about 1m to 5m, the upper limit being determined by attenuation. Other configurations of the position sensor, in which the moving part is a permanent magnet, and other applications of magnetostrictive delay lines are shown in [16].

12.1.6 Wiegand Sensors

In 1981, Wiegand patented a revolutionary sensor that generated high-voltage pulse when the magnetic field reaches some threshold value; the shape of the voltage pulse is highly independent of the rate of the field change and the device is passive, having just two terminals [17]. Wiegand

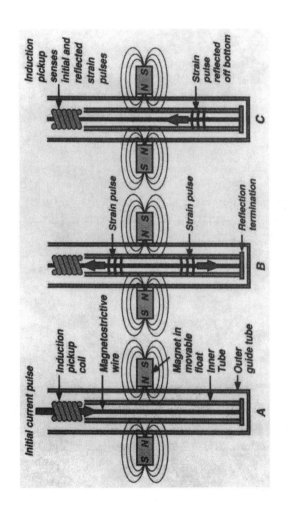

Figure 12.12 Magnetostrictive position transducer. The approach to magnetostrictive detectors shown here is patented technology assigned to AMETEK Patriot Sensors (U.S. Patent Number 5,017,867). The initial current pulse into the wire (a) induces torsial strain pulse in the position of permanent magnet (b). This pulse travels in both directions and it is reflected back from the sensor bottom. Both the initial and reflected pulses are detected by the induction pickup coil (c). The time delay between them is a measure of the magnet position.

made his sensors of 0.3-mm wire from Vicalloy (Co52Fe38V12), which was twisted to cause plastic deformation, resulting in higher coercivity in the outer shell and elastic stress in the central part. The central part forms a single domain. The pulse is caused by one large Barkhausen jump when the single domain reverses its magnetization. The pulse width is determined by eddy-current damping. The 30-mm-long wire with 1,000-turn coil may generate 2.5V pulses under optimum driving conditions (asymmetric sensed field), which have a frequency-independent amplitude between 1 mHz and 100 Hz [18]. Optimum working conditions are when the magnetization direction of the inner (magnetically soft) part reverses, while the magnetization of the (magnetically harder) outer shell is constant. Then the device characteristics are asymmetrical, and large pulse is generated in only one direction of the field change. The main disadvantage of Wiegand configuration is that the outer shell cannot be made really magnetically hard, so it can be unintentionally remagnetized by an external field. The main application field of Wiegand wires is not magnetic sensing but marking and security application. Wiegand wires are attached to access cards or antitheft labels; the sensing coil is part of the stationary detection device. The detected pulse has a characteristic shape, so it can be easily identified in noise. The presence of the switching field is necessary. Some of these devices may be deactivated by ac demagnetization.

Pulse wires were developed by VAC (Vacuumschmelze) Hanau, and they are produced by Siemens Electromechanical Components. They consist of magnetically soft wire under stress and a parallel wire of magnetically hard material, as shown in Figure 12.13. The magnetization of the latter wire has the same direction during the whole working cycle, while the magnetization direction of the magnetically soft wire is changing.

The simplified domain structure of the two states is shown in Figure 12.14. When the magnetization of both wires is parallel, both of them consist of a single domain. If the magnetization is antiparallel, closure domains appear. The switching process from the antiparallel state starts at the closure domains and is continuous so that the induced voltage is very low. The switching process from the single-domain parallel state is very fast and the pulse is high.

The typical shape of the voltage pulse is shown in Figure 12.15. The pulse energy is sufficient to supply an LED connected to optical fiber.

Application examples of pulse wire sensors are shown in Figure 12.16. Large mounting tolerances are permissible, and the air gap between the magnet and the sensor can be larger than 10 mm. Revolution counters using only one pulse wire can indicate the direction [Figure 12.16(a–c)]. In Figures

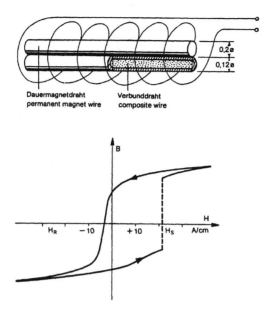

Figure 12.13 Construction of the pulse wire and typical hysteresis loop (from [12]).

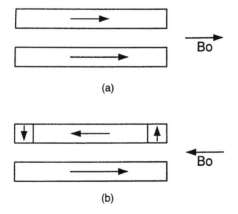

Figure 12.14 Domain structure of the pulse wire: (a) parallel state; and (b) antiparallel state.

12.16(d) and (e), clockwise rotation will drive only the sensor SE2 and counterclockwise only SE1. The effect is related to the homogeneity of the field at the switching moment; if the field is highly nonhomogeneous, the induced pulse will be very small.

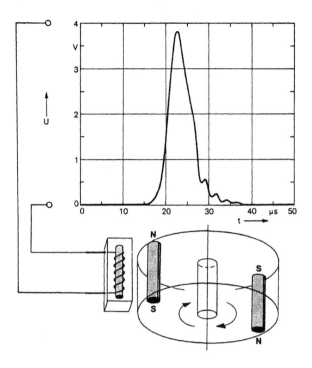

Figure 12.15 Pulse wire sensor: shape of the voltage pulse and typical application (from [12]).

A new candidate for devices based on large Barkhausen effect are amorphous wires, especially microwires covered by glass, which have built-in stress [19].

12.1.7 Magnetic Trackers

Trackers are devices that measure the location and relative orientation of the target. A complete tracker has 6 degrees of freedom (linear position in three axes and three rotation angles). The applications include body tracking in virtual reality, motion capture in animation and biomechanical measurement, and feedback systems. Miniaturized sensors are used to locate the position of probes and instruments such as biopsy needles inside the body. The target may be a source of signal (permanent magnet or transmitter coil) whose amplitude (and eventually phase) is sensed by receiving coils, or the sensor may be attached to the target.

The simplest type of transmitting target is a small permanent magnet. Such systems have been used for observing biomechanical movements [20].

Magnetic Sensors for Nonmagnetic Variables 443

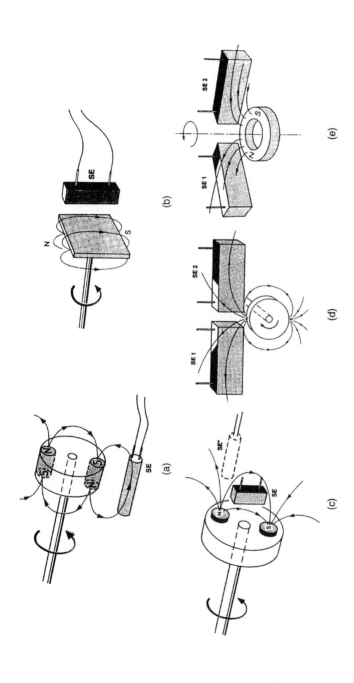

Figure 12.16 Recommended magnetic drives of pulse wires for large sensing distance or small magnets. A single sensor cannot recognize the direction (a–c). Using field nonhomogeneity, two sensors can be adjusted so that SE1 responds only to clockwise rotation and SE2 only to counterclockwise rotation (d, e). (From [12].)

Magnetic trackers with a sensing target consist of a transmitting coil (which may be flat) and miniature sensors attached to the target. The tracker resolution may be 0.7 mm and 0.1 degree, accuracy 2 mm and 0.5 degree with a translation range of up to 2.5m. Existing tracking systems may sample the location of each sensor 120 times per second for up to 120 sensors. The sensor signal may be wirelessly transmitted to the control unit. Some systems use a pulse dc magnetic field instead of an ac field and sample the position after the decay of eddy currents; this technique reduces the errors caused by conducting objects [21].

Another tracking system consists of three stationary large generator coils producing sequential pulses and one receiving coil for each target. The system was developed for three-dimensional imaging of the position and shape of long colonoscopes and endoscopes [22]. A simple three-dimensional positioning system with two flat coil sets allows determining the target position with 0.1-mm uncertainty in the range of ±20 mm [23].

12.2 Proximity and Rotation Detectors

These sensors have bistable (digital) output. They may be either activated by a moving magnet or biased by a fixed magnet and activated by a magnetically soft target. A typical example of a naturally bipolar sensing element is a reed contact, but any other magnetic sensor followed by a comparator or Schmidt trigger can be used for this purpose. The most popular are Hall sensors, AMR, and, recently, GMR magnetoresistors and also some semiconductor magnetoresistors.

Reed contacts are very cheap and totally passive devices. They consist of two magnetic strips of soft or semihard magnetic material sealed in glass pipe filled with inert gas. Normally open contacts are connected at a certain field by an attractive magnetic force between the free ends. Other contact types are normally closed. The reed contacts have hysteresis, and their switching zones have a complicated shape. However, they are still very popular because of their simplicity. Figure 12.17 shows that if the magnet is perpendicular to the contact and moves along it, there may be two switch-on zones. If the contact and the magnet have the same direction and the magnet moves along it, there may be three switching zones or one switching zone. This behavior is easy to explain by the shape of the magnet field lines.

High-security "balanced" switches use two reed contacts in the vicinity of the magnet; one of them normally is open and the other normally is closed. Any movement of the magnet activates one of the switches. The

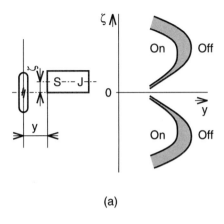

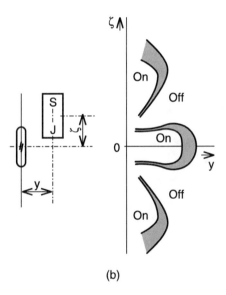

Figure 12.17 Switching zones of the reed contacts for driving the magnet perpendicular to the reed (a) and parallel with the reed (b). Sensor state is uncertain in gray regions due to the sensor hysteresis (from [11]).

normally open contacts usually are crossed by a resistor, which allows monitoring of the continuity of the wires.

Eddy-current sensors (for any conducting targets) and ac-excited variable gap sensors (only for magnetic targets) are also used for linear or angular gear position sensing. An example of an integrated inductive geartooth sensor is shown in Figure 12.18.

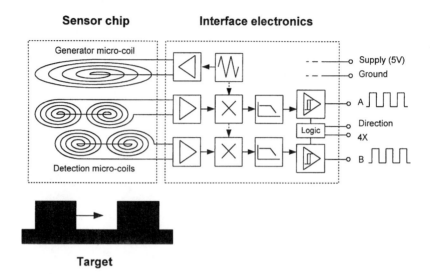

Figure 12.18 Block diagram of miniature inductive geartooth sensor (courtesy of CSEM [24].)

Pulse-output angular sensors are made using multipole magnets. Accuracy of 0.4 arc-sec and 160,000 pulses per revolution was achieved using a pair of magnetoresistive reading heads. The target was made of 3.5-inch rigid disk [25].

12.3 Force and Pressure

We will skip all the sensors that convert force, pressure, and torque into the displacement by using springs, diaphragms, columns, proving rings, and other mechanical converters [1].

A lot of the force, pressure, torque, and accelerometric magnetic sensors are based on inverse magnetostrictive (Villari) effect: The permeability of the sensing material changes due to applied strain [26]:

$$\mu = \frac{M_s^2}{(2K + 3\lambda E_0 \epsilon)} \quad (12.2)$$

where E_0 is the Young's modulus and ϵ is the strain (positive if material is compressed).

The device sensitivity is high if λ is high and K is low. Magnetostrictive materials are also referred to as piezomagnetic. The situation may be complicated by a change of E with an applied field (ΔE effect).

The stress $\sigma = E_0 \epsilon = F/A$ may be caused by the measured force F or indirectly by other variables.

Amorphous alloys are suitable for magnetoelastic sensors. They have excellent elastic properties: almost ideally linear stress-strain curves, even superior to spring alloys, with no plastic flow. Elastic strains of more than 1% are feasible, which is 10 times more than with crystalline materials [3]. The ΔE effect may cause nonlinearity at low stress levels. Negative magnetostriction materials are preferred, because they show linear increase of magnetoelastic anisotropy with mechanical stress [27]. A lot of force sensors using amorphous tapes and wires are described in scientific papers, but those devices are not widely used in industry.

The most popular industrial magnetic load cells are pressductors or torductors, which are based on variation of the flux distribution in the magnetic core (Figure 12.19). In an unloaded state, the mutual inductance between the perpendicular primary and secondary windings is zero. After loading, stress-induced anisotropy causes part of the primary coil flux to be coupled to the transformer secondary. These devices can measure forces up to 5 MN with linearity of 0.1% and hysteresis of 0.2%. They are well temperature compensated and withstand overloading.

A piezomagnetic pressure sensor prototype consists of a sputtered amorphous layer on GaAs membrane and microfabricated solenoid coils [28]. Strain gauges have been fabricated using similar technology.

A resonant pressure "vibrating cylinder" sensor with magnetic excitation and sensing are used for precise measurements, such as in aerometry [2],

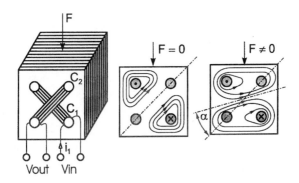

Figure 12.19 Magnetic load cell (Pressductor® or Torductor®) (from [11]).

but they probably will be pushed off the market by cheaper microfabricated capacitive sensors. Vibrating wire strain gauges also often use an electromagnetic driver and a variable reluctance detector.

12.4 Torque Sensors

Applications for sensing torque fall into two general categories: those in which the torque is transmitted by a rotating shaft and those in which the torque tends to twist a shaft having one end clamped. Devices for the latter category, often called reaction torque sensors, generally have simpler constructions, because there is no relative motion between the shaft and the electrical wiring associated with the actual torque detecting means. With rotating shafts, considerations of wear, friction, and reliability favor torque detection methods that require no physical contact with the shaft.

Three basic principles are available for measuring, without physical contact, the torque transmitted by a rotating shaft. The torque T can be determined from measurement of the twist angle θ, the surface strain ϵ, or of magnetic quantities related to the surface stress, σ. The relationships between T and the relevant variables are shown schematically in Figure 12.20 and are given quantitatively by (12.3) through (12.5), where G is the modulus of rigidity of the shaft material ($8.3 \cdot 10^{10}$ N/m^2 for most steels). It should be noted that in round shafts, tensile and compressive principal stresses—and hence principal strains—occur at ±45-degree angles to the rotational axis.

$$\sigma = \frac{16T}{\pi D^3} \quad (12.3)$$

$$\epsilon = \frac{\sigma}{2G} = \frac{8T}{\pi G D^3} \quad (12.4)$$

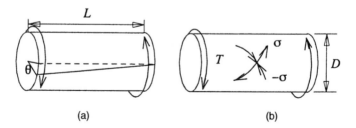

Figure 12.20 Twist angle (a) and principal stresses (b) induced by the applied torque.

$$\theta = \frac{2L(2\epsilon)}{D} = \frac{32LT}{\pi GD^4} \qquad (12.5)$$

Changes in surface strain can be measured by strain gauges attached to the shaft. The strains generally are too small (at most, a few parts of 10^3) to be accurately measured directly. Common practice is, therefore, to use four gauges arranged in a Wheatstone bridge circuit. With rotating shafts, coupling means, such as rotary transformers, are required to feed the excitation current to the gauges and to acquire the signal from the bridge circuit in a noncontacting manner.

The twist angle method of torque measurement generally requires a slender portion of the shaft to enhance the twist (2 to 3 degrees at most for $L/D = 5$) and a pair of identical toothed disks attached at opposite ends of the slender portion. The twist angle can be determined from the phase difference between magnetically or optically detected tooth/space patterns on each of the disks. This method generally requires the shaft to be rotating.

Determination of surface stress from the measurement of magnetic quantities provides an inherently noncontacting basis for measuring torque. Moreover, this method can be realized in a more compact construction than those required for either strain or twist angle methods. The rest of this section, therefore, is devoted to a more detailed explanation of the operation of torque sensors based on this magnetoelastic method.

The magnetoelastic effect provides a mutual interaction path between mechanical energy (elastic energy) and magnetic energy. The influence of the elastic energy is brought into the magnetic system through the stress-induced magnetic anisotropy, K_σ, given by (12.6), where λ is the magnetostriction constant of the shaft material.

$$K_\sigma = 3\lambda\sigma = \frac{48\lambda T}{\pi D^3} \qquad (12.6)$$

Magnetic anisotropy causes magnetic moments to incline toward its easy axis. When several sources of magnetic anisotropy are present at the same time, each having a different easy axis orientation, the magnetic moments will be inclined at that orientation where the anisotropies balance. Shafts generally are made of polycrystalline magnetic materials (steels) wherein, in the absence of torque, a magnetocrystalline anisotropy provides an easy axis orientation for the magnetic moments in each grain. The stress-induced magnetic anisotropy, K_σ, has its easy axis parallel to the line of tension if $\lambda > 0$, or parallel to the line of compression if $\lambda < 0$. The greater the applied torque, therefore, the more nearly will the equilibrium orientation of the magnetic moments be inclined toward the K_σ easy axis. Reversal of the direction of the torque will interchange the lines of tension and compression, resulting in a 90-degree rotation of the K_σ easy axis orientation. Because of that interaction, the permeability of a magnetic shaft material is related to the torque, showing larger values along the stress-induced easy axis and smaller values along the hard axis, which is perpendicular to the former.

There are two general ways to utilize the stress-induced magnetic anisotropy as the sensing mechanism for torque sensors. In type I, the permeability changes in the shaft surface, caused by the stress-induced magnetic anisotropy, affects the permeance of a magnetic flux path, which includes a magnetizing source and a pickup (sensing) coil. In type II, the stress-induced magnetic anisotropy causes a remanently magnetized magnetoelastically active member to generate magnetic flux. It is desirable for sensors of both types to have axisymmetric structures to avoid rotation-dependent outputs that degrade the attainable accuracy. It is also desirable that torque sensors work in a differential mode of operation, because that makes them robust against common mode types of disturbances, such as the ambient temperature.

Figure 12.21 is an example of a type I torque sensor constructed in accordance with fundamental principles and utilizing the magnetoelastic effect inherent in a shaft made principally of iron [29]. A pair of mutually orthogonal, U-shaped cores are combined, with their open ends facing toward the shaft but separated therefrom by small air gaps. The coils on the legs of the vertical core provide an ac excitation field to the shaft, and the coils on the legs of the horizontal core are used to pick up imbalances in the cyclically varying magnetic flux. The operating principle resembles that of ordinary bridge circuits if we consider the magnetic flux as the electric current. The permeability of the shaft becomes anisotropic under the stresses shown in Figure 12.21: larger in directions parallel to the line connecting

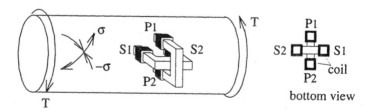

Figure 12.21 A torque sensor based on fundamental magnetoelastic principles.

points P1 and S1 and smaller in directions parallel to the line connecting points P1 and S2. Hence, the magnetic path, P1 ⇒ S1 ⇒ horizontal core (to the right) ⇒ S2 ⇒ P2 ⇒ vertical core ⇒ P1, has a larger permeance than the magnetic path, P1 ⇒ S2 ⇒ horizontal core (to the left) ⇒ S1 ⇒ P2 ⇒ vertical core ⇒ P1, thus yielding a net magnetic flux in the horizontal core. Because the excitation is ac, voltages reflecting the net flux will be induced in the pickup coils. It should be noted that the phase of the ac magnetic flux, hence the voltages induced in the pickup coils, corresponds to the direction of the applied torque. Another well-known excitation/pickup coil system, using a five-leg magnetic core, was proposed by Beth and Meeks [30].

The advantages of this type of torque sensor stem from its obviously simple and mechanically robust construction. However, local variations in magnetic properties of typical shaft surfaces limit their accuracy. Common practice to improve the accuracy is to use several circumferentially distributed excitation/pickup coils and average their outputs.

To allow multiple installations of the excitation/pickup coils within a small radial space, low-profile structures have been developed based on using a pair of figure-of-eight coils [31]. The operating principle of this arrangement is readily understood with the help of Figure 12.22.

The center branches of the two figure-of-eight coils shown in Figure 12.22 are aligned at +45-degree and −45-degree angles to the shaft axis. With the torque-induced stresses shown, the self-inductance of the left (a) figure-of-eight coil decreases, whereas that of the right one (b) increases. The difference in self-inductance between the two coils provides a measure of the torque. A small, simple, and mechanically robust construction is achieved by stacking the two coils shown in Figure 12.22 and embedding them both into a single ferrite core, as shown in Figure 12.23.

Another important group of type I torque sensors utilizes an axisymmetric construction, an example of which is shown in Figure 12.24. In the construction shown, oppositely directed helical grooves are machined or formed (typically along ±45° angles to the axis) on adjacent circumferential

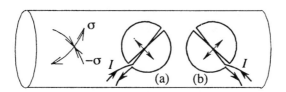

Figure 12.22 Operating principle of a pair of figure-of-eight coils to detect torque.

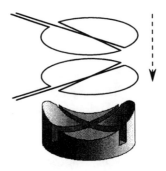

Figure 12.23 Structure of low-profile pickup coils.

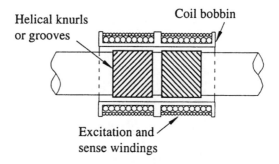

Figure 12.24 A torque sensor of axisymmetric structure.

regions of a steel shaft [32]. Solenoidal coils encircling those regions are used for excitation and sensing. The axial permeability of a grooved region increases when the easy axis of the stress-induced magnetic anisotropy occurs in parallel to the line of grooves, whereas it decreases otherwise. That results in different voltages being induced in the sense windings and that difference provides the measure of the torque. The advantage of this design is that the axisymmetric structure of the windings hides local variation of the magnetic properties of the shaft. Steels containing a small percentage of nickel are especially suitable for this torque sensor construction; indeed, nickel as an alloying element tends to enhance the performance of all types of magnetoelastic torque sensors because its presence both increases the magnetostriction constant and decreases the magnetocrystalline anisotropy.

Type II torque sensors are generally constructed with a thin ring of magnetoelastically active material (positively magnetostrictive) rigidly attached to the shaft [33]. The ring is expanded during installation on the shaft, thereby developing a magnetic anisotropy having the easy axis along the circumferential direction. A typical example is shown in Figure 12.25. The solid arrows in the figure indicate that the ring is magnetized in a way that each axial half is polarized in an opposite circumferential direction [33]. When torque is applied, the magnetizations tilt into helical directions (dashed arrows), causing magnetic poles to develop at the central domain wall and at the ring end surfaces. The polarity of the magnetic poles reverses when the applied torque changes its direction. Torque is determined by measuring magnetic flux with one or more magnetic field sensors. Two-region polarization can be accomplished by magnetizing the ring on the shaft with permanent magnets while rotating the shaft.

12.5 Magnetic Flowmeters

If the conducting fluid flows in the magnetic field, electric field **E** is generated. In an ideal case

$$\mathbf{E} = \mathbf{v} \times \mathbf{B} \tag{12.7}$$

In the case that v is perpendicular to B, voltage V is induced between two electrodes, which are at the distance of d perpendicularly to both v and B

$$V = k \cdot d \cdot v \cdot B \tag{12.8}$$

where k is a constant depending on fluid conductivity and the geometry. The magnetic field is either ac or pulse dc to avoid polarization effects. The coils are of the saddle shape. Induction flowmeters work for fluids with a

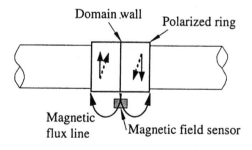

Figure 12.25 A torque sensor having a magnetoelastically active ring with double polarization.

conductivity higher than 1 μS/cm, which includes drinking water. The typical accuracy is 0.5%.

More complicated designs of contactless induction flowmeters use detection coil instead of electrodes. Although these devices are less accurate than contact flowmeters, they find applications in measurement of fluids, which would create sediments on electrodes.

Another industrial type of flowmeter uses bending of wire in the flow direction. Such a device for low flow rates that uses amorphous magnetic wire is described in [34].

12.6 Current Sensors

The current measurement using a shunt resistor is in some cases impractical or impossible [35]. Optical current sensors for high currents are being intensively developed (see Chapter 6). Optical fiber devices are suitable for high-voltage applications, but the reported errors are more than 1% even after temperature compensation [36].

Besides fulfilling the requirements common to magnetic field sensors, contactless current sensors should be geometrically selective, that is, sensitive to measured currents and resistant against interferences from other currents and external fields. The easiest way to guarantee that is to use a closed magnetic circuit with a measured conductor inside. If that is not possible, magnetic sensor arrays may be used [37].

Instrument current transformers have a primary winding with few turns (or a single conductor through the core opening) and a secondary winding, which ideally should be short-circuited [38].

Current comparators are described in a book by Miljanic and Moore [39]. An ac current comparator is in principle a three-winding device on the ring (toroidal) core. The ac comparators have errors below 1 ppm in amplitude and $3 \cdot 10^{-6}$ degrees in phase.

The dc current comparators are based on a fluxgate effect. They are usually feedback compensated, and the core consists of two detection toroids excited in opposite directions.

Fluxgate-like dc current sensor modules are similar to dc comparators, but of a much simpler design, and are manufactured by VAC Hanau for measurement ranges between 40A and 200A. The accuracy of a typical 40A module is 0.5%, linearity 0.1%, current temperature drift less than 30 μA (−25°C–70°C).

12.6.1 dc/ac Hall and MR Current Sensors

Traditional current sensors are based on the Hall element in the air gap of a magnetic yoke. To improve the linearity, the measured current may be compensated. These devices have a limited zero stability given by the Hall sensor offset: typical offset drift of a 50A sensor of 600 μA in the (0°C–70°C) range is 20 times worse than that of fluxgate-type current sensor modules. Hall current sensors also are more sensitive to external magnetic fields and close currents due to the magnetic leakage associated with the air gap. But especially for larger currents (>~10 A), the device precision typically is 1% for uncompensated and 0.5% for compensated type and is sufficient for most industrial applications.

Coreless MR current sensors are based on an AMR bridge, which is made insensitive to an external field but sensitive to measured current through the primary bus bar.

12.6.2 Current Clamps

The ac current clamps are based on current transformers made on an openable ferrite core. The measured conductor forms a primary winding; secondary winding is terminated by a small resistor or connected to current-to-voltage converter. Most of the available dc current clamps are based on the Hall sensor.

12.6.3 Magnetometric Measurement of Hidden Currents

The dc cables can be located and their current can be remotely monitored by measuring the magnetic field in several points. The magnetometric method is also used to measure the currents in constructions such as bridges and in pipelines.

12.7 Sensors Using Magnetic Liquids

One of the more exotic components of magnetism are magnetic liquids, usually suspensions that contain magnetic particles. Although in general they have problems with long-term stability, they have many possible sensor applications. The principle of inertial magnetohydrodynamic (MHD) angular rate sensor is shown in Figure 12.26. The rotation causes relative velocity between the dc magnetic field generated by a ring permanent magnet and the high conductive fluid acting as an inertial mass element. This relative

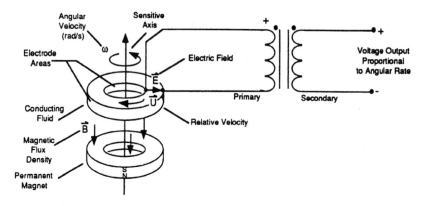

Figure 12.26 The principle of MHD angular rate sensor (from [40]).

velocity causes electric potential across the channel. The induced voltage is often amplified by a transformer integrated into the sensor case [40].

MHD angular rate sensors may have a 0.3 Hz to 1 kHz frequency band, so they can be used for shock sensing, image motion stabilization, and ride control. Precise MHD sensors may replace expensive gyroscopes in autonomous navigation and tracking systems.

References

[1] Norton, H. A., *Sensor and Analyzer Handbook*, London: Prentice Hall, 1982.

[2] Pallas-Areny, R., and J. G. Webster, *Sensors and Signal Conditioning*, New York: Wiley, 1991.

[3] Hinz, G., and H. Voigt, "Magnetoelastic Sensors," in R. Boll and K. J. Overshott (eds.), *Magnetic Sensors,* Veiden, Germany: VCH, 1989, pp. 98–151.

[4] Decker, W., and P. Kostka, "Inductive and Eddy Current Sensors," in R. Boll and K. J. Overshott (eds.), *Magnetic Sensors,* Veiden, Germany: VCH, 1989, pp. 255–313.

[5] Adelerhof, D. J., and W. Geven, "New Position Detectors Based on AMR Sensors," *Proc. Eurosensors XIII*, The Hague, 1999, pp. 421–424.

[6] Oudet, C., "A High-Linearity, Rotary Magnetic Sensor," *Sensors*, June 1995, pp. 28–31.

[7] Welsby, S. D., and T. Hitz, "True Position Measurement With Eddy Current Technology," *Sensors*, Nov. 1977, pp. 30–40.

[8] Anim-Appiah, K. D., and S. M. Riad, "Analysis and Design of Ferrite Cores for Eddy-Current-Killed Oscillator Inductive Proximity Sensors," *IEEE Trans. Magn.*, Vol. 33, No. 3, 1997, pp. 2274–2281.

[9] Tian, G. Y., Z. X. Zhao, and R. W. Baines, "The Research of Inhomogeneity in Eddy-Current Sensors," *Sensors and Actuators A*, Vol. 69, 1998, pp. 148–151.

[10] Passeraub, P. A., P. A. Besse, and R. S. Popovic, "Temperature Compensation of an Integrated Low Power Inductive Proximity Microsensor," *Sensors and Actuators A*, Vol. 82, 2000, pp. 62–68.

[11] Dadŏ, S., and M. Kreidl, "Sensors and Measurement Circuits," (in Czech), Prague, Czech Republic: CTU Publishers, 1996.

[12] Siemens Electromechanical Components, "Sensor-Applications for Your System Success," brochure, also on http://www.siemens.de/ec/eccs/sensors/magnetic.htm

[13] Ruhle, F., "Inductive Position Transducers," *Sensors*, Apr. 1995, pp. 19–27.

[14] Russel, J., "New Developments in Magnetostrictive Position Sensors," *Sensors*, June 1977, pp. 46.

[15] Hristoforou, E., and R. E. Reilly, "Displacement Sensors Using Soft Magnetostrictive Alloys," *IEEE Trans. Magn.*, Vol. 30, 1994, pp. 2728–2733.

[16] Hristoforou, E., "Magnetostrictive Delay Lines and Their Applications," *Sensors and Actuators A*, Vol. 59, 1977, 183–191.

[17] Wiegand, J. R., Switchable Magnetic Device, U.S. Patent 4247 601, 1981.

[18] Rauschner, G., and C. Radeloff, "Wiegand and Pulse-Wire Sensors," in R. Boll and K. J. Overshott (eds.), *Magnetic Sensors*, Veiden, Germany: VCH, 1989, pp. 98–151.

[19] Chiriac, H., et al., "Amorphous Glass-Covered Magnetic Wires for Sensing Applications," *Sensors and Actuators A*, Vol. 59, 1997, pp. 243–251.

[20] Sonoda, Y., "Applications of Magnetic Sensors to Observing Bio-Mechanical Movements," *IEEE Trans. Magn.*, Vol. 31, 1995, pp. 1283–1290.

[21] Ascension Technology Corp., "The Advantage of DC Magnetic Tracking," 1999, http://www.ascension-tech.com.

[22] Dogramadzi, S., C. R. Allen, and G. D. Bell, "Computer Controlled Colonoscopy," *Proc. IEEE Instumentation and Technology Conf.*, St. Paul, MN, May 18–21, 1998, pp. 210–213.

[23] Willenberg, G. D., and K. Weyand, "Three-Dimensional Positioning Setup for Magnetometer Sensors," *IEEE Trans. Magn.*, Vol. 46, 1997, pp. 621–623.

[24] MS 1200 Inductive position, speed and direction sensor, www.csem.ch.

[25] Tan, B., et al., "Demonstration of a Magnetic Angular Position Sensor," *Sensors and Actuators*, Vol. 81, 2000, pp. 332–335.

[26] Choudhary, P., and T. Meydan, "A Novel Accelerometer Design Using the Inverse Magnetostrictive Effect," *Sensors and Actuators A*, Vol. 59, 1977, pp. 51–55.

[27] Barandiaran, J. M., and J. Gutierrez, "Magnetoelastic Sensors Based on Soft Amorphous Magnetic Alloys," *Sensors and Actuators A*, Vol. 59, 1997, pp. 38–43.

[28] Gibbs, M. R. J., et al., "Microstructures Containing Piezomagnetic Elements," *Sensors and Actuators A*, Vol. 59, 1997, pp. 229–235.

[29] Dahle, O., "The Ring Torductor—A Torque-Gauge, Without Slip Rings, for Industrial Measurement and Control," *ASEA J.*, Vol. 33, No. 3, 1960, p. 23.

[30] Beth, B. A., and W. W. Meeks, "Magnetic Measurement of Torque in a Rotating Shaft," *Rev. Sci. Instrum.*, Vol. 25, 1954, p. 603.

[31] Sasada, I., "Magnetostrictive Methods of Detecting Torque From Case-Hardened Steel Shafts," *MRS Symp. Proc.*, Vol. 360, 1995, p. 231.

[32] Sendoh, I., and S. Kawahara, Torque meter, Japanese Patent 169326, 1945.

[33] Garshelis, I. J., and C. R. Conto, "A Torque Transducer Utilizing Two Oppositely Polarized Rings," *IEEE Trans. Magn.*, Vol. 30, 1994, p. 4629.

[34] Hristoforou E., I. N. Avaritsiotis, and H. Chiriac, "New Flowmeters Based on Amorphous Magnetic Wires," *Sensors and Actuators*, Vol. A59, 1997, pp. 94–96.

[35] Iwansson K., G. Sinapius, and W. Hoornaert, "Measuring Current, Voltage, and Power," New York: Elesevier, 1999.

[36] Cruden, A., et al., "Optical Current Measurement System for High-Voltage Applications," *Measurement*, Vol. 24, 1998, pp. 97–102.

[37] Bazzocchi, R., and L. Di Rienzo, "Interference Rejection Algorithm for Current Measurement Using Magnetic Sensor Arrays," *Sensors and Actuators A*, Vol. 85, 2000, pp. 38–41.

[38] Draxler, K., and R. Styblíková, "Use of Nanocrystalline Materials for Current Transformer Construction," *J. Magn. Magn. Mater.*, Vol. 157/158, 1996, pp. 447–448.

[39] Moore, W. J. M., and P.N. Miljanic, *The Current Comparator*, London: Peregrinus, 1988.

[40] Pewatron AG, "Inertial Angular Rate Sensors, Theory and Applications," brochure. Non-magnetic Final, May 21, 1999.

Magnetic Sensors, Magnetometers, and Calibration Equipment Manufacturers

Because the same instrument may be sold under several names, it is sometimes difficult to distinguish original manufacturers from resellers.

Applied Physics Systems
897 Independence Ave., Suite 1C
Mountain View, CA 94043
Phone: (650) 965-0500
Fax: (650) 965-0404
E-mail: aps@appliedphysics.com
http://www.appliedphysics.com/Magnetometers.html
fluxgates, SQUIDs

Bartington Instr., LTD
Spendlove Centre
Enstone Road, Charlbury
Oxford OX7 3PQ, England
http://www.bartington.com/
fluxgates, susceptibility meters

Billingsley Magnetics
2600 Brighton Dam Road
Brookeville, MD 20833

Lab phone: (301) 774-7707
Fax: (301) 774-0745
http://www.magnetometer.com/
fluxgates, Helmholz

Bruker Analyt. Messtech. Gmbh
Wikingerstrasse 13
D-7500 Karlsruhe 21
Germany
http://www.bruker.com/
NMR fax: (49) 721 5161-297
E-mail: ut@bruker.de
EPR fax: (49) 721 5161-237
E-mail: epr_sales@bruker.de
NMR spectrometers, electromagnets

Cryophysics SA
9, rue Dallery
F-78350 Jouy-en-Josas, France
Phone: +33-1-39 07 18 60
Fax: +33-1-39 56 42 56
E-mail: cryophysics_fr@compuserve.com
repr. LDJ + Lake Shore

Danfysik A/S
Mollehaven 31A
DK-4040 Jyllinge, Denmark
measurement benches, electromagnets, power converters

Dowty RFL Industries
353 Powerville Road
Boonton, NJ 07005-0239
gaussmeters, integrators, fluxgates, magnet chargers

Drusch GmbH
Altenbekner Damm 51
D-30173 Hannover, Germany
Phone: ++49 511 80 46 15
Fax: ++49 511 88 99 53
E-mail: info@drusch.com
http://www.drusch.com/
Hall generators, Gaussmeters, fluxmeters, search coils, NMR

F. W. Bell Inc.
6120 Hanging Moss Road
Orlando, FL 32807
Phone: (407) 678-6900
http://www.gaussmeter.com/index.html
gaussmeters, Hall probes, current sensors

Geofyzika a.s.
P.O. Box 90
Jecna 29a
62100 Brno, Czech Republic
Phone: 420-541634111
Fax: 420-541225089
http://geofyzika.com/
proton magnetometers

Geometrics
2190 Fortune Drive
San Jose, CA 95131
Phone: (408) 954-0522
Fax: (408) 954-0902
http://www.geometrics.com/
Cesium magnetometers

GMW Associates
629 Bair Island Road #113
Redwood City, CA 94063
http://www.gmw.com/
magnetometers, electromagnets, current meters
represents Danfysik, Group 3, Metrolab, Bartington, Resonance Research

Honeywell
Solid State Electronics Center
12001 State Highway 55
Plymouth, MN 55441
Phone: (800) 323-8295
Fax: (612) 954-2582
E-mail: clr@mn14.ssec.honeywell.com
http://www.honeywell.com/
AMR sensors and magnetometers, Hall sensors

Infineon Technologies (Siemens subsidiary)
1730 North First Street
San Jose, CA 95112
Phone: (408) 501-6000
Fax: (408) 501-2424
http://www.infineon.com/
Hall sensors

Institut Dr. Foerster
In Laisen 70
Postfach 1564
D-72766 Reutlingen 1
F. R. Germany
Phone: 07121-140-354
Fax: 07121-140-335
http://www.foerstergroup.com/
fluxgate magnetometers, bomb locators, nondestructive testing

Laboratorio Elettrofisico
via G. Ferrari 14
20014 Nerviano
Milano, Italy
Phone: 39-0331-589785
Fax: 39-0331-585760
E-mail: info@laboratorio.elettrofisico.com
http://www.laboratorio.elettrofisico.com/
fluxmeters, magnetizers, coercimeters

Lake Shore Cryotronics
575 McCorkle Blvd.
Westerville, OH 43082
Phone: (614) 891-2244
Fax: (614) 818-1600
E-mail: sales@lakeshore.com
http://www.lakeshore.com/
gaussmeters, fluxmeters, cryogenic equipment

LDJ Electronics Inc.
P.O. Box 219
Troy, MI 48099-0219
http://www.LDJ-electronics.com/
gaussmeters, integrators

Magnet-Physik Dr. Steingroever GmbH
Emil-Hoffmann-Strasse 3
D-5000 Köln 50, Germany
Phone: +49 2236 3919-0
Fax: +49 2236 391919
http://www.magnet-physics.com/
field measurement equipment, gaussmeters, integrators

Magnetic Instrumentation Inc.
8431 Castlewood Drive
Indianapolis, IN 46250
Phone: (317) 842-7500
Fax: (317) 849-7600
gaussmeters

MEDA, Inc.
485 Spring Park Place, Suite 350
Herndon, VA 20170
Phone: (703) 471-1445
Fax: (703) 471-9130
E-mail: bvayda@meda.com
http://www.meda.com/
fluxgates

Metrolab Instruments SA
110, ch. Pont-du-Centenaire
CH-1228 Geneva
Switzerland
Tel +41 22 884 33 11
Fax +41 22 884 33 10
E-mail: contact@metrolab.ch
http://www.metrolab.ch/
NMR, integrators, Hall

MμShield Company, Inc.
P.O. Box 439, 5 Springfield Road
Goffstown, NH 03045
Phone: (603) 666-4433
Fax: (603) 666-4013
http://www.mushield.com/design.html
shieldings

NVE (Nonvolatile Electronics)
11409 Valley View Road
Eden Prarie, MN 55344
Phone: (612) 829-9217
Fax: (612) 996-1600
http://www.nve.com/
GMR sensors

Oersted Technology
19480 SW Mohave Ct.
Tualatin, OR 97062
Phone: (503) 612-9860
Fax: (503) 692-3518
E-mail: oersted@oersted.com
http://www.oersted.com/
fluxmeters, Helmholz coils

Philips Semiconductors
P.O. Box 218
5600 MD Eindhoven
The Netherlands
Phone: 31-40-2791111
Fax: 31-40-2724825
http://www.semiconductors.philips.com/
AMR and GMR sensors

Quantum Design, Inc
11578 Sorrento Valley Road
San Diego, CA 92121-1311
Phone: (619) 481-4400
Fax: (619) 481-7410
E-mail: info@quandsn.com
http://www.quandsn.com/
SQUIDs

Redcliffe Magtronics Ltd.
20 Clothier Road
Brislington
Bristol
BS4 5PS
United Kingdom
Phone: 44-117-977-1404

Fax: 44-117-972-3013
E-mail: redcliffe@redmag.co.uk
http://www.redmag.co.uk/
magnetometers, scanning systems, magnetic diagnostics

Resonance Research Inc.
10 Cook Street, Pinehurst Business Park
Billerica, MA 01821
Phone: (978) 671-0811
Fax: (978) 663-0483
E-mail: rricorp@rricorp.com
http://www.rricorp.com/
MRI, EPR, and NMR instrumentation

RS Dynamics
Bocni II/1401 CZ-14301
Prague 4 Czech Republic
Phone: 42-0267103027
Fax: 42-0267103387
E-mail: info@rsdynamics.com
http://www.rsdynamics.com/
magnetometers (rotating coil and fluxgate), coil systems, micromagnetic instruments

Schonstedt Instrument Company
4 Edmond Road, P.O. Box 309
Kearneysville, WV 25430
Phone: (304) 725-1050
Phone: (800) 999-8280
Fax: (304) 725-1095
E-mail: info@schonstedt.com
http://www.schonstedt.com/
fluxgates

Scintrex Ltd.
222 Snidercroft Road
Concord, Ontario L4K 1B5
Canada
Phone: (905) 669-2280
Fax: (905) 669-6403
E-mail: scintrex@idsdetection.com
http://www.idsdetection.com/
magnetometers

Sentron AG
Baarerstr. 73
CH-6300 Zug
Switzerland
Phone: +41 41 711 21 70
Fax: +41 41 711 21 88
E-mail: info@sentron.ch
http://www.sentron.ch/
Hall sensors and magnetometers

Tristan Technologies, Inc.
6350 Nancy Ridge Drive, Suite 102
San Diego, CA 92121
Phone: (858) 550-2700
Fax: (858) 550-2799
E-mail: info@tristantech.com
http://www.tristantech.com/
SQUIDs

Walker Scientific
17 Rockdale Street
Worcester, MA 01606
Phone: (508) 852-3676
Fax: (508) 856-9931
E-mail: info@walkerscientific.com
http://www.walkerscientific.com/
gaussmeters, integrators, electromagnets, power converters, hysteresisgraphs, coercivity meters, Helmholtz coils

WUNTRONIC GmbH München
Heppstrasse 30
D-80995 München, Germany
Phone: ++49 89 313 30 07
Fax: ++49 89 314 67 06
E-mail: WUNTRONIC@wuntronic.de
http://www.wuntronic.de/
fluxgates, Hall effect

List of Symbols and Abbreviations

A	core cross-sectional area
ADC	analog-to-digital converter
AMR	anisotropic magnetoresistance
AU	astronomical unit
B	magnetic field (T)
B_N	magnetic noise
B_0	magnetic field (flux density) in the open space (in air)
B_{off}	offset induction
B_p	polarizing field
B_{res}	resultant field
C	capacitance
c	pitch
CMR	colossal magnetoresistance
d	diameter
d_c	core diameter
d_m	mean coil diameter
d_w	wire diameter
D	effective demagnetizing factor; dielectric displacement
DNP	dynamic nuclear polarization
DSP	digital signal processor
e	unit charge
e_{noise}	noise voltage

$\mathbf{e}_r$	unity vector in the direction of r
E	electrical field strength; energy; energy sensitivity of SQUID; Young's modulus
E_H	Hall electric field
ESR	electronic spin resonance
f	frequency; correction factor
F	Faraday rotation; force
FFT	Fast Fourier's transform
f_L, f_h, f_1, f_c	low frequency, higher frequency, lower frequency, corner frequency
g	acceleration of gravity
G	geometric correction factor
g_m	transconductance
GMI	giant magnetoimpedance
GMR	giant magnetoresistance, also sensitivity of GMR magnetotoresistor
h	height; Planck's constant
H	magnetic field intensity (A/m)
H_d	demagnetization field
H_{exc}	excitation field intensity
H_k	anisotropy field
H_m	maximum magnetic field intensity
H_o	characteristic field
H_0	intensity of the external field (in the open space)
i, I	current
I	intensity of light; moment of inertia
i_{EQ}	equivalent coil current
I_{exc}	excitation current
I_n, i_n	noise current
j	Coulomb's magnetic moment
J	polarization
J_c	current density
J_E	exchange coupling energy
j_z	longitudinal component of the current density
k	constant; wave factor

K		anisotropy constant
k_B		Boltzmann's constant
l, L		length
L		inductance; angular momentum
l_{eff}		effective coil length
L_0		arithmetic mean value $L_0 = <L(t)>$
LF		low frequency
L_{G0}		geometric mean value of the pickup coil inductance
L_s		serial inductance
m		mass; magnetic moment; length/diameter ratio
M		magnetization
m_e		mass of electron
M_s		spontaneous magnetization
m_w		mass of coil winding
MBE		mollecular beam epitaxy
n		multiple integer; carrier density
N		number of turns; number of atoms per cubic meter
NMR		nuclear magnetic resonance
N_{rms}		RMS level of the noise
N_Z		longitudinal demagnetization factor
$P(f)$		power spectrum density of the noise
q		elementary charge
Q		magnetic charge
Q_e		electric charge
q_m		Bohr magneton
r		distance
R		resistance; radius
r_{Cu}		dc wire resistance
RF		radio frequency
R_H		Hall coefficient
R_s		serial resistance
S		sensitivity; shielding factor
S_A, S_I, S_V		absolute sensitivity, supply-current sensitivity, supply voltage-related sensitivity
SDT		spin-dependent tunnelling

S/N	signal-to-noise ratio
t	time
T	period; torque; absolute temperature
T, t	thickness
T_1	spin-lattice relaxation constant
T_c	Curie temperature
U	magnetic voltage
v	speed
V	volume; voltage; Verdet constant
V_H	Hall voltage
V_I	induced voltage
v_n	noise voltage
w	width of the tape or strip; Weiss constant
W	energy per unit volume; equivalent volume
W_A	total anisotropy energy
W_C	magnetocrystalline anisotropy
W_D	shape anisotropy energy
Z	complex impedance; number of electrons per atom
Δ	change
Φ	magnetic flux
Φ_m	flux sensitivity
α	Gilbert's damping factor
β	linear birefringence
β	gradiometer balance
γ	density
γ	gyromagnetic ratio
γ_p	gyromagnetic ratio of proton
δ	skin depth
ϵ	relative permitivity
ϵ	strain
η_{max}	GMI factor
λ	magnetostriction
λ_s	saturation magnetostriction
λ	relative change of the length
λ	skin depth (also δ)

μ	permeability (usually relative μ_r)
μ_a	apparent permeability
μ_p, μ_e	proton, electron magnetic moment
μ_0	absolute permeability of open space $(= 4\pi \cdot 10^{-7}\text{ Hm}^{-1})$
μ_n, μ_{Hn}	drift mobility of electrons, Hall mobility
ρ	resistivity (specific resistance)
ρ_o, ρ_p	resistivity for J parallel respectively orthogonal to M_s
ρ_e	electrical charge density
ρ_s	sheet resistance
σ	conductivity
τ	transmission coefficient
ν	Larmor resonance frequency
ω	angular frequency
Ω	angular rotation rate

About the Authors

Mario H. Acuña received his MSEE degree from the University of Tucumán, Argentina, in 1967 and his Ph.D. in space physics from the Catholic University of America, Washington, D.C. in 1974. From 1963 to 1967 he worked for the Electrical Engineering Department and Ionospheric Research Laboratory at the University of Tucuman as well as the Argentine National Space Research Commission. In 1967 he moved permanently to the United States and joined the Fairchile-Hiller Corporation in Germantown, Maryland, to provide engineering and scientific support to NASA. Since 1969, he has been associated with NASA's Goddard Space Flight Center in Greenbelt, Maryland, where his research interests are focused on experimental investigations of the magnetic fields and plasmas in the solar system. As principal investigator, co-investigator, instrument scientist, and project scientist, he has been part of many NASA missions.

James R. Biard is a chief scientist, retired, from Honeywell Sensing & Control, in Richardson, Texas. He received his B.S., M.S., and Ph.D. degrees in electrical engineering from Texas A&M University, College Station, Texas. He is an inventor with 51 U.S. and foreign patents, including the GaAs LED, silicon MOS-ROM, Schottky clamped logic circuits, and a planar silicon vertical Hall element. He is a Fellow of IEEE and a member of the National Academy of Engineering. He also serves at Texas A&M University as an adjunct professor of electrical engineering. His major areas of interest are semiconductor device physics, semiconductor sensors, optoelectronic devices, and linear integrated circuit design.

Yuri S. Didosyan received a diploma in theoretical physics from Tbilisi State University. He received his Ph.D. from Moscow State University. Since

1976, he has worked as a senior research scientist for the Russian Institute of Metrology. From 1996–1998, he was a university lecturer, guest professor, and guest researcher at the Institute of Material Science in Electrical Engineering of the Technical University of Vienna, Austria.

Robert L. Fagaly received a B.S. in chemistry from San Jose State University in 1967, an M.S. in physics from San Jose State University in 1969, and a Ph.D. in physics from the University of Toledo in 1977. He also received an M.B.A. from the University of San Diego in 1992. After scientific fellowships at the Illinois Institute of Technology, Chicago, Illinois, and the Centre National de la Recherche Scientifique, Grenoble, France, he worked for Biomagnetic Technologies, General Atomics, and Conductus. He also lectured at the University of California at Los Angeles. Since 1997, he has been vice president and general manager of Tristan Technologies, Inc. in San Diego, California.

Ronald B. Foster is the design manager at Honeywell Sensing & Control in Richardson, Texas. He received his B.S. and M.S. degrees in physics from the University of Arkansas. He has worked for Texas Instruments. Since 1983 he has worked for Honeywell. His major areas of work have focused on the fabrication of a variety of sensors, including optical, magnetic, pressure, mass air flow, pH, RTD, humidity, and inertial measurement. He holds three patents related to sensing technology.

Ivan J. Garshelis is the president of Magnova, Inc. in Pittsfield, Massachusetts. He was also the president and founder of Magnetoelastic Devices, Inc. His research has focused on magnetism with emphasis on the magnetoelastic effects, primarily in areas relevant to sensor applications.

Hans Hauser received an Ing. degree in electrical engineering from the Technical University of Vienna. In 1988 he became a doctor of technical sciences, and in 1999 an IEEE senior member, and in 1997 he received his professor's degree. He is a professor and the head of the Industrial Electronics and Material Science Department at the Technical University of Vienna. His main areas of research are magnetism and magneto-optics.

Ludek Kraus works in the Institute of Physics at the Czech Academy of Sciences in Prague. He received an M.Sc. in solid state physics from Czech Technical University in Prague in 1968. In 1980 he received a Ph.D. in experimental physics and acoustics from the Institute of Physics at the Czech Academy of Sciences. His main area of interest is the magnetism of amorphous and nanocrystalline materials.

Eugene Paperno received an M.Sc. in electrical engineering from the Minsk Institute of Radio Engineering, Republic of Belarus, in 1983, and a Ph.D. in electrical engineering in 1997 from Ben-Gurion University of the Negev, Beer-Sheva, Israel. He has previously worked at the Institute of

Electronics, Belorussian Academy of Sciences, and at Kyushu University in Fukuoka, Japan. He works as a lecturer at Ben-Gurion University. His recent research involves instrumentation and measurements, and magnetic shielding including magnetic shaking, magnetic noise phenomena, magnetoresistive sensors, communications over existing power lines, and noise in oscillators.

Radjove S. Popovic is a professor of microtechnology systems in the Department of Microengineering at the Swiss Federal Institute of Technology at Lausanne (EPFL), in Switzerland. He received his Dipl. Ing. degree in applied physics from the University of Beograd, Yugoslavia, in 1969, and M.Sc. and doctorate of science degrees in electronics from the University of Nis, Yugoslavia, in 1974 and 1978, respectively. Dr. Popovic has previously worked at Elektronska Industrija Corp. in Nis, Yugoslavia, and at Landis & Gyr Corp. in Zug, Switzerland. His current research interests include sensors for magnetic, optical, and mechanical signals, the corresponding microsystems, physics of submicron devices, and noise phenomena.

Fritz Primdahl received an M.Sc. in electrical engineering and physics from the Technical University of Denmark in 1964. He worked in the Department of Geophysics at the Danish Meteorological Institute (1966–1980), and at the National Research Council of Canada (Post-Doctoral Fellow, 1968–1970). Now he is a senior scientist at the Danish Space Research Institute and also works in the Department of Automation at the Technical University of Denmark. His main research interests are space magnetometry and space plasma physics focusing on ionospheric plasma instabilities and field-aligned currents. He is presently involved in the vector magnetometers onboard several satellites, such as Oersted, ASTRID-2, CHAMP, and US SAC-C.

Pavel Ripka received his Ing. degree in 1984, a C.Sc. (equivalent to a Ph.D.) in 1989, and a Doc. degree in 1996 from the Czech Technical University in Prague. He currently is a lecturer at Czech Technical University, in the Department of Measurement, Faculty of Electrical Engineering, teaching courses in electrical measurements and instrumentation, engineering magnetism, and sensors. His main research interests are magnetic measurements and magnetic sensors, especially fluxgates. A stay at the Danish Technical University (1991–1992) was a milestone in his scientific career. Dr. Ripka is the author of more than 120 papers, and he holds five patents. He is a member of the Elektra Society, the Czech Metrological Society, the Czech National IMEKO Committee, and the Eurosensors Steering Committee and was elected a general chairman of the Eurosensors 2002 conference.

Ichiro Sasada received an M.S. in engineering in 1976 and doctorate of engineering degree in electronic engineering in 1986, both from Kyushu University. He is a professor in the Department of Applied Science for

Electronics and Materials at Kyushu University, in Fukuoka, Japan. Dr. Sasada's research fields involve the development of magnetostrictive torque sensors for automotive application, magnetic shielding system for biomagnetic field measurements using magnetic shaking method, magnetic imaging for nondestructive evaluation using eddy-current probes, and a new magnetic motion capturing system.

Christian Schott received his diploma degree in electrical engineering from the Technical University of Karlsruhe, Germany, in 1992. He obtained his Ph.D. in 1999 at the Swiss Federal Institute of Technology, Lausanne (EPFL). Dr. Schott has worked as a freelance inventor and engineer on industrial sewing machines and signal multiplexers, holding two patents in these fields. In 1995 he joined the Institute for Microsystems, where he worked on silicon Hall sensors for accurate magnetic field measurements. Dr. Schott is a systems and instrumentation manager for the Swiss company Sentron AG in Zug, Switzerland. His current research interests include accurate magnetic measurement in general, and single-chip sensors for 2D and 3D magnetic field measurement and their applications in particular.

Ichiro Shibasaki is a research fellow in the Corporate Research & Development Administration, Asahi Chemical Industry Co., Ltd., in Fuji-city, Shizuoka, Japan. He received a B.Sc. in physics from the Tokyo University of Science in 1966, and an M.Sc. in 1968 and a Ph.D. in 1971, both in theoretical physics, from the former Tokyo University of Education (now the University of Tsukuba) in Japan. Dr. Shibasaki has previously worked in the Department of Physics at the Tokyo University of Education. In 1975 he developed high sensitivity InSb thin film Hall sensors at Asahi Chemical Industry, for which he has received an award. Dr. Shibasaki is currently working on narrow bandgap semiconductor thin film elements and their application as practical magnetic sensors.

Mark Tondra received a B.S. in physics and mathematics, with honors, from the University of Wisconsin–Madison in 1989 and a Ph.D. in 1996 from the University of Minnesota in solid state physics. Since 1996, he has worked for NVE (Nonvolatile Electronics) as a physicist involved in the development of low field magnetic sensors and magnetoresistive memory. His research focuses on the magnetotransport properties of thin films, including GMR and spin dependent tunneling.

Index

1/f noise, 320–21
    performance, improving, 320
    power spectrum density, 41
    sources, 321
    See also Noise

Abbreviations, list of, 467–71
ac fluxgates, 114–15
ac measurements, 337–40
Air coils, 48–57
    current output, 55–57
    diameter, increasing, 49
    examples, 68
    frequency range, 56
    geometry, 49
    high-frequency, 50
    induction magnetometer with, 49
    influence of parasitic capacitances, 53–55
    PCB-fabricated, 50
    as pickup coil, 49
    short-circuited, 55
    thermal noise, 52–53
    voltage sensitivity, 51–52
    winding mass, 52
    See also Induction sensors
Alkali metal vapor self-oscillating magnetometers, 298–301
    development, 299
    photo cell output, 300
    principle of, 299
    symmetrical dual-cell system, 300
    VHF oscillator, 299
    See also Optically pumped magnetometers
Alloys, 32–33
Ambient field
    measurement, 396
    shielding of, 403
Amorphous alloys, 29, 447
Ampere's law, 251, 252
Amplifiers
    differential, 213, 214
    NMOS, 214
Analog trimming, 221
Angular shaft sensors, 426, 428
Animals, magnetic orientation in, 364–65
Anisotropic magnetoresistance (AMR), 129
    basis, 130
    geometric bias, 141–44
    linearization and stabilization, 136–44
    longitudinal bias, 140–41
    magnetoresistive films, 134–36
    perpendicular bias, 137–40
    planar Hall effect and, 130–34
    in soft magnetic thin films, 131
    theoretical analysis, 130

Anisotropic magnetoresistance (AMR)
    sensors, 77, 130–50
    design examples, 146–50
    general-purpose full bridge, 147–48, 149
    half bridge, 147, 148
    herringbone full bridge, 147
    layout, 144–50
    measuring range, 145–46
    sensitivity, 145
    for weak fields, 148–50
Anisotropy, 13–20
    exchange, 38–39
    forms, 13
    magnetocrystalline, 13–16
    shape, 19–20
    strain, 16–18
    uniaxial, 37
Annealing, 420–21
Antifuse trimming, 222
Antiperiodical, 51
Antitheft systems, 382–83
Applications
    automotive, 383
    auto-oscillation magnetometers, 97
    biomagnetic measurements, 369–71
    cylindrical Hall devices, 231–32
    geomagnetic measurements, 385–91
    GMR sensor, 166–69
    integrated Hall sensors (Hall IC), 222–23
    magnetic marking/labeling, 384–85
    military/security, 380–83
    miniature fluxgates, 112
    navigation, 371–79
    nondestructive testing, 383–84
    space research, 391–98
    SQUID, 337–45
Automotive applications, 383
Auto-oscillation magnetometers, 96–97
    applications, 97
    magnetic multivibrator, 96
    See also Fluxgate sensors
Axial shielding, 413–19
    with double-shell cylinders, 416
    with double-shell shields, 417–19
    factors, 417, 418–19
    with open-ended cylinders, 414
    with single cylinders, 415
    with single-shell shields, 413–17
    See also Magnetic shielding
Bandwidth, 43
Barber poles, 143–44
    full bridge with, 151
    resistivity of, 144
    structure illustration, 144
Bar codes, 385
Bardeen-Cooper-Schriefer (BCS) theory, 305
Barkhausen noise, 136, 140
Biological sensors, 362–65
    magnetic orientation in animals, 364–65
    magnetotactic bacteria, 363–64
Biomagnetism
    amplitudes/frequency ranges, 344
    applications, 345
    measurements, 369–71
Biot-Savart law, 3
Boltzmann's constant, 280
Broadband current output, 101–4

Calibration, 39–40, 403–22
    absolute, 39
    ac, 409
    equipment manufacturers, 459–65
    induction sensors with ferromagnetic cores, 61–62
    magnetometer, 405
    methods, 403
    NASA Goddard Space Flight Center, 408
    offset, 404
    shielding, 403–4, 410–22
Calibration coils, 39–40, 403, 405–10
    ac, 409
    Helmholz, 405–6
    ideal spherical, 408
    large, 409
    rectangular coil system, 407–8
    rules, 409
    size guideline, 409–10
    solenoids, 405
Canted ferromagnets, 260

Charges, 2–3
Chromatic modulation, 250
Closed-cycle refrigeration, 333–34
CMOS Hall elements, 205–6
CMOS Hall sensors, 203–4
CMOS N-well, 206
Colossal magnetoresistance (CMR), 129
Colpitts oscillator, 358
Compact spherical coil (CSC) magnetometer, 50, 116–17
   illustrated, 116
   in-flight calibration of, 117
   temperature sensitivity coefficient, 117
   uses, 117
Cored coils
   with current output, 63–64
   equivalent circuit for, 62–63
   thermal noise, 62
Cost, 44
Coulomb magnetic moment, 3
Crossfield effect, 110
CryoSQUID components, 333
Crystalline metals, 29
Curie temperature, 12
Current output, air coils, 55
Current-output fluxgate, 100–105
   basic equation, 103
   broadband, 101–4
   circuit diagram, 101
   defined, 100–101
   resonance condition, 105
   sensitivity, 101
   total pickup coil flux, 102
   turning, 104–5
   waveforms, 104, 106
   See also Fluxgate sensors
Current sensors, 454–55
   current clamps, 455
   dc/ac Hall and MR, 455
   dc comparators, 454
   fluxgate-like dc current sensor modules, 454
   magnetometric measurement of hidden currents, 455
Cylindrical Hall devices, 230–32
   applications, 231–32
   illustrated, 232
   photograph, 233
   sensitivity, 231
   sensor schematic, 233
   structure, 230–31
   See also Nonplatelike Hall magnetic sensors

dc/ac Hall and MR current sensors, 455
dc comparators, 454
dc SQUIDs, 311, 313, 316–17
   block diagram, 317
   electronics complexity, 314
   feedback current, 317
   LTS, 314
   operation, 316–17
   RF SQUIDs vs., 313–14
   white noise, 319
   See also SQUIDs
Deep space/planetary magnetometry, 391–92
Demagnetization, 421
   by application of decaying ac field, 25
   effective, factor, 58
   fluxgate sensors and, 85–88
   methods, 24–25
   results, 24
   thermal, of rock samples, 410
Demagnetizing fields, 19–20
Detection coils, 326–27
   design factors, 326
   illustrated, 327
   imbalance of, 330
   physical size of, 325
   sensitivity/noise level calculation, 325
   spatial resolution, 325
Dewars, 331–33
   fiberglass, 332
   metallic, 331–32
   with removable sections, 331
   See also Refrigeration
Diamagnetism, 10
Differential permeability, 26
Differential variable inductance transducer (DVRT), 434
Digital magnetometers, 93–94
   analog-to-digital conversion, 93
   feedback, 94

noise level, 94
phase-sensitive detection, 93
reference signal, 93
See also Fluxgate sensors
Digital signal processors (DSPs), 93, 288
Digital trimming, 221
Diode zap, 221
Direct registration of domain wall position (DRDWP), 260–63
   domain wall motion velocity, 263
   illustrated, 262
   scheme, 262
   See also Magneto-optical current transformers (MOCTs)
Domain nucleation, 31–32
Domain structures, 20–22
   complex nature, 22
   Faraday rotation of, 258
   Wiegand sensor pulse wire, 440, 441
Domain wall pinning, 31
Double-rod sensors, 80–81
Dual magnetometer technique, 395–97
   advantages, 397
   ambient field measurement, 396
   defined, 395–96
   illustrated, 397
   See also Spacecraft magnetic field measurement

Earth magnetic field, 386–87
   illustrated, 386
   variations, 387
Eddy-current sensors, 429–33
   based on differential relaxation oscillator, 434
   ferrite core shapes, 432
   ferromagnetic targets, 430
   at high frequencies, 431
   "killed oscillator," 432
   measurements, 429
   miniature, 433
   output frequency, 435
   pot core for, 431
   proximity, 433
   sensing coils, 430
   sensing diameter, 430
   target, 429

Eddy-current shield, 336
Eddy-current testing, 342–43, 390
Electric current, 3–4
Electromagnetic compatibility (EMC), 426
Electron spin resonance (ESR), 267–68
   continuous signal, 268
   defined, 267
   excitation, 268
   magnetometer, 11
   polarization saturation degree vs., 293
   RF, 289
Epitaxial garnet films, 258
Equivalent circuits, cored coils, 62–63
Exchange anisotropy, 38–39
Exchange coupled films
   longitudinal bias, 141, 142
   perpendicular bias, 138
Excitation, 97
   current, 98
   electron spin resonance (ESR), 268
   integrated Hall sensors (Hall IC), 210–13
   second-harmonic analog magnetometer, 90

Faraday effect, 244–47
   defined, 245
   scheme, 244
Faraday law, 47
Faraday rotation, 244–47
   angles, 251
   integral, 251–52
   larger than linear birefringence, 250
   magnetic field dependence, 249
   of multidomain structure, 258
   ratio to magnetic field, 259
   senses by adjacent domains, 261
   sensitivity of detection, 249
   smaller than linear birefringence, 251
   transparent ferromagnets and, 256–57, 259
Fast Fourier transforms (FFTs), 288
Ferromagnetic conductors
   magnetoinductive effects in, 350–51
   Maxwell equations for, 354
Ferromagnetic-core coils, 58, 68

Ferromagnetic shielding, 403
Ferromagnetism/ferrimagnetism, 11–28
    domain structure, 20–22
    magnetic anisotropy, 13–20
    magnetization curve, 24–28
    magnetization process, 22–24
    spontaneous magnetization, 12–13
Fiber-optic magnetostriction field sensors, 359–61
    block diagram, 360
    sensing element, 360
    for underwater detection, 360–61
Fiber-type MOCT, 252–53
    defined, 252
    scheme illustration, 253
    sensor head, 252
    See also Magneto-optical current transformers (MOCTs)
Field compensation systems, 410
Finite element modeling (FEM), 220–21
Finite permeability, 58
Flux density, 5–6
    contributions, 6
    maximum, dependence of ratio of, 420
    measurement, 6
Flux distribution, 419–20
Fluxgate compasses, 375–79
Fluxgate gradiometers, 119–20
    baseline, 119
    dual-sensor, 120
    uses, 119–20
Fluxgate sensors, 75–120
    ac, 114–15
    amorphous magnetic materials, 89
    auto-oscillation magnetometers, 96–97
    core materials, 88–90
    crossfield effect, 110
    current-output, 100–105
    defined, 75
    designs of, 111–12
    digital magnetometers, 93–94
    effect of demagnetization, 85–88
    excitation, 97
    gradiometers, 119–20
    material properties requirements, 88–89
    miniature, 112–14
    multiaxis magnetometers, 115–19
    nanotesla range resolution and, 76
    noise, 105–8
    noise level, 76
    noise spectrum, 107, 108
    nonselective detection methods, 94–96
    offset stability, 108–10
    operation theory, 83–88
    orthogonal-type, 78–79
    output voltage tuning, 97–100
    parallel-type, 79–83
    parametric amplification, 98
    portable and low-power instruments, 111
    principle, 76
    principles, 90–97
    second-harmonic analog magnetometers, 90–93
    station magnetometers, 111–12
    temperature range and, 76
    waveforms, 77
Flux quantization, 308
Force and pressure sensors, 446–48
Full-bridge sensors
    with barber-pole bias, 151
    herringbone, 147
    industrial, with conductor strips, 152
Fuse trimming, 222

GEM-19 magnetometer, 291
General-purpose full bridge, 147–48
    illustrated, 149
    offset, 148
    sensitivity, 147–48
    See also Anisotropic magnetoresistance (AMR) sensors
Geomagnetic measurements, 385–91
    airborne magnetic surveys, 388
    based on eddy currents, 390
    depth/amplitude behavior of dipole anomalies, 390
    Earth's magnetic field and, 386–87
    instruments, 388, 390
    Intermag, 390
    surface magnetic mapping, 387–88
    See also Applications

Geometric bias, 141–44
  barber poles, 143–44
  herringbones, 142–43
  See also Anisotropic magnetoresistance (AMR)
Giant magnetoimpedance (GMI), 351–58
  of amorphous CoFeSiB wire, 352
  asymmetrical effect, 357
  behavior, 356
  defined, 351
  effect origin, 351
  efficiency, 355
  materials, 355–57
  sensitivity, 357
  sensors, 357–58
  theoretical magnitude of, 354
Giant magnetoresistance (GMR) sensors, 129, 150–69
  applications. *See* GMR sensor applications
  bipolar material, 163–65
  bipolar response with biasing coils, 163–66
  changing state of, 151–52
  chips, 163
  construction, 163–66
  cross axis sensitivity, 166
  effect discovery, 162
  film deposition, 162
  gradiometer, 165
  introduction, 150–52
  material figure of merit, 157
  material processing techniques, 162
  micromagnetic design, 159–62
  multilayer, 158, 159
  nonmagnetic layer thickness, 157
  output, 152
  packaging, 166
  patterning techniques, 162
  sandwich, 158
  schematic top view, 163
  spin dependent scattering, 155–58
  spin dependent tunneling, 158–59
  spin valve effect basics, 152–62
  structures, 153
  temperature characteristics, 165–66
  trim sites, 166

  in Wheatstone bridge configuration, 163
GMR sensor applications, 166–69
  currency detection, 168
  cylinder position sensing, 167
  gear tooth sensing, 167
  geomagnetic, 168
  hard disk drive read-heads, 169
  linear and rotary position/motion sensors, 168–69
  memory, 169
  nondestructive evaluation, 167
  signal isolator/current sensors, 169
  unexploded ordnance, 167–68
  vehicle detection, 166
  See also Giant magnetoresistance (GMR) sensors
Gradiometers
  first-order, with three noise cancellation channels, 330
  fluxgate, 119–20
  GMR, 165
  for noise reduction, 334–35
  perfect, 328
  single-axial, 330
  SQUID input circuits, 327–29
  symmetric, 328

Half-bridge sensor
  Anisotropic magnetoresistance (AMR) sensors
  defined, 147
  illustrated, 148
Hall effect, 175–79
  coefficient, 177
  defined, 175
  in germanium, 184
  illustrated, 175
  voltage, 178, 180
Hall-effect magnetic sensors, 173–239
  absolute sensitivity, 181
  applied, 174
  characteristics, 180–83
  CMOS, 203–4
  current contacts (CCs), 179
  current-related sensitivity, 181
  direct-coupled, 205

early design used in keyboard switches, 202
geometry, 179–80
importance of, 173
integrated, 174, 201–23
long-term stability, 183
noise, 183
nonplatelike, 223–39
offset, 182–83
performance, 173–74
physics, 175–84
in plate shape, 178, 179
problems, 183–84
sense contacts (SCs), 179
sensitivity, 181–82
structure, 179–80
supply voltage-related sensitivity, 181
temperature cross-sensitivity, 183–84
thin-film Hall elements, 184–201
world market for, 174

Hall elements
biasing, with constant current, 212
CMOS, 205–6
commutated with CMOS switches, 217
cross-connected, 203
depletion layers, 211
development history, 207
dual, 203
epitaxial, 211, 212
fabrication, 201–2
geometries, 215
InAs, 192–200
InSb, 185–92
junction-isolated, 217
low drift requirement, 205
magnetic responsivity of, 205
N-type epitaxial, 211
resistance of, 211
silicon, biasing of, 210
thin-film, 184–201
vertical, 218–19

Hall mobility, 212

Hall plate
defined, 174
geometry of, 180
strongly extrinsic, 182

Hall probe, 174
Hard ferrites, 33
Hard magnetic films
longitudinal bias, 141
perpendicular bias, 137
Hard magnetic materials, 31–33
H-coils, 70–72
Rogowski-Chattock potentiometer, 70–71
straight potentiometer, 71–72
Helmholz coils, 405–6
Herringbones, 142–43
High-temperature superconductors (HTS), 307
$1/f$ noise, 320
characteristics, 311
crossovers, 312
intrinsic/step-edge grain boundary, 309
three-dimensional coil structures, 326
See also SQUIDs

Hysteresis
illustrated, 25
loops, 136
reducing effects of, 27
sensor specification, 39–40

Impedance, 353
InAs deep quantum well (DQW) structure, 198–200
cross section, 200
defined, 199
typical characteristics, 200
typical properties, 200

InAs Hall elements, 192–200
by MBE, 192–98
chip photograph, 196
deep quantum wells, 198–200
design and fabrication of, 195
electrical field characteristics, 197
electron density, 193
electron mobility, 193
fabrication process, 196
Hall output voltage, 195
magnetic field characteristics, 197
offset voltage, 197–98
properties, 192–95
resistance, 195–97

temperature characteristics, 193
temperature dependence, 193, 194
thickness, 193
typical characteristics of, 195–98
See also Thin-film Hall elements
Inclination compass, 364
Induction magnetometers, 48, 49
Induction sensors, 47–72
   air coils, 48–57
   coils for measurement of H, 70–72
   current-output, frequency
      characteristics, 56
   description, 47
   design examples, 65, 66–68
   EMI, noise of, 65
   with ferromagnetic core, 58, 61
   geartooth, 446
   general equation, 47
   moving-coils, extraction method, 69
   noise matching to amplifier, 64–65
   position, 427–29
   rotating coil magnetometers, 65, 69
   search coils, 57–64
   vibrating coils, 69–70
Inductosyn, 437
Initial permeability, 26
Input circuits, 323–31
   black box, 324
   detection coils, 326–27
   detection coil size, 325
   electronic noise cancellation, 329–31
   gradiometers, 327–29
   packaging, 323–24
   schematic diagram, 324
   sensitivity, 324–26
   See also SQUIDs
InSb Hall elements, 185–92
   absolute maximum ratings, 190
   by vacuum deposition, 185–86
   chip photograph, 189
   commercial success, 192
   constant current driving, 188–89
   cross section, 188
   device structure, 186
   electrical characteristics, 190
   magnetic field characteristics, 190
   problems, 192

   production process, 189
   temperature dependence, 187–88, 191
   typical characteristics, 186–92
   See also Thin-film Hall elements
Integrated Hall sensors (Hall IC), 201–23
   amplification, 213–14
   applications, 222–23
   CMOS Hall elements, 205–6
   defined, 174
   excitation, 210–13
   fabrication variables, 208
   geometry considerations, 215–18
   Hall offsets, 206–10
   historical perspective, 201–5
   junction-isolated Hall elements, 217
   N-type epitaxial resistors, 213
   offset voltage, 208
   packaging, 219–21
   stress effects, 208–9
   systematic thermal variables, 208
   trends, 222–23
   trimming methods and limitations,
      221–22
   vane switch, 222
   vertical Hall elements, 218–19
   zero-temperature-coefficient thin-film
      resistors, 213
   See also Hall-effect magnetic sensors
Intermag, 390
Irreversible Barkhausen jumps, 24
Irreversible domain wall displacements,
   23–24
Irreversible magnetization rotation, 32

Josephson effect, 308–9
   defined, 308–9
   junction types, 310
   tunnel junction illustration, 309
JPL two-cell He4 scalar sensors, 297
Junction-field effect transistor (JFET), 228

"Killed oscillator," 432

Laboratory applications, 337–42
   ac measurements, 337–40
   magnetometers/susceptometers, 340–42
   See also SQUIDs
Landau-Lifshitz equation, 354

LETI Overhauser magnetometer, 291, 292
Linear birefringence, 246
  Faraday rotation larger than, 250
  Faraday rotation smaller than, 251
Linear resolvers, 438
Linear transformer sensors, 433–37
  Inductosyn, 437
  LVDT, 433–34
  PLCD, 436–37
  variable gap, 434–36
  See also Position sensors
Linear variable differential transformer
      (LVDT) sensors, 433–34
  "half-bridge," 434
  illustrated, 436
  practical resolution, 434
  reliability, 434
Longitudinal bias, 140–41
  exchange tabs, 141, 142
  hard magnetic films, 141
  use of, 140
  See also Anisotropic magnetoresistance
      (AMR)
Longitudinal Kerr effect, 247, 248
Low-temperature superconductors (LTS),
      307
  advantages/disadvantages, 312–13
  characteristics, 311
  material forms, 313
  stable in air, 313
  tunnel junction weak links, 309
  See also LTS SQUIDs; SQUIDs
LTS SQUIDs
  dc, 314
  field dependence, 320
  loops, 323
  RF, 314
  temperature dependence, 319–20

MACOR, 118
Magnetically shielded rooms (MSRs),
      336–37
Magnetic anisotropy, 13–20
Magnetic dipoles, 6
Magnetic fields
  constant, 6
  Earth, 386–87

  energy per unit volume of, 6
  matter and, 4–8
  measurement reproducibility, 270–71
  planetary, origin of, 9
  spacecraft, 394–98
  units and magnitudes of, 8–9
  visualization, 4
Magnetic flowmeters, 453–54
Magnetic force, 2–3
  lines, 4
  microscope (MFM), 384
Magnetic imaging, 384
Magnetic liquids, 455–56
Magnetic mapping payload (MMP), 269
Magnetic materials, 28–39
  alloys, 32–33
  amorphous metals, 29
  classification, 28–29
  crystalline metals, 29
  hard, 31–33
  hard ferrites, 33
  magnetostrictive, 33
  nanocrystalline metals, 30
  sintered magnets, 33
  soft, 29–31
  soft ferrites, 30–31
  thin films, 33–39
Magnetic resonance, 267–71
  electron spin (ESR), 267–68
  historical overview, 268–70
  magnetic field reproducibility, 270–71
  nuclear (NMR), 267
Magnetic sensors. *See* Sensors
Magnetic Sensors and Magnetometers
  approaches, xv–xvi
  audience, xvi
Magnetic shaking, 421–22
  effective, 421
  efficiency, 421–22
Magnetic shielding, 50, 336–37, 403–4,
      410–22
  ambient field, 403
  annealing, 420–21
  axial, 413–19
  custom-made, 404
  demagnetizing, 421
  eddy-current shield, 336

effectiveness, 411–12
enhancement by magnetic shaking,
    421–22
ferromagnetic, 403
flux distribution, 419–20
shielded rooms, 336–37
superconducting, 404
theory, 410–12
transverse, 411, 412–13
See also Calibration; Noise reduction
Magnetic states, 9–28
Magnetic trackers, 369, 442–44
defined, 442
generator coils, 444
with sensing target, 444
Magnetism, 1–4
Magnetization
applied field and, 132–34
bulk, 23
coherent rotation, 35–38
curves, 24–28
macroscopic, 22
process, 22–24
remanent, 374
resistance and, 131–32
rotation, 24, 32
saturation, 277
spontaneous, 12–13
Magnetocrystalline anisotropy, 13–16
constants, 14
distinct, 16
energy areas, 18
Magnetoelastic field sensors, 358–62
fiber-optic magnetostriction, 359–61
magnetostrictive-piezoelectric, 361
shear-wave magnetometers, 362
Magnetohydrodynamic (MHD) angular
    rate sensor, 455–56
defined, 455–56
principle, 456
uses, 456
Magnetoimpedance. *See* Giant
    magnetoimpedance (GMI)
Magnetoinductive effect, 350–51
Magnetometers
auto-oscillation, 96–97
calibration, inflight, 405

coil, 47
CSC fluxgate, 50, 116–17
defined, xvii
digital, 93–94
ESR, 11
fluxgate, 50, 76, 90–97, 111–12
induction, 48, 49
manufacturers, 459–65
multiaxis, 115–19
optically pumped, 294–301
Overhauser effect proton, 268
proton precession, 271–94
relaxing-type, 96
resonance, 43, 267–301
rotating coil, 65, 69
scalar, 393
second-harmonic analog, 90–93
shear-wave, 362
spinning coil, 69
SQUID, 340–42
station, 111–12
three-axial, 48, 111
vector, 393–94
vehicle, readings, 376–78
Magneto-optical current transformers
    (MOCTs), 248–63
advantages, 248
based on diamagnets, 251–56
based on transparent ferromagnets,
    256–60
birefringence reduction and, 256
with direct registration of domain wall
    positions, 260–63
fiber-type, 252–53
field concentrator arrangement, 254
with glass SF6 sensor, 256
multipass sensing element, 254
reflection coatings and, 255
sensitivity, increasing, 253–54
sensitivity ratio, as function of gap size,
    255
Magneto-optical Kerr effect, 247–48
defined, 247
illustrated, 248
longitudinal, 247, 248
polar, 247, 248
transverse, 247, 248
types of, 247

Magneto-optical sensors, 243–64
  Faraday and Kerr effects, 244–48
  geometric measurements, 263–64
  MOCTs, 248–63
Magnetopneumography, 369–71
  defined, 369–70
  gradient distribution, 370, 373
  homogeneously deposited dust, 370, 372
  magnetic field distribution, 370, 371–72
  system elements, 370
Magnetoresistive films, 134–36
  film processing, 135
  materials, 134–35
  measurements, 135–36
Magnetoresistors, 129–69
  AMR, 130–50
  GMR, 150–69
  uses, 129
Magnetostatic energy density, 8
Magnetostriction, 16–19
  constants, 17
  isotropic saturation, 17
Magnetostrictive sensors
  materials, 33
  piezoelectric, 361
  position, 438, 439
Magnetotactic bacteria, 363–64
Manufacturers, 459–65
Marking/labeling applications, 384–85
Matter
  magnetic fields and, 4–8
  magnetic states of, 9–28
Maximum permeability, 26
Mechanical gyroscopes, 271–74
  illustrated, 272
  motion, 271–72
  spinning and coning, 274
  spinning wheel in perfect balance, 272
  wheel rotation around skew axis, 272–73
Medical applications, 343–45
Meissner effect, 307–8, 404
Meissner-Ochsenfeld effect, 10
Metal-oxide semiconductor (MOS) technology, 203

Metastable He4 magnetometers, 294–98
  frequency determination, 295
  JPL, 297
  modulation sweep range, 296
  Mz-mode, 295–96
  omnidirectional, 298
  sealed glass vessel, 294
  sweep frequency, 296
  See also Optically pumped magnetometers
Military/security applications, 380–83
  antitheft systems, 382–83
  target detection/tracking, 382
  UXO, 380–82
  See also Applications
Miniature fluxgates, 112–14
  advantages, 113–14
  applications, 112
  compensation sensor, 112
  orthogonal, with flat excitation and pickup coil, 113
  planar technology coils, 112–13
Molecular beam epitaxy (MBE), 185
  defined, 192
  InAs thin-film Hall elements by, 192–98
MOS-based Hall effect circuits, 203
Multiaxis magnetometers, 115–19
  individually compensated sensors, 117–19
  three-axial compensation systems, 116–17
Multilayer (GMR), 158, 159
Mumetal shielding, 320
Mz-mode optically pumped magnetometer, 295–96

Nanocrystalline metals, 30
Navigation, 371–79
  fluxgate compasses, 375–79
  magnetometer readings, 376–78
  mobile robots, 379
  pivoting needles, 374–75
  precision limitations, 372
  relaxation oscillator, 379
  remanent magnetization, 374

strapdown/electronically gimbaled
    compass, 379
  See also Applications
Noise, 41–42
  $1/f$, 320–21
  Barkhausen, 136, 140
  defined, 41
  density, 52
  electronic cancellation, 329–31
  EMI induction sensor, 65
  fluxgate sensor, 105–8
  Hall-effect magnetic sensors, 183
  matching to amplifier, 64–65
  spectrum density, 42
  SQUIDs, 319
  thermal, 52–53
  white, 41–42, 319
Noise reduction, 334–37
  gradiometers for, 334–35
  magnetic shielding for, 336–37
  See also SQUIDs
Nondestructive evaluation (NDE), 343
Nondestructive testing applications,
    383–84
  eddy-current methods, 384
  magnetic imaging, 384
Nonplatelike Hall magnetic sensors,
    223–39
  criteria, 224–25
  cylindrical Hall devices, 230–32
  introduction, 224–25
  three-axis Hall devices, 237–39
  two-axis vertical Hall devices, 232–36
  vertical Hall devices, 225–30
Nuclear magnetic resonance (NMR), 267
  coherent ac-magnetic signal, 267
  defined, 267
  historical overview, 268

Offset, 40
  calibration, inflight, 404
  fluxgate sensor, 108–10
  general-purpose full bridge, 148
  Hall-effect magnetic sensors, 182–83
  from magnetometer electronics,
    109–10
  of processing circuits, 109

  sources, 109–10
  vertical hall devices, 228
  zero, 108–9
Offset temperature
  coefficient, 40
  stability for amorphous sensors, 109
Optically pumped magnetometers,
    294–301
  alkali metal vapor self-oscillating,
    298–301
  metastable He4, 294–98
  See also Resonance magnetometers
Orthoferrites, 260
Orthogonal-type fluxgates, 78–79
Overhauser-effect proton magnetometers,
    268, 289–94
  GEM-19, 291
  LETI, 291, 292
  liquid proton sample, 291
  proton free precession rate, 291
  RF ESR, 289
  Tempone nitroxide free radical, 290,
    292
  Trityl-group, 293
  See also Proton precession
    magnetometers

Packaging
  integrated Hall sensor, 219–21
  SQUID input circuit, 323–24
Parallel-type fluxgates, 79–83
  core shapes, 79–83
  double-rod sensors, 80–81
  race-track sensors, 82–83
  ring-core sensors, 81–82
  single-rod sensors, 80
Paramagnetic glasses, 248
Paramagnetism, 10–11
Parametric amplification, 98–100
Parasitic capacitances, 54
PCB-fabricated coils, 50
Permanent magnetic linear contactless
    displacement (PLCD) sensor,
    436–37
Permanent magnet sensors, 426–29
  angular shaft, 426, 428
  defined, 426

induction position sensors, 427–29
nonlinearity of field dependence,
  426–27
variable reluctance, 429
See also Position sensors
Permeabilities
  apparent, 60
  differential, 26
  finite, 58
  initial, 26
  maximum, 26
  minor loops and, 27
  reversible, 26
  total, 25–26
Perming, 40–41
Perpendicular bias, 137–40
  exchange coupled films, 138
  field generation, 137
  permanent magnetic films, 137
  shunt bias, 139–40
  use of, 137
  See also Anisotropic magnetoresistance
    (AMR)
Phase-delay method, 94–95
Phase-sensitive detector (PSD), 90, 109
Piezomagnetic pressure sensors, 447
Planar Hall effect, 130–34, 229
Planck's constant, 7
PMOS N-well, 206
Polarimetric measurements, 249–51
Polar Kerr effect, 247, 248
Position sensors, 425–44
  eddy-current, 429–33
  induction, 427–29
  linear transformer, 433–37
  magnetic trackers, 442–44
  magnetostrictive, 438, 439
  with permanent magnet, 426–29
  rotation transformer, 437–38
  types of, 425
  Wiegand, 438–42
Power, 43
Proton precession magnetometers, 271–94
  amplified precession signal strength
    dependence, 286
  classic, 274–89
  continuously oscillating, 289

cross-sectional view, 279
electronic switching circuit, 283, 284
field gradient effect, 287
fraction of aligned protons, 277
magnetization decay, 286
mechanical gyroscopes, 271–74
omnidirectional toroid, 279
Overhauser, 289–94
polarization, 281–82
polarizing field, 282–83
proton angular momentum, 275
proton magnetic moment, 275, 276
proton precession frequency
  measurement, 287
toroid coil, 280
See also Resonance magnetometers
Proximity/rotation detectors, 444–46
Proximity sensor, 433

Racetrack sensors, 82–83
  defined, 82
  illustrated, 83
  noise level, 108
  problems, 82–83
  sensitivity, 100
  See also Parallel-type fluxgates
Reaction torque sensors, 448
Reed contacts, 444–45
Refrigeration, 331–34
  closed-cycle, 333–34
  dewars, 331–33
  See also SQUIDs
Relaxing-type magnetometer, 96
Resistance
  dependence on angle, 132
  dependence on applied field, 133
  magnetization and, 131–32
Resolvers, 437–38
Resonance magnetometers, 43, 267–301
  historical overview, 268–70
  magnetic field measurement
    reproducibility, 270–71
  optically pumped, 294–301
  proton precession, 271–94
Reversible permeability, 26
RF SQUIDs, 311, 314–16
  dc SQUIDs vs., 313–14

detected output voltage vs. flux, 316
locked-loop operation block diagram, 315
LTS, 314
operation, 314–16
white noise, 319
See also SQUIDs
Ring-core sensors, 81–82
advantages, 81
defined, 81
derivation of pickup coil flux, 85
illustrated, 80, 83
magnetization characteristics, 84
noise level, 108
sensitivity, 100
single-core dual-axis, 115
waveforms, 86
See also Parallel-type fluxgates
Rogowski-Chattock potentiometer, 70–71
Room temperature vulcanizing (RTV), 220
Rotating coil magnetometers, 65, 69
Rotation transformer sensors, 437–38
resolvers, 437–38
synchros, 437
See also Position sensors

Sandwich
cross section, 153
field-biased bridge sensor, 164
structure output, 158
unpinned, 158
See also Giant magnetoresistance (GMR) sensors
Saturation magnetization, 277
Scalar magnetometers, 393
Scanning SQUID microscope (SSM), 384
Search coils, 57–64
with current output, 63–64
equivalent circuit, 62–63
examples, 66–67
length, 61
thermal noise, 62
voltage output sensitivity, 57–62
See also Induction sensors
Second-harmonic analog magnetometer, 90–93

block diagram, 91
excitation, 90
pickup coil, 92–93
reference signal generation, 90
sensitivity, increasing, 92
See also Fluxgate sensors
Sensors
AMR, 77, 144–50
applications of, 369–98
biological, 362–65
calibration of, 39–40
current, 454–55
defined, xvii
double-rod, 80–81
eddy-current, 429–33
fiber-optic magnetostriction, 359–61
fluxgate, 75–120
force and pressure, 446–48
GMI, 357–58
GMR, 150–69
Hall-effect, 173–239
induction, 47–72
linear transformer, 433–37
with magnetic liquids, 455–56
magnetoelastic field, 358–62
magneto-optical, 243–64
magnetostrictive-piezoelectric, 361
manufacturers, 459–65
for nonmagnetic variables, 425–56
position, 425–44
race-track, 82–83
rotation transformer, 437–38
shielding. See Magnetic shielding
single-core, 81–82
single-rod, 80
torque, 448–53
vector, 43
Wiegand, 438–42
Sensor specifications, 39–44
bandwidth, 43
cost, 44
full-scale range, 39–40
hysteresis, 39–40
linearity, 39–40
long-term stability, 40
noise, 41–42
offset, 40

offset temperature coefficient, 40
perming, 40–41
power, 43
resistance against environment, 42–43
resistance against perpendicular field/
   field gradient, 43
temperature coefficient of sensitivity,
   39–40
Shape anisotropy, 19–20
   demagnetizing field, 19–20
   effect visualization, 20
   energy of demagnetizing stray fields,
      36
Shear-wave magnetometers, 362
   ac phase modulation, 363
   defined, 362
   illustrated, 363
   principle, 362
Shielding. *See* Magnetic shielding
Short-circuited induction coil, 55
Shunt bias, 139–40
Single-rod sensors, 80
Sintered magnets, 33
Soft ferrites, 30–31
Soft magnetic materials, 29–31
Solenoids, 405
   for calibration, 405
   self-capacitance in, 54
Space applications, 391–98
   deep-space/planetary magnetometry,
      391–92
   magnetic field measurement onboard
      spacecraft, 394–98
   magnetic instrumentation, 392–94
   See also Applications
Spacecraft magnetic field measurement,
   394–98
   alignments, 397–98
   booms for magnetic sensors, 395
   coordinate systems, 397–98
   dual magnetometer technique, 395–97
   magnetic interference, 394
   power spectrum, 397–98
   zero levels, 397–98
Space magnetic instrumentation, 392–94
   scalar magnetometers, 393
   vector magnetometers, 393–94
   See also Space applications

Spin dependent scattering, 155–58
   conditions, 156
   defined, 155
   multilayer, 158
   unpinning sandwich, 158
   See also Giant magnetoresistance
      (GMR) sensors
Spin dependent tunneling (SDT), 158–59
   biased resistor, 165
   devices, 158–59
   materials, 159
   structure cross section, 160
Spinning coil magnetometers, 69
Spin valves
   cross section, 153
   dependent tunneling, 158–59
   effects, 152–62
   elements, 152–53
   high field response of, 154
   low field response of, 154
   micromagnetic design, 159–62
   resistance of, 155
   scattering origin, 155–58
   simple, 152–55
   See also Giant magnetoresistance
      (GMR) sensors
Spontaneous magnetization, 12–13
SQUID applications, 337–45
   ac measurements, 337–40
   geophysical, 342
   laboratory applications, 337–42
   magnetometers/susceptometers, 340–42
   medical, 343–45
   nondestructive test and evaluation,
      342–43
SQUIDs
   $1/f$ noise, 320–21
   bandwidth, 322
   as black box, 324
   control electronics, 321–22
   dc, 305–45, 311, 313, 316–17
   design factors, 321
   energy sensitivity, 318
   fabricated as planar devices, 323
   field dependence, 320
   "flux-locked," 322
   humidity and, 43

input circuits, 323–31
introduction, 305–11
Josephson effect and, 308–9
"lost lock," 321
materials, 307, 312–13
noise, 317–18
noise reduction, 334–37
operation, 313–23
refrigeration, 331–34
RF, 311, 314–16
sensitivity, 317–19
sensor output, 321
slew rates, 321
technology limitations, 322–23
temperature dependence, 319–20
as vector magnetometers, 322–23
white noise, 319
See also SQUID applications
Station magnetometers, 111–12
Stoner-Wohlfarth astroid, 36–37
Straight potentiometer, 71–72
Strain anisotropy, 16–19
    energy areas, 18
    energy density, 17
    See also Anisotropy
Superconducting quantum interference device (SQUID) sensors. *See* SQUIDs
Superconducting shield, 404
Superconductivity, 10, 305–7
Susceptometers, 340–42
Symbols, list of, 467–71
Synchros, 437
Système International d'Unités (SI), 270

Target detection/tracking, 382
Temperature coefficient of expansion (TCE), 220
Temperature coefficient of resistance (TCR), 222
Temperature cross-sensitivity, 229
Temperature range, 42
Tempone nitroxide free radical, 290
    double positive/negative proton polarization structure, 292
    illustrated, 290
Testing, 403–22

Thermal noise
    air coils, 52–53
    calculating, 52–53
    cored induction sensor, 62
    See also Noise
Thin-film Hall elements, 184–201
    conclusion, 200–201
    Hall output voltage temperature dependence, 201
    InAs, 192–200
    InSb, 185–92
    introduction to, 184–85
    See also Hall-effect magnetic sensors
Thin films, 33–39
    amorphous iron-/cobalt-based, 34
    coherent rotation of magnetization, 35–38, 133
    magnetic structure, 34–35
    magnetization curves of, 37
    magnetization process, 34
    mechanically hard, 34
    Permalloy, 34
    sandwich and multilayer structures, 34
    See also Magnetic materials
Three-axis Hall devices, 237–39
    defined, 237
    Hall voltages, 237, 238
    magnetic field components, 237–39
    photograph, 237
    See also Hall-effect magnetic sensors
Three-axis magnetometers, 48
    compensation system, 116–17
    crosstalk problems, 117
    fluxgate, 111
    sensor head, 118
Toroid proton magnetometers, 279
    coil, 280
    coil impedance matching input stage, 281
    cross-sectional view, 279
    illustrated, 279
Torque sensors, 448–53
    of axisymmetric structure, 452
    based on fundamental magnetoelastic principles, 450
    figure-of-eight coils, 451

low-profile pickup coils, 452
with magnetoelastically active ring, 453
reaction, 448
strain gauges, 449
stress-induced magnetic anisotropy, 450
twist angle method, 448, 449
type I, 450–51
type II, 452–53
Total permeability, 25–26
Transparent ferromagnets
Faraday rotation, 257, 259
magnetic fields action on, 258
measurement sensitivity, 256
MOCTs based on, 256–60
polarization rotation, 257
sensitivity, 257–58
Transverse Kerr effect, 247, 248
Transverse shielding, 412–13
factor for double-shell shield, 412
factor for finite-length cylindrical shield, 412
factor for individual shell, 412
factors for set of elliptical cylinders, 413
illustrated, 411
with thin multiple shells, 413
See also Magnetic shielding
Trimming, 221–22
analog, 221
antifuse, 222
defined, 221
digital, 221
diode zap, 221
fuse, 222
laser, 221
limitations, 222
methods, 221–22
Two-axis vertical Hall devices, 232–36
angular position sensor, 235
defined, 232–35
illustrated, 234
three-phase angular position sensor, 236
See also Hall-effect magnetic sensors; Vertical Hall devices

Unexploded ordnance (UXO), 380–82
fieldwork techniques for locating, 380–82
as hazards, 380
See also Military/security applications
Vane switch, 222
Variable gap sensors, 434–36
Variable reluctance sensors, 429
Vector sensors, 43, 393–94
Vehicle magnetometer readings, 376–78
Verdet constant, 244, 248, 252, 253, 257
Vertical Hall devices, 225–30
accurate magnetic field measurement, 229–30
cut through illustration, 227
defined, 225
features, 227
illustrated, 226
matched sensors in probe head, 230
nonlinearity in magnetic field, 228–29
offset, 228
planar Hall effect, 229
technological structure, 225, 226
temperature cross-sensitivity, 229
two-axis, 232–36
See also Nonplatelike Hall magnetic sensors
Vertical Hall elements, 218–19
biasing, 219
compatibility, 218
defined, 218
illustrated, 219
orientation, 218–19
Vibrating coils, 69–70
Voltage sensitivity
air coils, 51–52
search coils, 57–62

Waveforms
broadband current-output fluxgate, 104
fluxgate, 77
ring-core fluxgate, 86
in tuned current-output fluxgate, 106
Weakly magnetic materials, 10–11
diamagnetism and superconductivity, 10
paramagnetism, 10–11

Wheatstone bridge, 143, 163
  gauges arranged in, 449
  unbalanced, 211
White noise, 41–42
  power spectrum density, 41
  SQUIDs, 319
  See also Noise
Wiegand sensors, 438–42

  defined, 438–40
  pulse wire construction, 440, 441
  pulse wire domain structure, 440, 441
  recommended magnetic drives, 443
  voltage pulse, 440, 442
  See also Position sensors

Zero offset, 108–9

Related Titles from Artech House

Chemical and Biochemical Sensing with Optical Fibers and Waveguides, Gilbert Boisdé and Alan Harmer

Optical Fiber Sensors, Volume III: Components and Subsystems, Brian Culshaw and John Dakin, editors

Optical Fiber Sensors, Volume IV: Applications, Analysis, and Future Trends, Brian Culshaw and John Dakin, editors

Sensor Technology and Devices, Ljubisa Ristic, editor

Smart Structures and Materials, Brian Culshaw

Understanding Smart Sensors, Second Edition, Randy Frank

For further information on these and other Artech House titles, including previously considered out-of-print books now available through our In-Print-Forever® (IPF®) program, contact:

Artech House
685 Canton Street
Norwood, MA 02062
Phone: 781-769-9750
Fax: 781-769-6334
e-mail: artech@artechhouse.com

Artech House
46 Gillingham Street
London SW1V 1AH UK
Phone: +44 (0)171-973-8077
Fax: +44 (0)171-630-0166
e-mail: artech-uk@artechhouse.com

Find us on the World Wide Web at:
www.artechhouse.com